Reading Essent

GLENCOE

PHYSICAL SCIENCE

An Interactive Student Textbook

McGraw Hill Education

To the Student

In today's world, knowing science is important for thinking critically, solving problems, and making decisions. But understanding science sometimes can be a challenge.

Reading Essentials takes the stress out of reading, learning, and understanding science. This book covers important concepts in science, offers ideas for how to learn the information, and helps you review what you have learned.

In each chapter:

- **Before You Read** sparks your interest in what you'll learn and relates it to your world.
- **Read to Learn** describes important science concepts with words and graphics. Next to the text you can find a variety of study tips and ideas for organizing and learning information:
    - The **Study Coach** offers tips for getting the main ideas out of the text.
    - **Foldables™ Study Organizers** help you divide the information into smaller, easier-to-remember concepts.
    - **Reading Checks** ask questions about key concepts. The questions are placed so you know whether you understand the material.
    - **Think It Over** elements help you consider the material in-depth, giving you an opportunity to use your critical-thinking skills.
    - **Picture This** questions specifically relate to the art and graphics used with the text. You'll find questions to get you actively involved in illustrating the concepts you read about.
    - **Applying Math** reinforces the connection between math and science.
- Use **After You Read** to review key terms and answer questions about what you have learned. The **Mini Glossary** can assist you with science vocabulary. Review questions focus on the key concepts to help you evaluate your learning.

See for yourself. *Reading Essentials* makes science easy to understand and enjoyable.

Credits

COV Koji Kitagawa/SuperStock; **59** simplytheyu/iStock/Getty Images Plus/Getty Images; **214** Ralph C. Eagle, Jr./Photo Researchers; **221** Andrew Syred/Science Photo Library/Photo Researchers; **260** file photo.

The McGraw·Hill Companies

 Education

Send all inquiries to:
McGraw-Hill Education
8787 Orion Place
Columbus, OH 43240-4027

ISBN: 978-0-07-896315-5
MHID: 0-07-896315-X

Printed in the United States of America.

15 16 17 18 19 QTN 20 19 18 17

Table of Contents

The Nature of Science

section ❶ The Methods of Science

What You'll Learn
- how scientists solve problems
- why scientists use variables
- how to compare and contrast science laws and theory

Mark the Text

Identify Highlight each heading that is a question. Use a different color of marker to highlight the answers to the questions.

FOLDABLES™

Ⓐ Build Vocabulary
Make the following Foldable to help you learn key terms from this section. You will need to make more than one Foldable.

| Variable |
| Dependent Variable |
| Independent Variable |
| Constant |
| Control |
| Scientific Method |
| Hypothesis |
| Experiment |

● Before You Read

When you hear the word *scientist*, what comes to mind? Brainstorm some words that describe a scientist. Write them on the lines below.

_____

_____

● Read to Learn

What is science?

Science is not just a subject in school. Science is a way of studying the world. The word *science* comes from a Latin word that means "knowledge." Science is a way to learn or gain knowledge by observing and investigating.

Nature follows a set of rules. The rules for understanding how the human body works are complicated. The rules for understanding the pattern of Earth spinning once every 24 h on its axis are simpler. Scientists ask questions to learn about the natural world.

What are the major categories of science?

Science covers many different topics. These topics fall under three main categories: life science, Earth science, and physical science. Life science is the study of living things. Earth science is the study of Earth and space. Physical science is the study of matter and energy. In this book, you will study mainly physical science. You also will learn how these three main categories sometimes overlap.

How does science explain nature?

Scientific explanations help explain the natural world. Scientists investigate and use technology to get new information. Sometimes, this new information causes scientific explanations to change. In the 1700s, most scientists thought heat was a fluid. After many studies, they concluded heat is not a fluid.

Scientific Methods

A **<u>scientific method</u>** is a set of steps used in an investigation. Scientists follow steps similar to those listed below when doing an investigation. These steps guide the scientist. Some steps may be repeated. Other steps may be skipped. The flowchart below shows six common steps found in various scientific methods.

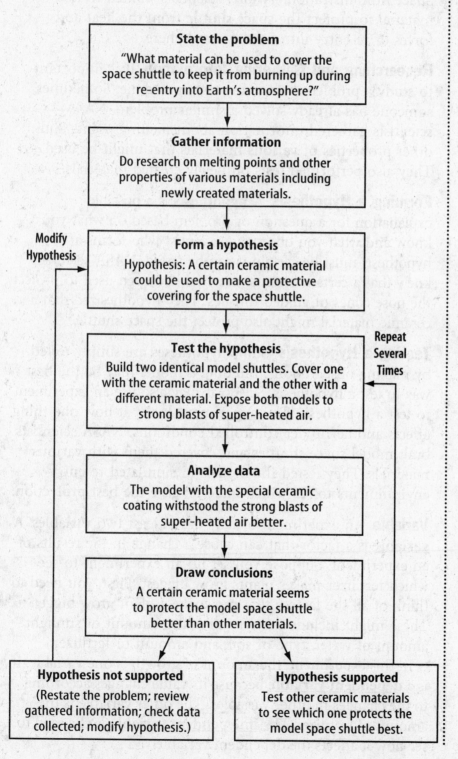

State the problem

"What material can be used to cover the space shuttle to keep it from burning up during re-entry into Earth's atmosphere?"

Gather information

Do research on melting points and other properties of various materials including newly created materials.

Modify Hypothesis

Form a hypothesis

Hypothesis: A certain ceramic material could be used to make a protective covering for the space shuttle.

Repeat Several Times

Test the hypothesis

Build two identical model shuttles. Cover one with the ceramic material and the other with a different material. Expose both models to strong blasts of super-heated air.

Analyze data

The model with the special ceramic coating withstood the strong blasts of super-heated air better.

Draw conclusions

A certain ceramic material seems to protect the model space shuttle better than other materials.

Hypothesis not supported

(Restate the problem; review gathered information; check data collected; modify hypothesis.)

Hypothesis supported

Test other ceramic materials to see which one protects the model space shuttle best.

Think it Over

1. **Draw Conclusions** Why might steps in scientific methods be skipped or changed?

_____

_____

_____

Picture This

2. **Interpret** Why are there two arrows leading to different parts of the chart at the bottom?

_____

_____

_____

Stating a Problem Many scientific investigations begin with a question about how or why something happens in nature. The problem is stated as a "how" or "why" question. Sometimes a question is asked because scientists want to improve upon something. Early work on guided missiles showed that the nose cone instruments needed a material to protect them from heat. Later, National Aeronautic and Space Administration (NASA) scientists wanted to find a material to protect the space shuttle from the heat and forces of re-entry into Earth's atmosphere.

Researching and Gathering Information It is important to study a problem before any testing is done. Sometimes someone has already solved a similar problem. NASA scientists gathered information about melting points and other properties of various materials that might be used. They also performed tests on newly created materials.

Forming a Hypothesis A <u>hypothesis</u> is a possible explanation for a question or problem based on what you know and what you observe. A scientist who forms a hypothesis must be certain it can be tested. NASA scientists knew that a certain ceramic material had been used to protect the nose cones of guided missiles. They hypothesized that a ceramic material might also protect the space shuttle.

Testing a Hypothesis Some hypotheses are simply tested by making observations. Building a model may be the best way to test a hypothesis. Scientists may also use an experiment to test a hypothesis. The **experiment** looks at how one thing affects another under controlled conditions. NASA scientists built model space shuttles and covered them with various materials. They tested the models in simulated re-entry environments to see which material gave the best protection.

Variable An experiment usually has at least two variables. A <u>variable</u> is a factor that can cause a change in the results of an experiment. Suppose you set up an experiment to see which fertilizer makes plants grow biggest. First, you need to think of all the factors that can make a plant grow bigger. These might include the type of plant, amount of sunlight, amount of water, type of soil, and amount of fertilizer.

In this experiment, the amount of growth is one variable. It is a **dependent variable** because its value changes according to the changes in other variables. The other variable is the amount of fertilizer. The **independent variable** is changed to see how it affects the dependent variable. ☑

☑ **Think it Over**

3. Research List two places where you might find information on the development of the space shuttle.

_____

_____

☑ **Reading Check**

4. Define What are the two kinds of variables that are tested in experiments?

_____

_____

Constants and Controls To keep an investigation fair, all other factors must be the same. A **constant** is a factor in an experiment that does not change. In the fertilizer experiment, the constants are the amount of water and sunlight the plants get and the temperature at which the plants are kept. These are the same for all plants tested. Three plants get different amounts of fertilizer, which is the independent variable.

A **control** is used to compare the results of the experiment. One plant is a control and does not get any fertilizer, but does get the same sunlight, water, and temperature as the other plants. Suppose after several days that the control plants grow between 2 and 3 cm. If the unfertilized plant grows only 1.5 cm, you might infer that the greater growth of the fertilized plants was caused by the fertilizer.

Analyze the Data An important part of any experiment is recording observations and organizing information. All results and observations should be recorded during an experiment. Many important discoveries have been made from unexpected results. The information or data should be organized into an easy-to-read table or graph. Later in this chapter, you will learn how to show your data.

Understanding what the data and observations mean is also important. The data must be organized logically. Poorly organized data may lead to a wrong conclusion. Scientists share their data through reports and meetings. Scientists may disagree about certain data, no matter how well the data is presented. ☑

Drawing Conclusions Scientists look at their data and decide if the data support the hypothesis. If the data is the same after many experiments, the hypothesis is supported. If the hypothesis isn't supported, scientists may change it or the experiment.

Sometimes others don't agree with the conclusions, so they design new experiments to test the hypothesis. In time, data will support a valid hypothesis.

Being Objective Scientists must avoid a bias. A **bias** occurs in an experiment when a scientist expects something to happen and lets this influence how the results are viewed. Scientists try to reduce bias whenever they can by doing the experiment many times and keeping careful notes about observations. Also, other scientists repeat the same experiment to see if they get the same results.

Think it Over

5. **Classify** Is the plant that receives no fertilizer in the plant experiment a variable, a constant, or a control?

_____

Reading Check

6. **List** two reasons why data must be organized.

_____

_____

_____

Visualizing with Models

Sometimes, scientists cannot see everything they are testing. They might be studying something too large or too small to see. It might even take too much time to see completely. In these cases, scientists use models. A **model** represents an idea, event, or object to help people better understand it. A model may be a physical object such as a scaled-down version of the space shuttle. A model can also be represented by a drawing on paper or by a computer program.

Models in History Models have been used throughout history. Lord Kelvin, a scientist who lived in England in the 1800s, was famous for making models. To model his idea of how light moves through space, he put balls in a bowl of jelly. He asked people to move the balls with their hands. Kelvin's work explaining heat and temperature is still used today. ☑

What are high-tech models?

Scientific models don't have to be something you can touch. Many scientists use computers to build models. Computer models are used to solve difficult mathematical equations. NASA uses computers in experiments with space flights to solve equations that are too hard or would take too long to solve by hand.

Another type of model is a simulator. A simulator can create the conditions found in real life. For example, a flight simulator is a model of an airplane. It can help a pilot pretend to be flying a plane. The pilot can test different ways to solve problems. The simulator reacts the same way a real plane does when it flies, but there is no danger to either the pilot or a plane.

Scientific Theories and Laws

A scientific **theory** is a way of explaining things or events based on what has been learned from many observations and investigations. When these observations and investigations have been repeated many times and support the hypothesis, then the hypothesis becomes a theory. New information in the future may change the theory.

Copyright © Glencoe/McGraw-Hill, a division of The McGraw-Hill Companies, Inc.

✔ Reading Check

7. Explain What are models?

_____

_____

_____

💡 Think it Over

8. List three differences between a flight simulator and a real airplane.

_____

_____

_____

A <u>scientific</u> <u>law</u> is a statement about what happens in nature and seems to be true all the time. A law explains what will happen under certain conditions, but it does not explain why or how it happens. Theories are used to explain how and why laws work. Gravity is an example of a scientific law. The law of gravity says that any one mass will attract another mass. To date, no experiments have been done that prove this law is not true. ☑

The Limitations of Science

Science is used to explain many things about the world. However, science cannot explain everything. Questions about emotions or values are not questions science can answer. A survey of peoples' opinions would not prove that these opinions are true for everyone. Scientists make predictions when they perform experiments. Then these predictions are tested and verified by using a scientific method.

Using Science—Technology

The words *science* and *technology* often are used in place of each other. However, the two words mean different things. Technology is the application of science to help people. For example, science methods are used when a chemist develops a new material. When this new material is used on the space shuttle, technology is applied.

Sometimes technology comes before science. For example, when the steam engine was invented, no one knew exactly how it worked. Scientists studied it and learned about the steam engine. This led them to discover new ideas about the nature of heat.

Not all technology produces something good. Some people question the benefits of some technology, such as nuclear technology. Learning more about science can help society make decisions about these issues.

☑ **Reading Check**

9. **Restate** What is a scientific law?

_____

_____

_____

_____

💡 **Think it Over**

10. **Think Critically** Give an example of when technology is an application of science.

_____

_____

_____

● After You Read

Mini Glossary

bias: what is expected changes how the results are viewed

constant: a condition in an experiment that does not change

control: a standard in an experiment against which the results are compared

dependent variable: the condition in an experiment that results from the changes made to the independent variable

experiment: an investigation that tests a hypothesis by collecting information under controlled conditions

hypothesis: an explanation for a question or a problem that can be tested

independent variable: in an experiment, the condition that is tested

model: anything that represents an idea, event, or object to improve understanding

scientific law: a statement about what happens in nature that seems true all the time

scientific method: the steps a scientist follows when performing an investigation

theory: an explanation that is supported by a large body of scientific evidence obtained from many different investigations and observations

variable: a quantity that can have more than a single value

1. Review the terms and their definitions in the Mini Glossary. Write a sentence explaining how bias can influence an experiment.

_____

_____

2. Complete the chart below to organize the information you have learned in this section. Put the following steps for scientific methods in order.

Analyze the data, Test the hypothesis, State the problem, Draw conclusions, Gather information, Form a hypothesis

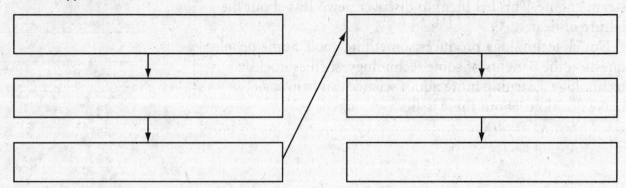

End of Section

The Nature of Science

section ② Standards of Measurement

● Before You Read

If someone asked you how wide your desk is, how would you measure it? Would you measure using inches, centimeters, feet, yards, or meters? Write why you selected this unit of measure.

_____

_____

_____

● Read to Learn

Units and Standards

A **standard** is an exact quantity that people agree to use to compare measurements. A standard is always exactly the same quantity when it is used anywhere in the world. Without standards, it is difficult to compare things that can be measured. Suppose you and a friend want to measure your desk but do not have a ruler. Instead, you use your hands as tools to measure the desk. If you each measure the desk using your own hands, will you both get the same measure? You can't be sure, because you don't know if your hands are the same size.

Precision and Accuracy

Imagine you are watching two people shoot arrows at a target. The first person shoots five bull's-eyes in a row. The second person never hits the bull's-eye, but all the arrows are close to one spot. What can you say about these two archers? The first person's aim was both precise and accurate. The second person's aim was only precise. Precision describes how closely measurements are to one another and how carefully measurements were made. Accuracy compares a measurement to a real or accepted standard. When you take measurements, the goal is to be both accurate and precise.

What You'll Learn

- the SI units and symbols for length, volume, mass, density, time, and temperature
- how to convert related SI units

Study Coach

Make an Outline Make an outline of the information in this section. Use each of the headings as part of the outline.

FOLDABLES™

🅑 Organize Information As you read this section, make the following Foldable to organize information about different types of measurements and units.

Distance

Volume

Matter

Time & Temperature

International System of Units

In 1960, an improvement was made to the metric system. This improvement is known as the International System of Units. This system is often abbreviated **SI** from the French *Le Systeme Internationale d'Unites*. The SI standards are accepted and used by scientists all over the world. Each type of SI measurement has a base unit. The base unit for length is the meter. The names and symbols for the seven base units are in the table below. All other SI units come from these seven base units.

1. Recognize Circle the base units that you have seen before.

SI Base Units		
Quantity Measured	Unit	Symbol
Length	meter	m
Mass	kilogram	kg
Time	second	s
Electric current	ampere	A
Temperature	kelvin	K
Amount of substance	mole	mol
Intensity of light	candela	cd

What are SI prefixes?

The SI system is easy to use because it is based on multiples of ten. A prefix is added to the name of the base unit to indicate how many multiples of ten it should include. For example, the prefix *kilo-* means 1,000. That means that one kilometer is equal to 1,000 meters. This also means that one kilogram equals 1,000 grams. The most commonly used prefixes are shown in the table below.

Picture This

2. Identify Which of the following is the smallest? (Circle your choice.)

a. decigram
b. nanogram
c. milligram
d. kilogram

Common SI Prefixes		
Prefix	Symbol	Multiplying Factor
kilo-	k	1,000
deci-	d	0.1
centi-	c	0.01
milli-	m	0.001
micro-	μ	0.000 001
nano-	n	0.000 000 001

How do you convert between SI units?

Sometimes quantities are measured using different units. Suppose a teacher has 1.3 L of water for a class experiment. She needs 125 mL to conduct the experiment. To determine if she has enough water, she must first find out how many mL of water she has.

Conversion Factors A conversion factor is used to change measurements from one unit to another. A conversion factor is a ratio that equals one. For a ratio to equal one, the numerator and denominator must have the same value. The numerator of a conversion factor should be the new unit. The denominator should be the old unit. For example, if you are converting liters to milliliters, use the following conversion factor.

$$\frac{\text{new unit}}{\text{old unit}} = \frac{1000 \text{ mL}}{1 \text{ L}}$$

To find out how much water she has in mL, the teacher multiplies the amount of water she has by the conversion factor.

$$1.3 \text{ L} \times \frac{1000 \text{ mL}}{1 \text{ L}}$$

$$1.3 \text{ L} \times \frac{1000 \text{ mL}}{1 \text{ L}}$$

$$1.3 \times 1000 \text{ mL} = 1{,}300 \text{ mL}$$

The teacher has 1,300 mL of water. That is enough for her experiment!

Measuring Distance

In science, the word *length* is used to describe the distance between two points. The SI base unit of length is the meter, m. A baseball bat is about 1 m long. Metric rulers and metersticks are commonly used to measure length. A meter is slightly longer than 1 yard, as shown by the meter-stick and the yardstick in the figure below.

Yard

Meter

Applying Math

3. **Convert Units** A length of rope measures 3,000 millimeters. How long is it in meters?

$$3{,}000 \text{ mm} \times \frac{1 \text{ m}}{1{,}000 \text{ mm}}$$

$$\frac{3{,}000}{1} \times \frac{1 \text{ m}}{1{,}000}$$

$$\frac{3{,}000 \text{ m}}{1{,}000} = \boxed{}$$

Picture This

4. **Circle** Circle the length by which the meterstick is longer than the yardstick.

How do you choose a unit of length?

When measuring distance, it is important to choose the proper unit. The unit you choose will depend on the object being measured. For example, you would measure the length of a pencil in centimeters (cm). The length of your classroom would be measured in meters. The distance from school to your house would be measured in kilometers (km). By choosing the best unit, you can avoid very large or very small numbers. It is easier to say something is 21 km rather than saying it is 21,000 m.

Measuring Volume

<u>Volume</u> is the amount of space an object fills. The volume of a rectangular solid, such as a brick, is found by multiplying its length, width, and height ($V = l \times w \times h$). If the sides of the brick were measured in centimeters, cm, the volume would be expressed in cubic centimeters, cm^3. When you multiply all three measurements, you multiply "cm" three times, once with each measurement. The result is cm^3. If you were trying to find out how much space there is in a moving van, you would measure the van using meters. Its volume would be expressed in cubic meters, m^3. Let's find the volume of this van.

First find the length, width, and height of the van.

> Length = 4 m
> Width = 2 m
> Height = 3 m

Substitute these values into the formula for finding volume.

$$V = l \times w \times h$$
$$= 4\ m \times 2\ m \times 3\ m$$
$$= (4 \times 2 \times 3)(m \times m \times m)$$
$$= 24\ m^3$$

The volume of the moving van is 24 m^3.

Copyright © Glencoe/McGraw-Hill, a division of The McGraw-Hill Companies, Inc.

Applying Math

5. **Define** In the calculations for finding the volume of the van, (m × m × m) is rewritten as m^3. The 3 in m^3 is called an exponent. What does an exponent represent?

_____

_____

_____

Applying Math

6. **Calculate** What is the volume of a brick that has a length of 20 cm, a width of 6 cm, and a height of 5 cm? Show your work.

_____

_____

_____

How do you measure the volume of a liquid?

Measuring the volume of a liquid in a container is different from measuring a solid object because the liquid does not have sides. To measure the volume of a liquid, you must use a container with a known capacity. Its measuring units should be marked. The most common units for expressing the volume of liquids are liters (L) and milliliters (mL). A milliliter is equal in volume to 1 cm³. So, the volume of 1 L equals 1,000 cm³. Look at food cans and bottles to see how these measurements are used. ☑

Measuring Matter

Mass is the measure of how much matter is in an object. **Matter** is anything that has mass and takes up space. A golf ball and a table-tennis ball are about the same size. The golf ball has more matter and mass, than the table-tennis ball.

What is density?

Another property of matter is density. The **density** of an object is the amount of mass in one cubic unit of volume of the object. You can find density by dividing an object's mass by its volume. Suppose an object weighs 10 g and has a volume of 2 cm³. The density of the object is 5 g/cm³.

If two objects are the same size and one object has a greater mass, it also has a greater density. This is because the more dense object has more mass in one cubic unit of volume than the other object has in one cubic unit of volume. The golf ball and the table-tennis ball have about the same volume. However, the golf ball has a greater mass. This means that the golf ball also has a greater density.

What are derived units?

The measurement for density, g/cm³, is a combination of SI units. A unit made by combining different SI units is called a derived unit. An SI unit multiplied by itself is also called a derived unit. For example, the liter, which is based on the cubic decimeter, and a meter cubed, m³, are both derived.

Measuring Time and Temperature

Sometimes scientists need to keep track of how long it takes something to happen or whether something heats up or cools down. These are measurements of time and temperature.

Time is the interval between two events. The unit of time in the SI system is the second (s). Seconds are usually measured with a clock or stopwatch.

✔ **Reading Check**

7. **Measure** What are the most common units for expressing the volume of liquids?

_____

_____

Applying Math

8. **Calculate** Suppose an object weighs 15 g and has a volume of 5 cm³. What is the density of the object?

_____

What's hot and what's not?

Sometimes scientists need to measure how much something heats up or cools down. Temperature is a measure of how much energy something has. Later, you will learn the scientific meaning of temperature.

What is Celsius?

Scientists use the Celsius (C) scale to measure temperature. This scale was designed to show that 0°C is the freezing point of water, and 100°C is the boiling point of water. The scale is divided into 100 equal divisions, or degrees, between the freezing point and the boiling point of water.

What is Fahrenheit?

The temperature measurement you are probably most familiar with is the Fahrenheit (F) scale. On this scale, water freezes at 32°F and boils at 212°F.

What is Kelvin?

The SI unit of temperature is kelvin (K). On the Kelvin scale, 0 K is called absolute zero. This is the coldest possible temperature. Absolute zero is equal to –273°C, which is 273° below the freezing point of water. The divisions on the Kelvin and Celsius scales are the same size. This makes it easy to convert between the two scales. Water freezes at 0°C. To convert to Kelvin, you add 273 to the Celsius temperature. So, on the Kelvin scale, water freezes at 273 K. Water boils at 100°C. So, on the Kelvin scale, water boils at 373 K. ☑

✔ Reading Check

9. Restate At what temperature does water freeze, and at what temperature does water boil on the Celsius scale?

_____

_____

_____

Picture This

10. Label each thermometer in the diagram with the name of its temperature scale.

● After You Read

Mini Glossary

density: the mass per unit volume of a material
mass: the measurement of the quantity of matter in an object
matter: anything that has mass and takes up space

standard: an exact quantity people agree to use to compare measurements
SI: International system of units; includes metrics
volume: the amount of space occupied by an object

1. Review the terms and their definitions in the Mini Glossary. In your own words, write a sentence that explains how mass affects an object's density.

_____

_____

2. Complete the chart below to organize the information from this section. For each unit include the unit's name, what it measures, and its symbol.

SI Base Units

3. **Study Coach** Think about what you have learned. Write a way to help you remember the meaning of volume.

_____

_____

End of Section

The Nature of Science

section ❸ Communicating with Graphs

What You'll Learn

- identify three types of graphs and tell how they are used
- how to distinguish between dependent and independent variables

◀ **Mark the Text**

Identify the Main Point
As you read through this section, highlight two main points about a line graph, a bar graph, and a circle graph.

FOLDABLES™

❻ **Compare and Contrast** Make the following Foldable. Use each section to draw an example of one type of graph.

Line Graphs

Bar Graphs

Circle Graphs

● Before You Read

Think about graphs that you have seen. Why do you think graphs are a useful way to display information?

_____

_____

_____

● Read to Learn

A Visual Display

A **graph** is a visual display of information or data. Often graphs are used to show results of experiments. Patterns in data show better in a graph than when data are in a table. The most common types of graphs are line graphs, bar graphs, and circle graphs. Each type of graph is used to show different types of data. Graphs make data easy to compare.

Line Graphs

Line graphs show how a relationship between variables changes over time. A line graph can show more than one event on the same graph. For example, a builder had a choice of three brands of thermostats to install in school classrooms. He tested each brand to find the most efficient one. The builder set each thermostat at 20°C. He checked the classroom temperature every five minutes for 25 minutes and recorded the data.

The graph on the next page shows the results of the test. The break in the vertical axis between 0 and 15 means that the numbers 1 to 14 have been left out. This is done so there is more room to spread the scale and the graph is easier to read. The horizontal line represents the 20°C setting and is the control in this experiment. The three lines (A, B, and C) show how quickly each thermostat reached the setting of 20°C.

Classroom Temperature

Picture This

1. **Interpret** Which thermostat reached its temperature setting the quickest?

_____

How do you construct line graphs?

There are several steps that you must follow to make a line graph. Always use the horizontal axis, or *x*-axis, for the independent variable. The vertical axis, or *y*-axis, is always used for the dependent variable. You should select a scale that makes your graph sensible and readable. For example, the scale on the *x*-axis would not make sense or be very readable if the units were by hours instead of minutes. All of lines A, B, and C that you see on this graph would be in the first tenth of the section ending with 5.

The data on a line graph are related. To draw the lines, you first plot points that show the relationship of the variables. For example, in line B, after five minutes the temperature was 17°C. You plot a point that aligns with 5 on the *x*-axis and 17 on the *y*-axis. Then plot the points representing the other data. Then you draw a line to connect the points.

You must also make sure that all the data use the same units. If some measurements were made in the Celsius scale and others were made in the Fahrenheit scale, the units must be converted to the same unit of measurement.

Bar Graphs

Bar graphs are useful for comparing data collected by counting. Each bar shows a number counted at a particular time. As on a line graph, the independent variable is plotted on the *x*-axis, and the dependent variable is plotted on the *y*-axis. The data are not related so the bars do not touch. ☑

✔ Reading Check

2. **Describe** What kind of data are bar graphs useful for?

_____

_____

Example The bar graph below shows classroom size on one day (January 20, 2004). The height of each bar shows the number of classrooms having a particular number of students. For example, three classrooms had 21 students. The number of students in one classroom is not related to the number of students in other classrooms. This is why the bars do not touch.

Classroom Size (January 20, 2004)

Copyright © Glencoe/McGraw-Hill, a division of The McGraw-Hill Companies, Inc.

Applying Math

3. Interpret Data How many classrooms had 26 students?

_____

Circle Graphs

A circle graph, or pie graph, is used to show how a certain quantity is divided into parts. The circle represents the whole and the segments, or slices, are the parts of the whole. The segments are usually given as percentages of the whole.

This circle graph below shows how much of each different kind of heating fuel is used in different buildings. According to the circle graph, gas is the heating fuel used the most.

When you use a graph, make sure that the conclusions are based on accurate information. Also select scales that help make the graph easy to read. In a circle graph, label the sections. In bar and line graphs, always label both the *x*-axis and the *y*-axis.

Applying Math

4. Use Models What do the percentages in a circle graph add up to? Why is this important?

_____

_____

_____

Heating Fuel Usage

Steam 25%

Gas 50%

Other 5%

Coal 10%

Electric 10%

● After You Read

Mini Glossary

graph: a visual display of information or data

1. Review the term *graph* and its definition in the Mini Glossary. Write a sentence that tells how a graph helps you understand data.

   _____

   _____

2. Complete the outline below. Tell what type of data each type of graph displays and then draw a small sample of each type of graph.

   Three Types of Graphs
   I.   Line Graph
        A. data type: _____
        B. example:

   II. Bar Graph
        A. data type: _____
        B. example:

   III. Circle Graph
        A. data type: _____
        B. example:

3. ◀ Mark the Text ▶ Think about what you have learned. Give an example of each type of data best suited for each type of graph.

   Line Graph: _____

   Bar Graph: _____

   Circle Graph: _____

End of Section

The Nature of Science

section ④ Science and Technology

What You'll Learn

- types of technology
- value of technology
- consumer effect on technology

● Before You Read

You use different kinds of technology every day. What do you think of when you hear the word *technology*?

_____

_____

FOLDABLES™

ⓓ Classify Make a two-tab Foldable like the one below. Take notes on how different forces shape technology.

Technology | Society

● Read to Learn

What is technology?

The terms *science* and *technology* mean very different things. Science is a process of exploration. Scientific processes are used to gain knowledge of and predict events in the natural world. Scientists often search for answers to human needs and problems. Scientists also search for scientific knowledge for the sake of learning new information.

Technology When scientific knowledge is used to solve a human need or problem, the result is called technology. **Technology** is the use of scientific knowledge of materials and processes to benefit people. An aspirin pill, a car, and a national highway system may not seem to have much in common. These things are all very different, but they are all examples of technology. Technology can be:

- any human-made object (such as a radio, computer or hammer),
- methods or techniques for making any object or tool (such as the method for making glass),
- skills needed to operate a human-made object (such as the skills needed to drive a car), or
- a system of people and objects used to do a task (such as the Internet, which is a system to share information). ☑

☑ Reading Check

1. **Define** What is technology?

_____

_____

_____

What are technological methods or techniques?

Just as writing instruments have changed over time, so have methods or techniques for doing many tasks. The methods for printing books have changed over time. Long ago people had to copy each page of a book by hand. This took many hours. Books were expensive, and only the very rich could pay for them. Now, books are created in many ways. Modern printing presses create most of the books today. Printing press technology printed your textbook. Books can be written on computers and are sometimes printed on computer printers. The printing press and computer technology save countless hours in producing books.

What are technological knowledge and skills?

Technology is not just new machines and new techniques. It is also the knowledge and the skills needed to do a task. For example, you need knowledge and special skills to use most computers. Any skill used to operate a complex machine is a technological skill.

What are technological systems?

A network of people and objects working together to do a job also is technology. The network is technology, the objects are technology, and the skills of the people are technology.

The Internet is a good example of this kind of technology. The Internet is a collection of computers and software that people with skills use to send and receive information. The airline industry is another example of a technological system. This industry is a collection of objects, methods, systems, and rules. An airport's planes, pilots, and computers create a technology system that moves people and goods.

Global Technological Needs

You just read how technology has changed over time. Today, not all parts of the world value the same technologies. The technology valued in the United States may not be valued in other parts of the world. Industrialized nations have different technology needs than developing nations.

Think it Over

2. **Infer** What is a technological system that you use each day?

_____

_____

Picture This

3. **Apply** Name two technologies that people in developing nations could benefit from.

_____

_____

What are the needs of developing countries?

In some parts of the world, people work hard for basic needs such as food, shelter, clean water, and basic health care. Consider a family living in rural Kenya. Most rural Kenyan families live without electricity or running water in their homes. Droughts or small amounts of rain can cause food shortages and famine. Rural Kenyans also have limited health care. Because of limited health care infections and diseases, such as HIV/AIDS, are severe problems in many developing nations. The life expectancy of rural Kenyans is mid-forties.

The most important technologies in developing countries help take care of basic needs. Technologies that provide enough clean water and food are very important. Basic health care technology such as clean bandages, the right medicine, and skilled doctors are also valued. Generally, rural people value different technologies than urban people because cities are more industrialized.

What technology is used in industrialized countries?

The United States is an industrialized country. Because of technology, industrialized countries have access to safe, clean water and food supplies. Most homes have running water and electricity. Good health care is available to treat people. The life expectancy of Americans is the late-seventies.

People in industrialized nations often value technology that helps provide a higher quality of life. Most homes in the United States have computers, telephones, and televisions. The people of industrialized nations value technologies like faster computers, compact discs, and safer cars.

Advances in medicine improve people's lives by curing illnesses and diseases. Some people spend money on medical procedures to improve their quality of life. Plastic surgery can remove wrinkles from a person's face. Eye surgery can improve a person's vision so he or she may not have to wear glasses.

How do developing nations contrast?

Both developing and industrialized nations value technology that helps supply basic human needs: clean water, enough food, and basic health care. However, the technology for basic needs may be very different in developing nations. For example, people in rural areas in some countries could get clean water if hand-pumps were installed in villages.

Think it Over

4. **Explain** What are two technologies that are valued in industrialized nations?

_____

_____

Social Forces that Shape Technology

Science and society are closely connected. <u>Society</u> is a group of people that shares similar values and beliefs. Discoveries in science and technology change society, and likewise society affects how new technologies develop. The politics, values, and economics of a society affect what new technologies develop.

How did attitudes toward cars change?

An example of the connection between a society's values and its technology is the development of the car in the United States. Over the last 100 years, the people of the United States have changed their attitudes toward cars.

When they were first invented, cars were expensive. Then technology was developed to mass-produce affordable cars. Many people were able to own cars because of the changes in technology and manufacturing. This increased the demand for fossil fuels like gasoline. Technology developed that made gas and oil more accessible and affordable. As more cars were sold, the demand for gas and oil increased.

Now gas prices are high and oil is becoming scarce, so people are beginning to buy more fuel-efficient cars. The automobile industry is responding with new technology, such as hybrid cars that use both electricity and gasoline.

Do consumers affect technology?

If people do not want a technology, they will not buy it. Companies are not likely to spend additional money to develop technology that consumers will not buy. On the other hand, if consumers buy a technology, companies will spend additional money to improve the technology. For example, if consumers continue to buy fuel-efficient cars, more money will be spent on improving that technology.

Do personal values affect technology?

People support new technologies that agree with their personal needs and values, directly and indirectly. For example, people support the development of technology indirectly when they choose to vote for a political candidate. Support for a candidate translates into support for the projects, technological and otherwise, that that candidate favors. People support the development of technology directly when they give money to organizations committed to a specific project, such as cancer research.

Think it Over

5. **Explain** How does buying a certain brand of CD player affect the development of CD players in general?

_____

_____

_____

Economic Forces that Shape Technology

Many factors influence whether money is spent on a technology, and many questions must be answered before the money is spent. The questions include who will buy the product, what the benefits of the technology are, and how much it will cost to make.

What is the influence of the federal government?

One way technology research and development is funded is through the federal government. Every year, Congress and the president place large amounts of money in the federal budget for scientific research and development. For example, Congress and the president decide how much money will be spent on new technologies in agriculture, defense, energy, and transportation. The money is given to companies and institutions in the forms of contracts and grants. ☑

What is the influence of private foundations?

Some scientific and technological research is supported with money from private foundations. Foundations raise money for many types of research, such as cancer and muscular dystrophy. Foundations raise money in many ways, including charity races, telethons, and benefit concerts. Many private foundations focus on research for specific scientific issues, such as treatment for Alzheimer's disease.

What is the influence of private industries?

Private industries also fund research and development of new technology. Like the federal government, private industries use a portion of their budget on research and development. Investing in new technology can make money for the company. Selling the new products or technologies they develop to consumers is one way companies make profits.

Responsible Technology

Humans can invent technology that may have an impact on other living things. Humans are part of many ecosystems on Earth, and it is important that new technology does not destroy the environment. Humans must think of both the positive and negative effects of their inventions.

How can technology affect the environment?

When the effects of technology are known, the benefits of the technology often are more important than the negative effects. For example, gasoline-powered cars give people safe, reliable transportation. The negative effects are the environmental problems created by the use of cars.

✔ **Reading Check**

6. **Identify** three things the federal government pays to develop.

_____

_____

_____

💡 **Think it Over**

7. **Conclude** Why is it important for humans to think about the positive and negative effects when they create technology?

_____

_____

_____

● After You Read

Mini Glossary

society: a group of people that share similar values and beliefs

technology: application of scientific knowledge to benefit people

1. Write a sentence describing one value or belief of your society.

_____

_____

2. Complete the outline to help you organize what you learned about technology.

I. Technology
    A. _____ objects
    B. Technological _____ or techniques
    C. Technological knowledge or _____
    D. _____ systems

II. Global Technological Needs
    A. Needs of developing _____
    B. Industrialized countries
    C. Differences in _____ nations
    D. Bioengineering and _____ supply

3. Complete the table below to help you organize the information you learned about forces that shape technology.

Force	How It Shapes Technology
Government	
Private Foundations	
Private Companies	
Consumers	
Citizens	

End of Section

section ❶ Describing Motion

What You'll Learn

- the difference between displacement and distance
- how to calculate an object's speed
- how to graph motion

Mark the Text

Identify the Main Point

Highlight the main point in each paragraph in this section. Highlight in a different color a detail or example that helps explain a point.

FOLDABLES™

Ⓐ Find Main Ideas As you read this section, make the following Foldable to help you identify the main ideas about motion.

● Before You Read

Have you ever been on a roller coaster? You can feel the steep drops and quick turns in your body. Write how it feels to travel up a steep hill slowly and then to go down the other side quickly.

_____

_____

● Read to Learn

Motion and Position

Distance and time are important in describing a race. The winner covers the distance in the shortest amount of time. It takes more time to run a 10-km race than to run a 5-km race because 10-km is a longer distance.

How are motion and position related?

You don't need to see something move to know that it has moved. Suppose you see a mail truck stopped next to a mailbox. When you look again later, the truck is farther down the street by a tree. You didn't see the truck move, but you know that motion has taken place because of the truck's new position.

What is reference point?

A reference point is needed to determine the position of an object. Motion occurs when an object goes from one reference point to another. In the example above, the mailbox was the first reference point for the mail truck. The tree was the second reference point.

After a reference point is chosen, a frame of reference can be created. A frame of reference is a coordinate system in which the position of the objects is measured. A coordinate system is like a map. The reference point is at the origin and each object's position can be described with its coordinates.

How are distance and displacement different?

Distance is how far something has moved. It is an important part of describing motion. In the 50-m dash, a runner travels 50 m between the start line and the finish line. The distance of the race is 50 m.

The SI unit of length or distance is the meter (m). Long distances are measured in kilometers (km). One kilometer is equal to 1,000 meters. Short distances are measured in centimeters (cm). One meter is equal to 100 centimeters.

Not all motion is in a straight line. In the figure, the runner jogged 50 m to the north. Then she turned around and jogged 30 m to the south. The total distance she jogged is 80 m. She is 20 m from the starting point. **Displacement** is the distance and direction of an object's position relative to the starting point. The runner's displacement is 20 m north.

Displacement = 20 m north of starting line
Distance traveled = 50 m + 30 m = 80 m

Speed

So far, motion has been described by the distance something has moved and by displacement from the starting point. You might also want to tell how fast something is moving. To do this, you need to know how far it travels in a certain amount of time. **Speed** is the distance an object travels per unit of time.

How is speed calculated?

The SI unit of distance is the meter (m). The SI unit of time is the second (s). So, in SI, speed is measured in meters per second (m/s). Sometimes it is easier to express speed in other units so that the numbers will not be very large or very small. Something that moves very quickly, such as a rocket, can be measured in kilometers per second (km/s). Very low speeds, such as geological plate movements, can be measured in centimeters per year (cm/y).

To calculate the speed of an object, divide the distance it traveled by the time it took to travel the distance. Here is a formula for calculating speed.

$$\text{speed} = \frac{\text{distance}}{\text{time}}, \text{ or } s = \frac{d}{t}$$

Picture This

1. **Describe** another way a runner could have a displacement of 20 m north.

_____

_____

_____

Applying Math

2. **Calculate** A train traveling at a constant speed covers a distance of 960 meters in 30 s. What is the train's speed? Show your work.

_____

_____

_____

What is motion with constant speed?

A speedometer measures the speed of a car. Suppose you look at the speedometer when you are riding on a freeway. The car's speed hardly changes. If the car is not speeding up or slowing down, it is moving at a constant speed. If you are traveling at a constant speed, you can measure your speed over any distance from millimeters to light years.

What is changing speed?

Usually, speed is not constant. The graph below shows how the speed of a cyclist changes during a 5-km ride. Follow the graph as the ride is described. As the cyclist starts off, his speed increases from 0 km/h to 20 km/h. Then he comes to a steep hill. He slows down to 10 km/h as he pedals up the hill. He speeds up to 30 km/h going down the other side of the hill. At the bottom, he stops for a red light. He speeds up when the light turns green. At the end of the ride, he slows down and then stops. The ride took 15 min.

What is average speed?

Look at the graph of speed for the bicycle trip. Sometimes the bicycle was moving fast, sometimes it was moving slowly, and sometimes it was stopped. How could you describe the speed of the whole bike ride? Would you use the fastest speed or the slowest?

Average speed describes the speed of motion when speed is changing. **Average speed** is the total distance traveled divided by the total time of travel. It is calculated using the relationships among speed, distance, and time.

The total distance the cyclist traveled was 5 km. The total time was 15 minutes, or $\frac{1}{4}$ h. You can write $\frac{1}{4}$ h as 0.25 h. The average speed for the bicycle trip can be found using a mathematical equation.

$$\text{average speed} = \frac{\text{total distance}}{\text{total time}} = \frac{5 \text{ km}}{0.25 \text{ h}} = 20 \text{ km/h}$$

Applying Math

3. **Describe** what is happening when the line on the graph is horizontal.

_____

_____

_____

Applying Math

4. **Find** the average speed in kilometers per hour of a race car that travels 260 km in 2 h.

$s = \frac{260}{2}$

$s =$ _____ km/h

What is instantaneous speed?

The speed shown on a car's speedometer is the speed at one point in time, or one instant. **Instantaneous speed** is the speed at one point in time. If an object is moving with constant speed, the instantaneous speed doesn't change. The speed is the same at every point in time. However, when a car speeds up or slows down, its instantaneous speed is changing. The speed is different at every point in time. ☑

Graphing Motion

A distance-time graph shows the motion of an object over time. The graph below shows the motion of three swimmers during a 30-min workout. The line for Mary is straight. This means that she swam at a constant speed. Her speed was 80 m/min for the whole 30 min. The line for Kathy is also straight. Kathy also swam at a constant speed.

Her speed was 60 m/min for the whole 30 min. Notice that the line for Mary is steeper than the line for Kathy. This is because Mary swam faster than Kathy. Mary was swimming faster, so she swam farther than Kathy in the same amount of time. The steepness of the line is called the slope. The slope of a line on the graph is the speed. A steeper slope means a greater speed. Mary was swimming faster than Kathy, so the slope of Mary's line is steeper than the slope of Kathy's line.

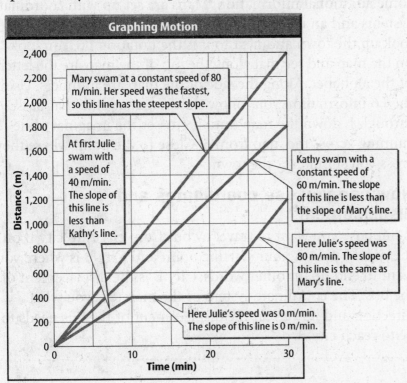

☑ Reading Check

5. **Describe** a car's instantaneous speed as it stops for a red light.

_____

_____

_____

FOLDABLES™

B Making Graphs Make the following Foldable using graph paper. Draw and explain a motion graph on the inside.

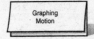

Applying Math

6. **Explain** why the slopes are different for each swimmer.

_____

_____

_____

How is changing speed graphed?

Julie's line is not straight. Julie did not swim at a constant speed for the entire 30-min practice. Julie swam at a constant speed for the first 10 min. Her rate was 40 m/min. Then she rested for 10 min. For this part of the practice, the line is horizontal. Her speed was 0 m/min, and the line has zero slope. She swam 800 m during the last 10 min. During this part of the practice, she swam as fast as Mary, so that part of her line has the same slope as Mary's line.

How do you draw a distance-time graph?

A distance-time graph plots data for distance and time. The distance traveled is plotted on the vertical axis. Time is plotted on the horizontal axis. Each axis has a scale, or a series of numbers, that covers the range of the data.

Mary swam the farthest during the practice, 2,400 m. So, the vertical scale must go to 2,400. The practice was 30 min long, so the horizontal axis must go to 30. Each axis is divided into equal parts. When each axis is finished, the data points are plotted on the graph. Then the data points for each swimmer are connected with a line.

What is a coordinate system?

When you use a map to locate a small town in a different state, you may not know where to look at first unless you have some additional information. Maps are set up with coordinate systems and an index of the towns on the map. Suppose you look up the town and next to it is the notation L6. You look on the map and see that along the top of the map are the letters of the alphabet. Along the side of the map are numbers. Using the L6 information you follow the imaginary line that passes through L down the map until it crosses the imaginary line running across the map from 6. Close to the point where they cross you should see the town.

How can you use coordinates and displacement?

Suppose you want to travel by bus from city A to city B on the map. City A is your starting point and City B is where you want to end up. In other words, City B is the displacement of the bus. The coordinate system of the map tells you what direction and how great the displacement of the bus needs to be to reach city B.

Copyright © Glencoe/McGraw-Hill, a division of The McGraw-Hill Companies, Inc.

Think it Over

7. Recall The vertical axis is called the *y*-axis. What is the horizontal axis called?

● After You Read

Mini Glossary

average speed: the total distance traveled in a unit of time

displacement: the distance and direction that something moved from a starting point

distance: a measure of how far an object has moved

instantaneous speed: the speed of an object at a one point in time

speed: the distance an object travels in an amount of time

1. Use the data below to make a distance-time graph. Be sure to include labels and scales.

Bike Ride

Time (min)	Distance (m)
15	150
30	300
45	450
60	600

End of Section

Motion

section ❷ Velocity and Momentum

What You'll Learn

- the difference between speed and velocity
- how to describe the motion of two objects relative to each other
- how to calculate an object's momentum

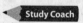

Study Coach

Outlining As you read the section, make an outline of the important information in each paragraph.

💡 Think it Over

1. **Describe** how the velocity of an escalator that is going up is different from the velocity of one that is going down at the same speed.

_____

_____

_____

● Before You Read

If an object is moving at 2 m/s towards a porcelain vase, do you think it will have enough momentum to break the vase? Explain.

_____

_____

_____

● Read to Learn

Velocity

Suppose you hear that there is a storm nearby. The storm is traveling at a speed of 20 km/h and is 100 km east of your location. Do you have enough information to know whether the storm will reach you?

Knowing only the speed of the storm is not enough to find the answer. The speed tells you how fast the storm is moving. To find the answer, you also need to know the direction the storm is moving. In other words, you need to know the velocity of the storm. **Velocity** is the speed of an object and the direction it is moving.

How do speed and direction affect velocity?

To help you understand velocity, think about two escalators. Some escalators go up and others go down. One escalator is moving upward at the same speed that another escalator is moving downward. The two escalators are going the same speed, but they are going in different directions. They each have a different velocity. If the second escalator were moving upward, both elevators would have the same velocity.

Velocity depends on both speed and direction. Because of this, an object moving at a constant speed will have a changing velocity if it changes direction. A race car on an oval track has a constant speed. But, as the race car goes around the track, the direction in which the car is moving changes. This means that the velocity of the car is changing. An object has constant velocity if neither the speed nor direction it is moving changes. The light from a laser beam travels at a constant velocity.

Motion of Earth's Crust

Some motion is so slow that it is hard to see. The surface of Earth doesn't seem to change from year to year. But if you look at geological evidence of Earth over 250 million years, you will see that large changes have occurred.

According to the theory of plate tectonics, the continents are moving constantly over Earth's surface. The movement is shown in the figure below. The changes are so slow that we do not notice them.

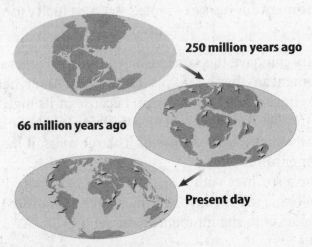

250 million years ago

66 million years ago

Present day

How do continents move?

Earth is made of layers. The top layer is the crust. The layer below the crust is the upper mantle. Together, the crust and the top part of the upper mantle are the lithosphere. The lithosphere is broken into huge sections, called plates.

Below the lithosphere, the layers are like putty. The plates slide slowly on these soft layers. The moving plates cause geological changes on Earth. These changes include the formation of mountain ranges, earthquakes, and volcanic eruptions.

The plates move very slowly. The speed of the plates is measured in centimeters per year. In California, two plates are sliding past each other along the San Andreas Fault. The average relative speed of the two plates is about 1 cm per year.

Picture This

2. **Describe** the movement of the continents over the past 250 million years.

_____

_____

_____

Think it Over

3. **Explain** Why is the speed of the movement of Earth's plates measured in centimeters per year instead of in meters per second?

_____

_____

_____

Think it Over

4. **Compare** Which has more momentum, a car traveling at 12 km/h or a bicycle traveling at the same speed? Explain why.

_____

_____

_____

Applying Math

5. **Apply** What is the momentum of a bicycle with a mass of 18 kg traveling at 20 m/s?

_____

_____

_____

Relative motion

Not all motion is as obvious as the mail truck's motion. When you are sitting still in a desk, you appear not to be moving. However, you are moving. You are not moving in relation to your desk or school building. You are moving in relation to the Sun because you are sitting on Earth.

Relative motion means that one thing moves in relation to another thing. Earth is moving in space in relation to the Sun. The Sun is the reference point for Earth's motion.

Momentum

A moving object has a property called momentum. Momentum is related to how much force is needed to change an object's motion. The **momentum** of an object is the product of its mass and its velocity. Momentum can be found using the following equation. The symbol p represents momentum. The unit for momentum is kg•m/s.

$$\textbf{momentum (kg•m/s)} = \textbf{mass (kg)} \times \textbf{velocity (m/s)}$$
$$p = mv$$

Two cars can have the same velocity. But the bigger car has more momentum, because it has more mass. An archer's arrow can have a large momentum because of its high velocity, even though its mass is small. A walking elephant may have a low velocity, but because of its large mass, it has a large momentum.

Suppose a sprinter with a mass of 80 kg has a speed of 10 m/s. What is the sprinter's momentum? Substitute the known values into the momentum equation.

$$p = mv$$
$$= (80 \text{ kg})(10 \text{ m/s})$$
$$= 800 \text{ kg•m/s}$$

The sprinter's momentum is 800 kg•m/s.

How are force and momentum related?

Recall that acceleration is the difference between final and initial velocity, divided by the time. Also recall that the net force on an object is its mass times its acceleration. When you combine these two relationships, you get the following equation.

$$F = \frac{(mv_f - mv_i)}{t}$$

● After You Read

Mini Glossary

momentum: the product of an object's mass and velocity, represented by the formula $p = mv$

velocity: the speed of an object and the direction of its motion

1. Review the terms and their definitions in the Mini Glossary. Explain how speed is related to an object's momentum.

_____

_____

_____

_____

2. Choose one of the question headings in the Read to Learn section. Write the question in the space below. Then write your answer to the question.

Question:

Answer:

_____

_____

3. ◀ **Study Coach** Think about what you have learned in this section. How did identifying the main point and supporting details of each paragraph help you learn the new material?

_____

_____

_____

End of Section

chapter 2 Motion

section ❸ Acceleration

What You'll Learn

- how acceleration, time, and velocity are related
- the different ways an object can accelerate
- how to calculate acceleration
- the similarities and differences between straight line motion, projectile motion, and circular motion

Outlining As you read the section, make an outline of the important information in each paragraph.

ⅅ Construct a Venn Diagram Make the following trifold Foldable to compare and contrast the characteristics of acceleration, speed, and velocity.

● Before You Read

Describe what happens to the speed of a bicycle as it goes uphill and downhill.

_____

_____

_____

● Read to Learn

Velocity and Acceleration

A car sitting at a stoplight is not moving. When the light turns green, the driver presses the gas pedal and the car starts moving. The car moves faster and faster. Speed is the rate of change of position. **Acceleration** is the rate of change of velocity. When the velocity of an object changes, the object is accelerating.

Remember that velocity is a measure that includes both speed and direction. Because of this, a change in velocity can be either a change in how fast something is moving or a change in the direction it is moving. Acceleration means that an object changes it speed, its direction, or both.

How are speeding up and slowing down described?

When you think of something accelerating, you probably think of it as speeding up. But an object that is slowing down is also accelerating. Remember that acceleration is a change in speed. A car that is slowing down is decreasing its speed. It is also accelerating, because its speed is changing.

Imagine a car being driven down a road. If the speed is increasing, the car has positive acceleration. When the car slows down, the speed decreases. The decreasing speed is called negative acceleration. In both cases, the car is accelerating, but one acceleration is positive and one is negative.

Acceleration has direction, just like velocity. In the figure below, both cars are accelerating because their speeds are changing. When a car's acceleration and velocity are in the same direction, the speed increases and the acceleration is positive. Car A has positive acceleration. When a car is slowing down, the acceleration and velocity are in opposite directions. The acceleration is negative. Car B has negative acceleration.

Car A

Acceleration

Velocity

Car B

Acceleration

Velocity

Does changing direction affect acceleration?

A change in velocity is either a change in an object's speed or its direction. When a moving object changes direction, its velocity changes and it is accelerating. The speed of a horse moving around on a carousel remains constant, but it is constantly changing direction. So, the horse is accelerating.

How do you calculate acceleration?

Acceleration is the rate of change in velocity. To calculate acceleration, you first find the change in velocity. To find change in velocity subtract the beginning velocity of an object from the velocity at the end of its movement. Beginning velocity is called the initial velocity, or v_i. Velocity at the end is called the final velocity, or v_f. ✓

$$\text{change in velocity} = \text{final velocity} - \text{initial velocity}$$
$$= v_f - v_i$$

If motion is in a straight line, the change in speed can be used to calculate the change in velocity. The change in speed is the final speed minus the initial speed.

To find acceleration, divide the change in velocity by the length of time during which the velocity changed.

$$\text{acceleration } (a) = \frac{\text{change in velocity}}{\text{time } (t)} \quad \text{or} \quad a = \frac{(v_f - v_i)}{t}$$

The SI unit for velocity is meters per second (m/s). To find acceleration, velocity is divided by the time in seconds (s). So, the unit for acceleration is m/s^2.

Picture This

1. **Describe** the acceleration of the cars in each figure.

_____

_____

✔ Reading Check

2. **Use Variables** Write what v_f and v_i mean.

_____

_____

_____

Applying Math

3. **Calculate** Suppose a bird takes off from a tree and flies in a straight line. It reaches a speed of 10 m/s. What is the change in the bird's velocity?

_____

_____

How is positive acceleration calculated?

How is the acceleration of an object that is speeding up different from that of an object that is slowing down? The acceleration of an object that is speeding up is always positive. The acceleration of an object that is slowing down is always negative.

Suppose an airplane is sitting at the end of a runway. The plane takes off and moves down the runway. It takes 20 s for the plane to travel from one end of the runway to the other. When the airplane reaches the end of the runway, it is traveling 80 m/s. The airplane is traveling in a straight line. The initial velocity of the plane is 0 m/s and the final velocity of the plane is 80 m/s. The time is 20 seconds. The acceleration for the plane can be calculated as follows:

$$a = \frac{(v_f - v_i)}{t} = \frac{(80 \text{ m/s} - 0 \text{ m/s})}{20 \text{ s}} = 4 \text{ m/s}^2$$

The airplane is speeding up as it goes down the runway. The final speed is greater than the initial speed. The acceleration is positive.

How is negative acceleration calculated?

Now imagine a skateboarder moving in a straight line. The skateboarder is moving at a speed of 3 m/s. It takes the person 2 s to come to a stop. The initial velocity is 3 m/s and the final velocity is 0 m/s. The total time is 2 seconds. The calculation for the skateboarder's acceleration is as follows:

$$a = \frac{(v_f - v_i)}{t} = \frac{(0 \text{ m/s} - 3 \text{ m/s})}{2 \text{ s}} = -1.5 \text{ m/s}^2$$

The skateboarder is slowing down. The final speed is less than the initial speed. The acceleration has a negative value.

Amusement Park Acceleration

Roller coasters are exciting rides. People who design roller coasters use the laws of physics. The steep drops and loops of steel roller coasters give the rider large accelerations. When riders move down a steep hill, gravity will cause them to accelerate toward the ground. When riders go around a sharp turn, they are also accelerated. This acceleration makes them feel as if a force were pushing them toward the side of the car.

Copyright © Glencoe/McGraw-Hill, a division of The McGraw-Hill Companies, Inc.

---

4. **Explain** Why is the acceleration of an object moving at a constant velocity always 0?

_____

_____

_____

 Think it Over

5. **Think Critically** A car that is slowing down is still moving forward. Why is this considered negative acceleration?

_____

_____

_____

● After You Read

Mini Glossary

acceleration: the rate of change in velocity

centripetal acceleration: acceleration toward the center of a curved or circular path

1. Review the term and its definition in the Mini Glossary. Explain why a change in velocity affects acceleration.

_____

_____

_____

2. Complete the chart to organize information about how acceleration is calculated.

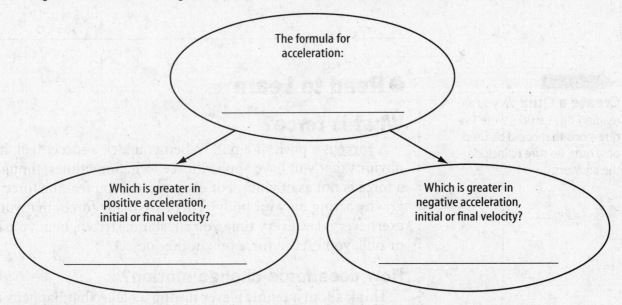

The formula for acceleration:

_____

Which is greater in positive acceleration, initial or final velocity?

_____

Which is greater in negative acceleration, initial or final velocity?

_____

3. ◀ Study Coach As you read the section, you made an outline describing the points covered in each paragraph. How did you decide what to write as the major points in your outline?

_____

_____

_____

End of Section

Forces and Newton's Laws

section ❶ Forces

What You'll Learn

- how force and motion are related
- what friction is between objects
- the difference between mass and weight

● Before You Read

When you hit a baseball with a bat, you apply a force that moves another object. Think of three more examples from sports in which force is used to move something. Write your examples on the lines below.

_____

_____

_____

Study Coach

Create a Quiz As you are reading this section, write five questions that could be used on a quiz. Be sure to include the answers.

FOLDABLES

A Find Main Idea As you read, use quarter or half sheets of paper to help you identify the main ideas about forces and friction.

Forces

Friction

● Read to Learn

What is force?

A **force** is a push or a pull. When you kick a soccer ball, it is obvious that you have applied force to it. Sometimes, though, a force is not as obvious. For example, do you feel the force of gravity acting on your body? There are many forces that you exert every day. Every time you sit, stand, stretch, bend, push, or pull, you exert a force on another object.

How does force change motion?

Think about a tennis player hitting a ball. What happens to the motion of the ball when the racket hits it? The force of the racket hitting the ball makes the ball stop. Then the force makes the ball move in a different direction.

What are balanced forces?

Not all forces change velocity. Suppose two students are pushing on opposite sides of a box. As shown in the figure, if both students are pushing with an equal force, the box does not move. When two or more forces act on an object at the same time, the forces combine. This is called a **net force**. When two students are pushing with the same force, but in opposite directions, the two forces cancel each other. The net force on the box is zero. Forces on an object that are equal in size and opposite in direction are called balanced forces.

$\longrightarrow$ + $\longleftarrow$ = 0

Net force = 0

Picture This

1. **Describe** Why are the forces in the figure said to be balanced?

_____

_____

_____

What is the result of unbalanced forces?

Not all forces in opposite directions balance each other. Think about two students pushing on the opposite sides of a box. What happens if one student pushes harder than the other as in the picture? The forces are not balanced, and the box will move in the direction of the larger force. The student who is pushing harder will move the box in the direction of the force. The net force that moves the box is the difference between the two forces.

$\longrightarrow$ + $\longleftarrow$ = $\longrightarrow$

Net force = $\longrightarrow$

$\longrightarrow$ + $\longrightarrow$ = $\longrightarrow$

Net force = $\longrightarrow$

Suppose both students push on the same side of the box as in the second figure. The students are both exerting force in the same direction. Again the forces are unbalanced. The forces combine, or add together, because they are exerted in the same direction. Again, net force is equal to both forces added together. ☑

Friction

Suppose you give a skateboard a push with your hand. If the net force on the skateboard is zero, it will continue to move with a constant speed. Does it move with constant speed after it leaves your hand? No, the skateboard slows down and stops. There must be a net force acting on it. So, what force is acting? The force that brings the skateboard to a stop is friction. **Friction** is the force that opposes the sliding motion of two surfaces that are touching each other. The amount of friction between two surfaces depends on two things. The first is the kind of surfaces that are touching. And the second is the amount of force that is pressing the surfaces together.

☑ Reading Check

2. **Define** What is the net force?

_____

_____

_____

_____

💡 Think it Over

3. **Infer** Which would have more friction, a car traveling down a street or a toy remote-control car traveling on the same street? Why?

_____

_____

_____

What causes friction?

Something that seems very smooth, like polished metal, may actually be rough. You can see the dips and bumps on the surface when you look at it under a microscope. If two surfaces are pushed tightly together, welding, or sticking, happens where the bumps touch each other. These places are called microwelds. *Micro-* means "very small." Friction is caused by the microwelds that form where the surfaces are in contact.

What makes things stick together?

The greater the force pushing two surfaces together, the stronger the microwelds will be. More of the surface bumps will be touching each other. You can see microwelds in the figure below. To move one surface over another, a force must be applied to break these microwelds.

Force

Surfaces

Microwelds form where bumps come into contact.

More force

Same two surfaces

More force presses the bumps closer together.

What is static friction?

Suppose you fill a cardboard box with books, but it is too heavy to lift. You try to push it, but you cannot get it to move. This means there must be a force that is working against you. The force is friction caused by microwelds between the bottom of the box and the floor. This is called static friction. Static friction is the force that keeps two surfaces at rest from sliding across each other. In this case, your push is not large enough to break the microwelds between the box and floor. So, the box will not start moving. ☑

Picture This

4. **Observe** Do surfaces touch more when there is less force or greater force?

_____

_____

_____

☑ Reading Check

5. **Complete** the sentence: The force that keeps two surfaces at rest from sliding over each other is

_____.

What is sliding friction?

Suppose you ask your friend to help move the box. As you and your friend push together, the box starts to move. The two of you have applied enough force to break the microwelds between the box and the floor. If you stop pushing, the box will stop moving. Another force, sliding friction, is opposing the motion of the box. Sliding friction is the force that works against the motion of two surfaces that are sliding across each other. Sliding friction is caused by the microwelds breaking and then forming again as the box slides along the floor.

What is rolling friction?

A wheel digs into the surface that it is rolling over. This causes both the wheel and the surface to be deformed. There is static friction in the deformed area where the wheel and surface are in contact. This makes a frictional force called rolling friction. Rolling friction is the frictional force between a rolling object and the surface it rolls on.

Have you ever seen a car's wheels spinning in the mud or snow? There is sliding friction between the spinning wheels and the surface. Sand or gravel on the surface increases the sliding friction, so the wheels will stop slipping.

What is gravity?

At this moment you are exerting an attractive force on everything around you—your desk, your classmates, and even the planet Jupiter, millions of kilometers away. This attractive force acts on all object with mass and is called gravity. **Gravity** is an attractive force between any two objects that depends on the masses of the objects and the distance between them. Gravity increases as the mass of either object increases, or as the objects move closer to each other. You can't feel the force between you and your desk because the force is weak. Earth is both close and has a large enough mass that you can feel its gravitational attraction. The Sun has more mass, but it is too far away to exert an attraction you can feel.

If the mass of either of the objects increases, the gravitational force between them increases.

If the objects are closer together, the gravitational force between them increases.

6. **Compare and Contrast** How do static friction and sliding friction differ?

_____

_____

_____

Picture This

7. **Interpret Illustrations** What two things can cause the gravitational force between objects to increase?

_____

_____

8. Determine What do the 1 and 2 in m_1m_2 mean?

_____

_____

_____

The Law of Universal Gravitation

The law of universal gravitation lets us find the force of gravity between any two objects if their masses and the distance between them are known. Here is an equation for the law:

$$\textbf{gravitational force} = (\text{constant}) \times \frac{(\text{mass 1}) \times (\text{mass 2})}{\textbf{distance}^2}$$

$$F = G\frac{m_1m_2}{d^2}$$

In this equation, G is the constant called the universal gravitational constant. A constant is an amount that never changes. The variable d is the distance between the two masses, m_1 and m_2.

Why is gravity called a long-range force?

The law of universal gravitation states that the gravitational force between two masses decreases as the distance between them increases. Suppose the distance between two objects increases from 1 m to 2 m. Then the gravitational force between them becomes one fourth as large. If the distance increases from 1 m to 10 m, the gravitational force between the objects is one hundredth as large. No matter how far apart two objects are, the gravitational force between them never completely goes to zero. Gravity is called a long-range force. No matter how long the distance is between two objects, gravity never disappears.

Because contact between objects is not required, gravity is sometimes discussed as a field. A **field** is a region of space that has a physical quantity (such as a force) at every point. All objects are surrounded by a gravitational field. Earth's gravitational field is strongest near Earth and becomes weaker as the distance from Earth increases.

Is there a difference between weight and mass?

Weight and mass are not the same. Weight is a force. Mass is a measure of the amount of matter in an object. But weight and mass are related. The greater an object's mass, the stronger the gravitational force between the object and Earth. **Weight** is the gravitational force on an object. Your weight on Earth is greater than your weight on the Moon because Earth has a greater mass than the Moon.

Weight is found using the following equation. The letter g represents the acceleration of gravity.

$$\textbf{weight (N)} = \textbf{mass (kg)} \times \textbf{acceleration of gravity (m/s}^2\textbf{)}$$

$$W = mg$$

● After You Read

Mini Glossary

field: a region of space that has a physical gravity (such as force) at every point

force: a push or pull

friction: the force that opposes the sliding motion of two surfaces that are touching each other

gravity: an attractive force between any two objects that depends on the mass of the objects and the distance between them

net force: when two or more forces act on an object at the same time, the forces combine

weight: the gravitational force exerted on an object

1. Review the definition of air resistance in the Mini Glossary. Give an example of air resistance.

_____

_____

2. Complete the chart below to organize the different kinds of friction.

Describe static friction.	Give an example of static friction.
Describe sliding friction.	Give an example of sliding friction.
Describe rolling friction.	Give an example of rolling friction.

3. ◀ **Study Coach** As you read this section, you wrote the main idea of each paragraph in your own words. How did you decide what the main idea of each paragraph was?

_____

_____

Forces and Newton's Laws

section 2 Newton's Laws of Motion

What You'll Learn

- how inertia is related to Newton's first law
- how acceleration is calculated using Newton's second law
- how forces relate to Newton's third law

Mark the Text

Identify the Main Point

Highlight the main point in each paragraph. Use a different color to highlight examples of the main point.

FOLDABLES

B Find Main Idea As you read, use quarter or half sheets of paper to help you identify the main ideas about inertia and Newton's laws of motion.

Inertia

Newton's First Law of Motion

The Second Law of Motion

The Third Law of Motion

Before You Read

Imagine stepping off the bed of a pick up truck onto a platform. As you step you push against the truck bed. Now imagine stepping off of a skateboard with the same effort. What happens to the truck and the skateboard as you step off of them?

_____

_____

Read to Learn

Newton's Laws of Motion

You have seen many examples of how forces change the motion of objects. Forces change the motion of objects in specific ways. The British scientist Sir Isaac Newton stated rules that describe the effects of forces on the motion of objects. These rules are known as Newton's laws of motion. They apply to all objects in motion, from billiard balls to planets orbiting the Sun.

What is Newton's first law of motion?

Newton's first law of motion states that an object moving at a constant velocity keeps moving at that velocity unless an unbalanced net force acts on it. An object at rest stays at rest unless a net force acts on it. Newton's first law of motion is sometimes called the law of inertia.

Inertia and Mass

Inertia is the tendency of an object to resist any change in its motion. If an object is moving, it will keep moving in a straight line at a constant speed until a force changes its direction or speed. Inertia causes an object to resist changes in direction and speed. A dirt bike will move in a straight line with a constant speed unless a force acts on it. A force can turn the wheel and change the direction. Another force, friction, can slow the speed of the bike. An object that is not moving also has inertia. It will remain motionless until an unbalanced force causes it to move.

Why can different objects have different inertia?

The inertia of an object is related to its mass. The greater the mass is of an object, the greater its inertia is. Remember, mass is the amount of matter in an object. A bowling ball has much greater mass than a table-tennis ball does. Therefore, a bowling ball has greater inertia than a table-tennis ball. Imagine hitting a bowling ball with a table-tennis paddle. The bowling ball would not move very much. Imagine hitting a table-tennis ball with the same amount of force. The table-tennis ball would move quite easily. ☑

Newton's Second Law

<u>Newton's second law of motion</u> states that the net force acting on an object causes the object to accelerate in the direction of the net force. To find the acceleration of an object, you need to know its mass and the net force that is acting on it. The acceleration of an object can be found using the following equation. ☑

$$\text{acceleration (meters/second}^2) = \frac{\text{net force (in newtons)}}{\text{mass (in kilograms)}}$$

$$a = \frac{F_{net}}{m}$$

Suppose you are pushing a friend on a sled. The mass of your friend and the sled together is 70 kg. The net force on the sled is 35 N (kg m/s^2). Find the acceleration of the sled.

$$a = \frac{35N}{70\,kg} = 0.5 \text{ m/s}^2$$

The acceleration of the sled is 0.5 m/s^2.

Force, Mass, and Acceleration

You read about Newton's first law of motion. The law says that the motion of an object changes only if an unbalanced force acts on it. Newton's second law of motion describes the relationship between the acceleration of an object, its mass, and the forces exerted on it.

How are force and acceleration related?

How are throwing a ball as hard as you can and tossing it gently different? When you throw the ball hard, you exert a much greater force on the ball than when gently tossing it. Therefore, the thrown ball has a greater velocity when it leaves your hand than it does when you toss it. The ball you throw hard has a greater change in velocity. And, the change in velocity happens over a shorter period of time.

✔ **Reading Check**

1. **Compare** the inertia of a car to the inertia of a bicycle.

_____

_____

_____

_____

Since acceleration is the change in velocity divided by the time it takes for the change to happen, throwing a ball harder has a greater acceleration because its velocity changes in a shorter time.

Are mass and acceleration related?

You would think that throwing a baseball and a softball as hard as you can, would give them the same speed. It does not because the softball has more mass. If the time to throw both balls is equal, the difference in mass affects the velocity, and therefore the acceleration. Thus mass and acceleration are related.

How can you find net force?

Using an object's mass and acceleration, you can use the formula for Newton's second law to calculate the net force. For example, a tennis ball touches the racket during a serve for only a few thousandths of a second. The ball accelerates in a very short time. Suppose the ball leaves the racket at a speed of 100 km/h and accelerates at 5,000 m/s². The ball's mass is 0.06 kg. Use this formula to find the force placed on the ball by the racket.

$$F_{net} = ma$$
$$F_{net} = (0.006 \text{ kg})(5,000 \text{ m/s}^2)$$
$$F_{net} = 300 \text{ kg m/s}^2 \quad F_{net} = 300 \text{ N}$$

Newton's Third Law

If you push against a wall while wearing in-line skates, you will roll backwards. The action of pushing against the wall produced a reaction—moving backward. This is an example of Newton's third law of motion.

Newton's third law of motion describes action and reaction. It states that when one object applies a force on a second object, the second object applies a force on the first object that is equal in strength and opposite in direction. In other words "to every action force there is an equal and opposite reaction force."

How do action and reaction happen?

When a force is applied in nature, a reaction force occurs at the same time. When you jump on a trampoline, you exert a downward force on the trampoline. At the same time, the trampoline exerts an equal upward force on you. ☑

How do action and reaction make you move?

The action and reaction forces are equal. So how do you move by walking if each time you push on the ground, Earth pushes back with an equal force? The forces are acting on objects that have different masses. Earth has more mass than you do. Even though the forces are equal, their net force is not equal. Unequal net forces determine the direction you move.

Copyright © Glencoe/McGraw-Hill, a division of The McGraw-Hill Companies, Inc.

Applying Math

3. Demonstrate Show how multiplying both sides of the acceleration equation, $a = \dfrac{F_{net}}{m}$, by m results in the net force equation $F_{net} = ma$.

_____

_____

_____

☑ Reading Check

4. Explain what is meant by action and reaction.

_____

_____

_____

● After You Read

Mini Glossary

inertia: the tendency of an object to resist any change in its motion

Newton's first law of motion: an object moving at a constant velocity keeps moving at that velocity until acted on by an unbalanced force

Newton's second law of motion: an object's acceleration is in the same direction as the net force and is equal to the net force divided by the mass

Newton's third law of motion: when an object exerts a force on a second object, the second object exerts an equal but opposite force on the first

1. Review the terms and their definitions in the Mini Glossary. Choose two terms that are related and write a sentence using both terms.

   _____

   _____

   _____

2. Place the number of the law of motion in the blank next to the example of the law given.

   1. _____ $F_{net} = ma$

   2. _____ the seven ball is struck by the cue ball

   3. _____ an air-filled balloon is released

   4. _____ you push on a box of books

3. ◀ Mark the Text You highlighted the main parts of each paragraph as you read. How did you decide what the main points were?

   _____

   _____

   _____

End of Section

Forces and Newton's Laws

section ❸ Using Newton's Laws

● Before You Read

Imagine riding on a sled, or in a wagon, or perhaps a school bus that stops quickly or suddenly. What happens to your body if you are not expecting the sudden stop? Describe the motion and feeling on the lines below.

_____

_____

_____

Study Coach

Make a List of 10 things that help you use Newton's laws.

FOLDABLES

❸ **Relate** Make this foldable to help you connect the ideas of air resistance, terminal velocity, and free fall.

● Read to Learn

What happens in a crash?

The law of inertia can explain what happens in a car crash. Imagine that a car traveling at 50 km/h crashes head-on into a stationary object. It crumples and comes to a stop within 0.1 s. Passengers who are not wearing safety belts continue to move forward at the same speed the car was traveling. Within about 0.02 s after the car stops, the unbelted passengers slam into the dashboard, steering wheel, windshield, or the backs of the front seats. They are still moving at the original speed of 50 km/h. This is about the same speed they would reach falling from a three-story building.

How do safety belts help?

A person who is wearing a safety belt will be attached to the car. The person will slow down as the car slows down. Safety belts also prevent people from being thrown out of cars. About half of the people who die in car crashes would survive if they wore safety belts. Thousands of others would suffer fewer serious injuries.

What about air bags?

Since the late1990s all new cars have air bags. Designed to protect the front seat occupants from injury during collisions, air bags inflate and then deflate again in one twenty-fifth of a second. Some newer cars even have air bags for side collisions. Air bags and seat belts save lives.

Earth's Gravitational Acceleration

If you dropped a marble and a bowling ball at the same time, which one would hit the ground first? Suppose the effects of air resistance were small enough to be ignored. When all forces except gravity can be ignored, a falling object is said to be in free fall. If there were no air resistance, all objects near Earth's surface would fall with the same acceleration. The marble and the bowling ball would hit the ground at the same time.

The acceleration of an object in free fall is about 9.8 m/s². This acceleration is sometimes called the acceleration of gravity. It is given the symbol g. The force of Earth's gravity on a falling object is the object's mass times the acceleration of gravity. This can be expressed by the following equation.

force of gravity (N) = **mass** (kg) × **acceleration of gravity** (m/s²)
$$F = mg$$

You can use this equation to find the gravitational force on a sky diver with a mass of 60 kg.

$$F = mg = (60 \text{ kg})(9.8 \text{ m/s}^2) = 588 \text{ N}$$

The gravitational force on the sky diver is 588 N.

Air Resistance

Objects falling toward Earth are being pulled downward by the force of gravity. **Air resistance** is a force that opposes the movement of objects through the air. Air resistance is similar to friction. You can feel air resistance on your face when you ride your bike very fast.

Like friction, air resistance acts in the direction opposite to the object's motion. In the case of a falling object, air resistance pushes up as gravity pulls down. If there were no air resistance, only gravity would affect falling objects. All objects would fall at the same rate.

Air resistance causes different objects to fall with different accelerations and different speeds. The amount of air resistance depends on an object's size and shape. Imagine dropping two identical plastic bags. One is crumpled into a ball and the other is spread out, resembling a parachute. When the bags are dropped, the crumpled bag falls faster than the spread-out bag. The downward force of gravity on both bags is the same. But, the upward force of air resistance on the crumpled bag is less. So, the net downward force on the crumpled bag is greater.

Applying Math

1. **Calculate** Find the weight of a 50-kg person on Earth. Remember, $g = 9.8$ m/s². Show your work.

_____

_____

_____

2. Compare Measure the length of each arrow beside the sky diver. Compare the length of the gravity arrow to the lengths of the air resistance and net force arrows.

_____

_____

_____

What is terminal velocity?

Imagine an object falling toward Earth. As the object falls, gravity causes it to accelerate. This causes the upward force of air resistance to increase. At some point, the upward force of air resistance becomes equal to the downward force of gravity. This means that the net force is zero. So, from this point on, the object will fall at a constant speed. This constant speed is called terminal velocity. **Terminal velocity** is the highest speed a falling object can reach. The terminal velocity of an object depends on its size, shape, and mass.

Look at the figure below. The air resistance force on an open parachute is much greater than the air resistance on the sky diver with a closed parachute. With the parachute open, air resistance increases. This makes the terminal velocity of the sky diver become small enough that the sky diver can land safely.

Gravity = air resistance + net force

Weightlessness and Free Fall

Suppose you are standing on a scale in an elevator that is not moving. The scale would record your normal weight. But, what would happen if you were standing on the scale and the elevator were falling rapidly? If gravity was the only force acting on you and the scale, you would be in **free fall**. The dial would say you have zero weight, even though your weight has not changed. You might say you are weightless.

You may have seen pictures of astronauts floating inside a space shuttle. They are experiencing weightlessness. The astronauts are not really weightless. However, they do weigh less because they are farther from Earth's gravitational pull. ☑

Reading Check

3. Explain Why do astronauts in the space shuttle weigh less?

_____

_____

_____

Centripetal Force

Look at the path of the ball as it travels through the curved tube in the figure below. When the ball enters a curve, even if its speed does not change, it is accelerating. This is because its direction is changing. When the ball goes through a curve, the change in its direction is toward the center of the curve. Acceleration toward the center of a curved path is called **centripetal acceleration**.

The ball has centripetal acceleration. So, according to the second law of motion, the direction of the net force on the ball must be toward the center of the curved path. The net force exerted toward the center of a curved path is called a **centripetal force**. An object that moves in a circle is doing so because a centripetal force is acting on it in a direction toward the center. The centripetal force is the force exerted by the walls of the tube on the ball.

How does centripetal force depend on traction?

When a car rounds a curve, a centripetal force must be acting on the car to keep it moving in a curved path. If it does not, the car will slide off the road. This centripetal force is the frictional force, or traction, between the tires and the road. The road may be slippery because of rain or ice. As tires get older, they get smoother and their traction decreases. If either or both of these situations occur, the car may slide in a straight line and not follow the curve. This is because the centripetal force, traction, is not strong enough to keep the car moving around the curve.

Can gravity be a centripetal force?

Imagine swinging a yo-yo on a string above your head. The string places centripetal force on the yo-yo. The string keeps the yo-yo moving in a circle. In the same way, Earth's gravitational pull places centripetal force on the Moon and keeps it moving in a nearly circular orbit.

Picture This

4. **Trace** the path of the ball at right with a pen. Try to move the pen at a constant speed.

Think it Over

5. **Describe** a real-world situation that involves centripetal force.

_____

_____

Momentum

A moving object has a property called momentum. Momentum is related to how much force is needed to change an object's motion. The **momentum** of an object is the product of its mass and its velocity.

What is the law of conservation of momentum?

Momentum can be passed from one object to another. When a cue ball hits a group of balls that are motionless, the cue ball slows down and the other balls move. The momentum that the group of balls gained equals the momentum that the cue ball lost. But the total momentum of all the balls before and after the collision is the same. Total momentum has not been lost, nor has new momentum been created. This is an example of the **law of conservation of momentum**. If a group of objects applies forces on each other, their total momentum does not change.

What happens when objects collide?

In a game of pool, suppose one ball is moving in one direction, and another ball moving the same direction strikes it from behind. The ball that is struck will continue to move in the same direction, but more quickly. The striking ball has given it more momentum in the same direction.

What if two balls of equal mass are moving toward each other with the same speed? They would have the same momentum, but in opposite directions. If the balls collided, each would reverse direction, and move with the same speed as before the collision.

What is rocket propulsion?

Suppose you are standing on skates holding a softball. You exert a force on the softball when you throw it. Newton's third law says the softball exerts a reaction force on you. This force pushes you in the direction opposite the softball's motion.

Rockets use this same principle to move. In a rocket engine, burning fuel produces hot gases. The rocket engine applies a force on the gases and causes them to escape out of the back of the rocket. By Newton's third law, the gases apply a reaction force on the rocket and push it in the opposite direction.

💡 **Think it Over**

6. **Recognize Cause and Effect** In a game of pool, why will the balls eventually stop after a collision?

_____

_____

_____

● After You Read

Mini Glossary

air resistance: a friction-like force opposing objects moving in air

centripetal force: a force exerted toward the center of a curved path

free fall: when gravity is the only force acting on an object

law of conservation of momentum: if a group of objects exerts forces only on each other, their total momentum does not change

terminal velocity: the maximum velocity an object will reach while falling through air

1. Review the terms and their definitions in the Mini Glossary. Describe a real-world example of the law of conservation of momentum.

_____

_____

_____

2. Match the terms with the correct statements. Put the letter of the statement in Column 2 on the line in front of the term it matches in Column 1.

Column 1

_____ 1. rocket propulsion

_____ 2. momentum

_____ 3. conservation of momentum

_____ 4. Newton's third law of motion

Column 2

a. To every action force there is an equal and opposite reaction.

b. Momentum cannot be created or destroyed.

c. the product of a moving object's mass and velocity

d. An engine applies a force on hot gases and the gases apply a force in the opposite direction.

3. ◄ **Study Coach** You created a list of things that help you use Newton's laws. How do these examples teach you Newton's laws?

_____

_____

_____

End of Section

Work and Energy

section ① Work and Machines

What You'll Learn

- about work
- how machines make work easier
- what mechanical advantage and efficiency are

Mark the Text

Selective Underlining As you read this section, underline the key idea in each paragraph.

✓ Reading Check

1. **Identify** What two things have to happen for work to be done?

_____

_____

_____

● Before You Read

Name two things in your life that require work. Name two things you would use to do work.

_____

_____

● Read to Learn

What is work?

When people talk about work, they might mean doing a job, earning money, or using a computer. But scientists have a different definition of work. <u>Work</u> is the energy transferred when a force makes an object move a certain distance.

How do you know if you are doing work?

Press your hand into the surface of your desk as hard as you can. Because you are not making the desk move, you are not doing work. If you push against any object and it doesn't move, then you haven't done any work.

How is motion related to work?

In order for work to be done on an object, two things have to happen. One is that the applied force must make the object move. The second is that the object must move in the direction of the applied force.

When you pick up a stack of books, you are doing work. Your arms apply an upward force, and the books move upward in the direction of the force. But if you hold the books in your arms without moving them, your arms are not doing work. You are still applying an upward force to keep the books from falling. However, because the books are not moving, no work is being done. ✓

How is direction related to work?

If you hold a stack of books in your arms and begin to walk forward, are your arms doing work? Your arms may get tired, but they are not doing work. The upward force exerted by your arms is not causing the book to move forward. The force exerted by your arms is upward, but the books are moving horizontally. Your body and legs move the books horizontally.

When you carry the books, the force your arms put on the books is perpendicular, or 90°, to the force of direction you move them. Therefore your arms do zero work on the books. When the force is perpendicular to the motion, the work done is zero. If force and motion are parallel, the work done is the force times the distance. But what if the force is not perpendicular or parallel? For these in between directions, the work done is less than the force times the distance, but more than zero. Work is done only when force is applied.

What is a machine?

A **machine** is a device that changes the force or increases the motion from work. In other words, it makes work easier. Imagine trying to lift a car by yourself. It might seem impossible, but with a machine called a lever you could arrange it to lift it all by yourself.

A lever increases the force on an object. Other machines can increase the speed at which you do work. You can move faster if you run instead of walk, but if you use a machine called a bicycle, you can travel faster still.

Machines might be very simple, like a knife or a doorknob. If you cut your food with a knife you are using a simple machine. A **simple machine** does work with just one movement of the machine.

Machines might be more complex like a bicycle, or a cuckoo clock. Pulling a pulley to wind the clock starts gears and levers moving inside the clock. A **compound machine** is a combination of two or more simple machines (the pulley, gears, and levers).

Making Work Easier

Machines make work easier in several ways. Machines can increase the force on an object. For example, a screwdriver increases the force you apply to turn a screw. Machines also can increase the distance over which a force is applied, as in a leaf rake. Other machines make work easier by changing the direction of an applied force. A pulley changes a downward force to an upward force. ☑

Copyright © Glencoe/McGraw-Hill, a division of The McGraw-Hill Companies, Inc.

☑ **Reading Check**

3. **Describe** How does a machine affect work?

_____

_____

_____

Efficiency

Some of the energy put into a machine is changed into thermal energy by friction. For this reason, the work produced by a machine, the output work, is always less than the work put into it. **Efficiency** is a measure of how much of the work put into a machine, the input work, is changed into useful output work by the machine. A high-efficiency machine produces less thermal energy from friction, so more of the input work is changed to useful output work. Work input and output is measured in joules. A joule (J) is one newton (1N) of force applied over one meter (1m).

How is efficiency calculated?

To calculate the efficiency of a machine, divide the output work by the input work. Efficiency is written as a percentage.

$$\text{efficiency (\%)} = \frac{\text{output work (in joules)}}{\text{input work (in joules)}} \times 100\%$$

$$\text{efficiency} = \frac{W_{\text{out}}}{W_{\text{in}}} \times 100\%$$

You can calculate the efficiency of a machine with W_{in} of 50 J and W_{out} of 40 J.

$$\text{efficiency (\%)} = \frac{40J}{50J} \times 100\%$$

$$\frac{40J}{50J} = 0.8 \times 100\% = 80\%$$

The efficiency of the machine is 80 percent.

An ideal machine does not produce friction, so the output work equals the input work. The efficiency of an ideal machine is 100 percent. Because a machine produces friction, the output work is less than the input work. Therefore, the efficiency of a machine is always less than 100 percent.

How can machines be made more efficient?

Machines can be made more efficient by reducing friction. Adding oil or grease to the surfaces that rub together fills the gaps between the surfaces. This allows the surfaces to slide across each other more easily.

Copyright © Glencoe/McGraw-Hill, a division of The McGraw-Hill Companies, Inc.

Applying Math

4. Determine What is the efficiency of a machine whose work output is 4 joules, and whose work input is 10 joules?

_____

_____

_____

 Think it Over

5. Analyze Explain to a classmate how oil makes a machine more efficient.

Your force

Distance jack moves

Force exerted by jack

Distance you push

Picture This

6. **Infer** Look at the picture. Color the arrow red for the distance the jack is pushed. Use blue to color the arrow for the distance the car moves. Why is one arrow longer than the other?

_____

_____

How are machines useful?

How are machines useful if a machine's output work is always less than its input work? Machines change the way work is done. They can increase speed, change the direction of a force, or increase force.

How do machines increase speed?

A bicycle is a machine that increases speed. By pushing on levers, a person can travel faster by bike than on foot. To increase speed the force must be reduced. Round wheels help reduce the friction and the force, but riding uphill still requires the person to apply a large force to overcome gravity.

Why change the direction?

Some machines change the direction of the force being used. A splitting maul changes a downward force into outward forces to split wood. A pulley changes a downward force into an upward force to lift an object. A screwdriver changes the force applied to it into a rotating force to tighten or loosen a screw. Think about how a crowbar works. How does it change direction?

How can you increase the force?

If you have ever changed a tire, you know that a car jack, like the one in the photo, is a useful machine. You could never lift the car by yourself, but the jack provides the force to do the job. It increases the force by decreasing the speed. The upward force of the jack is greater than the downward force that you apply to the jack handle. The upward speed of the jack is also slower than the downward speed of the handle.

We describe the effectiveness of a machine—increasing speed, changing direction of force, or increasing force—as the mechanical advantage of the machine. **Mechanical advantage** is the ratio of the output force to the input force. ☑

Think it Over

7. **Describe** an example of work that is made easier by increasing distance.

_____

_____

_____

_____

Reading Check

8. **Define** What is mechanical advantage?

_____

_____

_____

⬤ After You Read

Mini Glossary

compound machine: a combination of two or more simple machines

efficiency: measure of how much work put into a machine is changed into useful output by the machine

machine: a device that makes doing work easier

mechanical advantage: the ratio of the output force to the input force

simple machine: does work with just one movement of the machine

work: the application of a force through a distance

1. Review the terms in the Mini Glossary. Write one sentence using the terms *efficiency* and *mechanical advantage*.

   _____

   _____

   _____

2. Complete the chart below to organize information about the three ways machines make work easier.

3. ◄ **Mark the Text** How did underlining the key ideas help you to learn?

   _____

   _____

   _____

End of Section

Work and Energy

chapter **4**

section ② Describing Energy

● Before You Read

Energy comes in many forms. List as many types of energy as you can think of on the lines below.

_____

_____

What You'll Learn

■ the different forms of energy
■ how energy can be stored

● Read to Learn

What is energy?

Changes are taking place all around you all the time. For example, lightbulbs are heating the air around them, and the wind may be blowing leaves in the trees. Even you are changing as you breathe, blink, or move around in your desk.

Does change require energy?

Energy is the ability to cause change. Anything that causes change must have energy. For example, you use energy when you comb your hair. You also use energy when you move between classes, open a book, or write with a pen. You even use energy when you yawn or sleep. ☑

Because work may be done when change occurs, we can describe energy as the ability to do work. Energy then, like work, can be measured in joules. A tennis racket doing 250 joules (J) of work on a tennis ball transfers 250 J of energy to the ball.

What are systems?

Both the tennis racket and the tennis ball are systems. A **system** is anything you can draw a boundary around. A system may be a single object such as a tennis ball, or a group of objects such as the solar system of planets. When one system does work on a second system, energy is transferred from the first system to the second system. You are a system that shares energy with the tennis racket system and the solar system.

Study Coach

State the Main Ideas As you read this section, stop after each paragraph and write the main idea of what you read in your own words.

✔ **Reading Check**

1. **Define** What is energy?

_____

_____

Picture This

2. **Illustrate** In the space below, draw another representation of one of the types of energy shown in the figure.

What are some different forms of energy?

Turn on an electric light, and a dark room becomes brighter. Turn on your CD player, and sound comes through the headphones. In both of these cases, energy moves from one place to another. These changes are as different from each other as from the change caused by the racket hitting the ball. Energy has many different forms. Some forms are electrical, chemical, radiant, and thermal.

The figure above shows some forms of energy and some objects associated with these forms of energy.

How is energy like money?

Suppose you have $100. You could have one hundred-dollar bill, two fifty-dollar bills, 100 one-dollar bills, or 10,000 pennies. You could start with the $100 in one form and change it into another form. But, no matter what form it is in, it is still $100. Likewise energy is the same no matter what form it is in. Energy from the Sun that warms you and the energy from the food you eat are just different forms of the same thing—energy! ☑

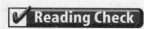

✔ Reading Check

3. **Explain** how energy and money are alike.

_____

_____

_____

Kinetic Energy

When you think of energy, you might think of moving objects. An object in motion, like the tennis ball, does have energy. **Kinetic energy** is the energy a moving object has because of its motion. The kinetic energy of a moving object depends on the object's mass and speed. You can find the kinetic energy of an object using the following equation.

$$\textbf{kinetic energy (joules)} = \frac{1}{2}\textbf{mass (kg)} \times [\textbf{speed (m/s)}]^2$$

$$KE = \frac{1}{2}mv^2$$

FOLDABLES™

Ⓐ Find Main Ideas Fold a piece of paper into 12 sections and label. Fill in the main ideas about kinetic and potential energy.

The Nature of Energy	Define	Examples of...	Calculate
Kinetic Energy			
Potential Energy			

A jogger whose mass is 60 kg is moving at a speed of 3 m/s. Use the equation to find the jogger's kinetic energy.

$$KE = \frac{1}{2}mv^2$$

$$= \frac{1}{2}(60 \text{ kg})(3 \text{ m/s})^2$$

$$= \frac{1}{2}(60 \text{ kg})(9 \text{ m}^2/\text{s}^2)$$

$$= \frac{1}{2}(540 \text{ kg} \cdot \text{m}^2/\text{s}^2)$$

$$= 270 \text{ kg} \cdot \text{m}^2/\text{s}^2$$

The kinetic energy of the jogger is 270 J.

Potential Energy

An object with energy does not have to be moving. Objects that are at rest have stored energy. For example, an apple that is hanging from a tree has stored energy. If the apple stays in the tree, it will keep its stored energy because of its position above the ground. If the apple falls to the ground, a change happens. Because the apple can cause change, it has energy. Stored energy due to position is called **potential energy**. The stored energy of position, potential energy, will change to energy of motion, kinetic energy, when the apple falls.

What is elastic potential energy?

Energy can be stored in other ways, too. Suppose you take a rubber band and stretch it. If you let the rubber band go, it will fly through the air. Where did this kinetic energy come from? The stretched rubber band had something called elastic potential energy. **Elastic potential energy** is energy stored by an object that can stretch or shrink, like a rubber band or a spring. When you let the rubber band fly through the air, its elastic potential energy becomes kinetic energy.

What is chemical potential energy?

Where does your body get the energy to make it move? The food that you eat each day has stored energy. To be more exact, food's energy is stored in chemical bonds between atoms. Natural gas stores energy in the same way. Energy stored in chemical bonds is called **chemical potential energy**. In natural gas, energy is stored in the bonds that hold the carbon and hydrogen atoms together. This energy is released when the gas is burned.

Applying Math

4. Explain Look at the problem about the jogger. Why does (3 m/s)² become (9 m²/s²)?

_____

_____

_____

Think it Over

5. Apply Describe a situation where potential energy becomes kinetic energy.

_____

_____

_____

FOLDABLES

B Compare and Contrast Make the following Foldable to compare and contrast the properties of different types of potential energy.

What is gravitational potential energy?

Anything that can fall has stored energy called gravitational potential energy. <u>**Gravitational potential energy**</u> (GPE) is energy that is stored by objects that are above Earth's surface. The apple in the tree has GPE. The GPE of an object depends on two things—the object's mass and its height above the ground. Gravitational potential energy can be found using the following equation. ☑

gravitational potential energy (J) = **mass** (kg) × **acceleration of gravity** (m/s²) × **height** (m)

$$GPE = mgh$$

Look at the vase near the bottom of the bookcase. If you fill the vase with water, you increase its GPE by increasing its mass. If you move the vase to a higher shelf, you also increase its GPE by increasing its height. The gravitational potential energy of an object can increase if you change its mass or move the object higher above the ground.

If two objects are at the same height, then the object with the greater mass will have more GPE. If two objects have the same mass, the one that is higher above the ground will have the greater GPE.

Applying Math

7. Interpret In the formula *GPE = mgh,* which symbols represent the constants and which symbols represent the variables?

_____

_____

Picture This

8. Reasoning Which probably has a greater GPE, a feather on a high shelf or a large book on the next shelf down? Explain your reasoning.

_____

_____

_____

● After You Read

Mini Glossary

chemical potential energy: energy stored in chemical bonds

elastic potential energy: energy stored by things that stretch or shrink

energy: the ability to cause change

gravitational potential energy: energy stored by things that are above Earth

kinetic energy: energy in the form of motion

potential energy: energy stored in a motionless object

system: anything around which you can imagine a boundary

1. Review the terms and their definitions in the Mini Glossary. Explain the difference between kinetic energy and potential energy.

_____

_____

2. Complete the chart below. Fill in the first column with the three kinds of potential energy. Fill in the second column with an example of something that stores each type of potential energy.

Kinds of Potential Energy   Examples

3. Study Coach Think about what you have learned. How did you decide what was the main idea of each paragraph?

_____

_____

End of Section

section ⊖ Conservation of Energy

What You'll Learn
- how power is related to energy
- how energy is conserved

● Before You Read

The motion of objects appear to change all the time. Imagine a person swinging on a swing. Explain the person's motion while swinging.

_____

_____

_____

Mark the Text

Identify the Main Point
Highlight the main point in each paragraph. Highlight in a different color a detail or example that helps explain the main point.

✔ Reading Check

1. **Explain** What does the law of conservation of energy state?

_____

_____

_____

● Read to Learn

The Law of Conservation of Energy

As a batted ball speeds up or slows down, its kinetic and potential energy are always changing. But the amount of mechanical energy always stays the same. The kinetic and potential energy continually change form back and forth and no energy is destroyed.

This is true for all forms of energy. Energy can change from one form to another, but the total amount of energy never changes. Another way to say this is that energy is conserved. The **law of conservation of energy** states that energy cannot be created or destroyed. This means that the total amount of energy in the universe is always the same. It just changes from one form to another. ✔

You might have heard the phrase *energy conservation* before. Conserving energy means reducing the need for energy so we use fewer energy resources such as coal and oil. The law of conservation of energy and conserving energy are not the same. The law of conservation of energy describes what happens to energy as it changes form or is transferred from one object to another.

Changes Between Kinetic and Potential Energy

Recall that a rubber band has elastic potential energy. When a stretched rubber band is let go, the potential energy is changed into kinetic energy. **Mechanical energy** is the total amount of potential and kinetic energy in a system. The mechanical energy of the rubber band is the total of its potential energy and its kinetic energy at any one time. Mechanical energy comes from where an object is and the movement of the object.

Does mechanical energy change?

What happens to the mechanical energy of an object as its potential energy is changed into kinetic energy? Look at the apple in the tree below. It has gravitational potential energy because Earth is pulling down on it. The apple does not have kinetic energy while it hangs in the tree. The apple's gravitational potential energy and its mechanical energy are the same.

GPE + Kinetic Energy = Mechanical Energy

Look at the apple in the second figure. As it falls, the apple loses height, so its gravitational potential energy becomes less. As the velocity of the apple increases, its kinetic energy increases. The potential energy of the apple is being changed to kinetic energy. However, the mechanical energy will not change. The form of the energy changes, but the total amount of the energy remains the same.

GPE + Kinetic Energy = Mechanical Energy

FOLDABLES

C Collect Information
Make note cards from two half-sheets of paper as shown. Write on each note card what you learn about mechanical energy and the law of conservation of energy.

💡 **Think it Over**

2. **Explain** Does the mechanical energy of an apple falling from a tree change? Explain.

_____

_____

_____

Picture This

3. **Interpret** What do the arrows in the figures mean?

_____

_____

_____

How does energy change in projectile motion?

Energy changes also occur during projectile motion. During projectile motion, an object moves through the air in a curved path. Look at the figure below. As the ball leaves the bat, it has mostly kinetic energy. As the ball rises, its gravitational potential energy becomes greater, but its kinetic energy becomes less due to decreasing speed. As the ball falls, its gravitational potential energy becomes less, but its kinetic energy becomes greater due to increasing speed. However, the total mechanical energy of the ball does not change as the ball moves through the air.

What happens to the energy of a swing?

When you ride on a swing, part of the fun is the way you feel just as you drop from the highest part of the swing's path. Energy is constantly changing during this ride on a swing. The push that gets you moving is kinetic energy. As the swing rises, you lose speed. This means the kinetic energy is changing into gravitational potential energy. At the top of the path, the GPE is its greatest and the kinetic energy is at its lowest. As the swing starts its downward path and its speed increases, the GPE changes into kinetic energy. At the bottom of the swing's path, the kinetic energy is greatest and the GPE is at its lowest. As you swing back and forth, kinetic and potential energy are constantly changing back and forth.

How does friction affect energy?

Suppose you are swinging on a swing. If you stop pumping and no one is pushing you, you will soon stop swinging. It would seem that the mechanical (kinetic and potential) energy of the swing is lost. Wouldn't this go against the law of conservation of energy?

If the energy of the swing decreases, then the energy of some other object must increase by the same amount. What object has an energy increase? With every motion, the swing's ropes or chains rub on their hooks, and air pushes on the rider. Friction and air resistance cause the temperature of the hooks and air to increase a little. The mechanical energy is not destroyed. It is changed into thermal energy. So, the total amount of energy stays the same—it just changes forms.

How can electrical energy change?

Every day you use items that change energy from electricity into light, sound, thermal energy, radiant energy, and motion. In the figure you can see that electricity is changed into light and thermal energy in a lightbulb.

Light energy out

Thermal energy out

Electrical energy in

How can chemical energy change?

Chemical energy can be changed into kinetic energy. A fuel, such as gasoline, stores energy in the form of chemical potential energy. Cars and buses usually run on gasoline. As shown in the figure, an electrical spark causes a small amount of fuel in the engine to burn. This changes the chemical potential energy stored in the gasoline molecules into thermal energy. The thermal energy heats up gases and they expand. The expanding gases cause parts of the car to move. The moving parts have kinetic energy. Chemical energy has been changed into thermal energy and then into kinetic energy. ☑

Spark plug fires   **Gases expand**

Picture This

6. Relate Name other items that change electrical energy into thermal energy.

_____

_____

☑ Reading Check

7. List three forms of energy that are involved in the running of a car engine.

_____

_____

_____

Picture This

8. Identify Circle the moving parts in the figure. What type of energy is this?

_____

How does your body store and use energy?

Your body requires energy to stand still. The law of conservation of energy applies to the chemical and physical changes that are going on in your body. Your body stores potential energy in the form of fat and other chemical compounds. The potential energy is the fuel for processes such as the beating of your heart, digesting of food, and moving muscles.

Green plants convert radiant energy into chemical potential energy and store it as sugar.

Power

Suppose you and a friend want to see who can push a box of books up a ramp faster. Each box of books weighs the same. You each push your box the same distance, but your friend pushes her box faster than you do. You both do the same amount of work on the boxes because the force and the distance are the same. The only difference is how long it takes to do the work. Your friend has more power than you because she can do the work faster than you can. **Power** is the rate at which energy is converted. Something has more power if it can do the same amount of work in less time.

How do you calculate power?

To calculate power, divide the amount of work done by the time it takes to do the work. Below is the power equation.

$$\textbf{Power (in watts)} = \frac{\textbf{energy (in joules)}}{\textbf{time (in seconds)}}$$

$$P = \frac{E}{t}$$

The SI unit for power is the watt (W). One watt equals one joule of work done in one second. The watt is a very small unit. Power usually is given in kilowatts. One kilowatt equals 1,000 watts.

Find the power of a machine that can do 5,000 joules of work in 20 seconds. Use the power equation.

$$P = \frac{W}{t}$$

$$P = \frac{5000 \text{ joules}}{20 \text{ seconds}}$$

$$P = 250 \text{ watts}$$

The power of the machine is 250 watts.

Copyright © Glencoe/McGraw-Hill, a division of The McGraw-Hill Companies, Inc.

Think it Over

9. **Apply** How are work and power related?

_____

_____

_____

After You Read

Mini Glossary

law of conservation of energy: energy may change from one form to another, but the total amount of energy never changes.

mechanical energy: the total amount of potential and kinetic energy in a system

power: the rate at which energy is converted

1. Review the terms and their definitions in the Mini Glossary. Describe a real-world example in which the amount of potential and kinetic energy change, but the total amount of mechanical energy stays the same.

_____

_____

_____

_____

2. A 20 watt lightbulb uses 20 joules of energy every second. A person expends 50 watts of energy per stair when climbing up stairs. How long could you light the lightbulb if the energy used to climb 20 stairs were converted to electricity?

Power Equation

$$\text{Power (in watts)} = \frac{\text{Energy (in joules)}}{\text{time (in seconds)}}$$

$$P = \frac{E}{t}$$

3. **Mark the Text** Think about what you have learned. How did highlighting the main points and details or examples help you learn the new material?

_____

_____

_____

End of Section

section ① Temperature and Heat

What You'll Learn

- what temperature is
- how thermal energy depends on temperature
- how thermal energy and heat are related
- calculate the change in thermal energy

Study Coach

Make an Outline Make an outline of the information you learn about in this section. Use the headings in the reading as a starting point. Include the boldface vocabulary terms in your outline.

☑ **Reading Check**

1. **Explain** Why is the temperature of hot tea higher than the temperature of iced tea?

_____

_____

_____

● Before You Read

You wake up in the morning and get out of bed. Does the floor feel cold or warm on your bare feet? On the lines below, write a sentence that compares how it feels to step on a bare floor and on a rug on a cold morning.

_____

_____

● Read to Learn

Temperature

The words *hot* and *cold* describe temperature. When you heat water on a stove, the temperature of the water increases. You've often heard the word *temperature*, but what is it?

What does motion have to do with matter?

All matter is made of tiny particles—atoms and molecules. The particles in all matter are constantly moving. The movement of the particles is random. This means that the particles move in all directions at different speeds. Because the particles are moving, they have kinetic energy. The faster they move, the more kinetic energy they have. When an object is hot, the particles move faster. As it cools, the particles move more slowly. Particles move faster in hot objects than in cooler objects.

What is temperature?

Temperature and kinetic energy are related. **Temperature** is a measure of the average value of the kinetic energy of the particles in an object. As the temperature of something increases, the average speed of its particles increases. The temperature of hot tea is higher than the temperature of iced tea because the particles in the hot tea are moving faster. Recall that in SI units, temperature is measured in kelvins (K). The Celsius scale is used more commonly than the Kelvin scale. One kelvin is the same size as one Celsius degree. ☑

Thermal Energy

When you take butter out of the refrigerator, it is cold and hard. If you let it sit at room temperature for a while, it gets warmer and softer. The air in the room is at a higher temperature than the butter. So, particles in the air have more kinetic energy than the butter particles. Air particles and butter particles bump into each other. The collisions transfer kinetic energy from the air to the butter. The butter particles begin moving faster and the temperature of the butter increases.

Potential Energy to Kinetic Energy Particles also have potential energy that can be changed into kinetic energy. How can particles have potential energy? Think about a ball. Earth exerts an attractive gravitational force on the ball. When you hold the ball above your head, the ball and Earth are separated. This gives the ball potential energy. The particles in a substance also exert attractive forces on each other. They have potential energy when they are separated. ☑

As particles move farther apart, their potential energy increases. As they move faster, their kinetic energy increases. <u>Thermal energy</u> is the sum of the kinetic and potential energy of all the particles in an object. The figure shows that if either potential or kinetic energy increases, thermal energy increases.

Kinetic energy increases

The kinetic energy increases as the particles move faster.

Speed increases

The potential energy increases as the particles move farther apart.

Potential energy increases

Separation increases

How are thermal energy and temperature related?

When the temperature of an object increases, the average kinetic energy of its particles increases. When the average kinetic energy of its particles increases, the object's thermal energy increases. Therefore, the thermal energy of an object increases as its temperature increases.

FOLDABLES™

A **Relate** Make the following Foldable to tell how temperature and mass are related to thermal energy.

Thermal Energy

Temp.   Mass

✔ Reading Check

2. Identify What gives particles in a material potential energy?

_____

_____

_____

_____

Picture This

3. Use Scientific Illustrations In the space below, sketch the particles in the figure if separation increased even more.

How are thermal energy and mass related?

Suppose you have a glass of water and a beaker of water. They are at the same temperature. The beaker contains twice as much water as the glass. The average kinetic energy of the water molecules is the same in both containers, since they are the same temperature. But there are twice as many water molecules in the beaker as there are in the glass. So the total kinetic energy of the water molecules in the beaker is twice as large as that of the water molecules in the glass. The water in the beaker has twice as much thermal energy as the water in the glass has. If the temperature doesn't change but the mass of the object increases, the thermal energy in the object increases.

Heat

Have you ever noticed that your chair felt warm when you sat down? You could tell that someone had been sitting in it recently. The chair felt warm because thermal energy from the person's body flowed to the chair and increased its temperature.

<u>Heat</u> is thermal energy that flows from something at a higher temperature to something at a lower temperature. Recall that joules are the units that energy is measured in. Heat is a form of energy, so it is measured in joules.

When you put ice in water, the ice seems to be cooling the water. Actually, the water is heating the ice. The water is transferring its thermal energy to the ice. This transfer of thermal energy does both, cool the water, and warm the ice. Even after the ice melts, heat continues to flow from the warmer water to the cooler water until all the water is the same temperature. Heat always flows from warmer materials to cooler materials.

Specific Heat

When you are at the beach in the summertime, the ocean probably feels much cooler than the air or the sand. Energy from the Sun is warming the air, sand, and water at the same rate. But the temperature of the water has changed less than the temperature of the air or sand has changed.

As a substance or material absorbs heat, how much its temperature changes depends on how much heat is added. But it also depends what the material is made of. For example, think about 1 kg of sand and 1 kg of water. It takes six times as much heat to raise the temperature of water 1°C than it takes to raise the temperature of sand 1°C. The ocean water at the beach would have to absorb six times as much heat as the sand to be at the same temperature.

Think it Over

4. **Communicate** Write in your own words the difference between temperature and heat.

_____

_____

_____

_____

Think it Over

5. **Infer** Read the Before You Read exercise on the first page of this section again. Use what you learned about specific heat. Why does a carpeted floor feel different to your bare feet than a bare floor feels on a cold morning?

_____

_____

_____

The **specific heat** of a substance is the amount of heat needed to raise the temperature of 1 kg of that substance by 1°C. Specific heat is measured in joules per kilogram degree Celsius [J/(kg • °C)]. The table shows the specific heat of some familiar materials.

Specific Heat of Some Common Materials	
Substance	Specific Heat [J/(kg • °C)]
Water	4,184
Wood	1,760
Carbon (graphite)	710
Glass	664
Iron	450

Applying Math

6. **Use a Table** How much heat is needed to raise the temperature of 1 kg of iron from 25°C to 30°C? Show your work.

_____

_____

How does water cool things?

Look at the table of specific heat above. Compared with the other common materials, water has the highest specific heat. The figure below shows why this is. Water molecules form strong bonds with each other. When heat is added, the strong bonds make it hard for the molecules to move. Some of the added heat has to break these bonds before the molecules can start moving faster. In metals, electrons can move freely. When heat is added, no strong bonds have to be broken before the electrons can start moving faster.

Strong bond

When heat is added to water, some of the added heat has to break some of these bonds before the molecules can start moving faster.

In metals, electrons can move freely. When heat is added, no strong bonds have to be broken before the electrons can start moving faster.

Picture This

7. **Interpret Scientific Illustrations** Explain how strong bonds give water a high specific heat.

_____

_____

_____

_____

A coolant is a substance that is used to absorb heat. Water can absorb heat without a large change in temperature, so it is useful as a coolant. Water is used in the cooling systems of automobile engines. As long as the water temperature is lower than the engine temperature, heat will flow from the engine to the water. Compared to other materials that could be used in an engine, the temperature of water will increase less.

How does thermal energy change?

The thermal energy of an object changes when heat flows into or out of the object. You can use the following equation to calculate the change in thermal energy.

change in thermal energy (J) =

mass (kg) × **change in temperature** (°C) × **specific heat** $\left(\dfrac{J}{kg \cdot °C}\right)$

$$Q = m(T_f - T_i)C$$

In the equation, Q stands for the change in thermal energy. C stands for the object's specific heat. T_f is the final temperature and T_i is the initial temperature.

Suppose a wooden chair with a mass of 20 kg is sitting in sunlight. As it sits, it warms from 15°C to 25°C. The specific heat of wood is 1,700 J/(kg • °C). What is the change in the thermal energy of the chair?

The mass of the chair is 20 kg, so m = 20 kg.
The temperature warms from 15°C to 25°C.
   So T_i = 15°C and T_f = 25°C.
The specific heat of wood is 1,700 J/(kg • °C), so
   C = 1,700 J/(kg • °C).

$$Q = m(T_f - T_i)C$$
$$= (20 \text{ kg})(25°C - 15°C)(1{,}700 \text{ J/kg} \cdot °C)$$
$$= (20 \text{ kg})(10°C)(1{,}700 \text{ J/kg} \cdot °C) = 340{,}000 \text{ kg } °C \text{ J/kg} \cdot °C$$
$$= 340{,}000 \text{ J}$$

The change in thermal energy of the chair is 340,000 J.

Measuring Specific Heat

A calorimeter is a device that can be used to find the specific heat of a material. The specific heat of a material can be determined if the mass of the material, its change in temperature, and the amount of thermal energy absorbed or released are known. In a calorimeter, an object that has been heated transfers heat to a known mass of water. The transfer of heat continues until the object and the water are the same temperature. The energy absorbed by the water can be calculated by measuring the water's temperature change. The thermal energy released by the object equals the thermal energy absorbed by the water. ☑

Applying Math

8. **Calculate** The specific heat of copper is 385 J/kg • °C. Find the change in thermal energy for a copper pipe with a mass of 8 kg when it is heated from 12°C to 21°C. Show your work.

_____

_____

_____

✔ Reading Check

9. **Explain** What does a calorimeter measure?

_____

_____

● After You Read

Mini Glossary

heat: thermal energy that flows from something at a higher temperature to something at a lower temperature

specific heat: the amount of heat needed to raise the temperature of 1 kg of a substance by 1°C

temperature: the average value of the kinetic energy of the particles in an object

thermal energy: the sum of the kinetic and potential energy of all the molecules in an object

1. Review the terms and their definitions in the Mini Glossary. What is the difference between temperature and thermal energy?

_____

_____

_____

2. Complete the chart below to organize information from this section about temperature.

3. **Study Coach** You made an outline as you read this section. How did that strategy help you learn the material?

_____

_____

Thermal Energy

section ② Conduction, Convection, and Radiation

What You'll Learn
- the three ways heat is transferred
- the difference between insulators and conductors
- how insulators control the transfer of thermal energy

● Before You Read

There is an old joke that says a sweater is something you put on when your mother is cold. On the lines below, write about a time when you felt warm, but someone else felt cold.

_____

_____

_____

Mark the Text

Identify Definitions As you read the section, highlight the definition of each word that appears in bold.

FOLDABLES

ⓑ Compare and Contrast
Make the Foldable shown below to help you understand how conduction, convection, and radiation are similar and different.

Conduction

Convection

Radiation

● Read to Learn

Conduction

The three ways that thermal energy can be transferred from place to place are conduction, convection, and radiation. In conduction, heat is transferred by direct contact. **Conduction** occurs when the particles within a substance or between two substances bump into each other and transfer kinetic energy. Conduction occurs because the particles in matter are in constant motion.

How does conduction transfer thermal energy?

The figure shows how thermal energy is transferred when one end of a metal spoon is heated by a flame. The kinetic energy of the particles near the flame increases. The heated particles bump into surrounding particles. In these collisions, thermal energy is transferred from particles with more kinetic energy to particles with less kinetic energy. The collisions continue as thermal energy is transferred to the other end of the spoon. When heat is transferred by conduction, thermal energy is transferred from place to place without transferring matter. Thermal energy is transferred by the collisions between particles, not by movement of matter.

What are thermal conductors?

Energy can be transferred by conduction in all materials. But energy moves more quickly in some materials than in others. Conduction is faster in solids and liquids than in gases. In gases, particles are farther apart. Collisions with other particles happen less often in gases than in solids or liquids.

The best thermal conductors are metals. Remember that an atom has a nucleus surrounded by electrons. In metals, there are electrons that are not bound to individual atoms. They move easily through the metal and bump into other electrons. This helps the electrons transfer thermal energy. Silver, copper, and aluminum are some of the best energy conductors. ☑

Convection

Thermal energy can also be transferred by a method called convection. **Convection** is the transfer of thermal energy in a fluid by the movement of warmer and cooler fluids from one place to another. A fluid is a substance that can flow. It can be a liquid or a gas. The movement of fluids from one place to another causes currents. These currents transfer thermal energy.

How is heat transferred by convection currents?

When a pot of water is heated on a stove, heat is transferred by convection currents. First, the water molecules at the bottom of the pot gain thermal energy from the stove. The water becomes less dense as it is heated. The water at the top of the pot is still cool, so it is more dense than the warm water. Since the warm water is less dense, it rises. As it rises, it is replaced at the bottom of the pot by the sinking, cooler water. The cycle continues until all the water in the pot is at the same temperature. This rising-and-sinking action is a convection current. Also, as the warmer water rises, it heats some of the cooler water that it comes in contact with by conduction. In a convection current, both conduction and convection transfer thermal energy.

How do convection and conduction differ?

Both convection and conduction transfer heat. However, convection is different from conduction. Convection transfers thermal energy by moving particles from one place to another. Warm particles change places with cooler particles. In conduction, thermal energy is transferred by particles bumping into surrounding particles. But, in conduction, no particles actually move from place to place as they do in convection.

FOLDABLES™

☾ Compare Make the following Foldables to compare conductors and insulators.

Conductors: Insulators:

☑ Reading Check

1. **Explain** why metals are good conductors.

_____

_____

_____

💡 Think it Over

2. **Explain** A convection oven has a fan to move the air inside it. Is the air in a regular oven also affected by convection? Why or why not?

_____

_____

_____

3. **Make a Drawing** In the first box, draw an ice-covered mountain. Then draw arrows to show how radiation is reflected and absorbed by an ice-covered mountain range. In the second box, draw trees to represent a rain forest. Then draw arrows to show how radiation is reflected and absorbed by a dark rain forest.

Radiation

How does heat travel through space? There is almost no matter between the Sun and Earth. So heat does not travel by conduction or convection. Instead, heat is transferred from the Sun to Earth by radiation.

Radiation is the transfer of energy by electromagnetic waves. They carry energy through solids, liquids, and gases. But they can also carry energy through empty space. Energy transferred by radiation is often called radiant energy. When you sit by a fireplace, you feel warm because heat is transferred by radiation from the fire to your skin.

How do radiant energy and matter interact?

When radiation strikes a material, three things can happen. The material absorbs some of the energy. It also reflects some of the energy. And some of the energy may be transmitted, or passed through, the material. The amount of energy absorbed, reflected, and transmitted depends on the type of material. Light-colored materials reflect more radiant energy, while dark-colored materials absorb more radiant energy. When material absorbs radiant energy, the thermal energy of the material increases. The figure above shows what happens to radiant energy from the Sun as it reaches Earth.

How is radiation different in solids, liquids, and gases?

Radiant energy can travel through the space between particles in a solid, liquid, or gas. Particles can absorb the radiation and emit some of the energy they absorbed. This energy then travels through the space between particles. It is absorbed and emitted by other particles. The transfer of energy by radiation works best with gases. The particles in gases are much farther apart than in solids or liquids. This extra space allows the radiant energy to travel more easily through gases.

Controlling Heat Flow

You probably do things every day to control the flow of heat. For example, you put on a jacket when you leave your house on a cool morning. And you might wear an oven mitt on your hand when you reach into the oven to pull out a hot pan. In both cases, you use different materials to help control the flow of heat. Your jacket reduces the flow of heat from your body to the surrounding air. The oven mitt reduces the flow of heat from the hot dish to your hand.

Animal Adaptations Almost all living things have special features that help them control the flow of heat. The antarctic fur seal has a coat that is about 10 cm thick. This coat helps keep the seal from losing heat. The emperor penguin has a thick layer of blubber and thick, closely spaced feathers. This helps the penguin reduce the loss of body heat. These adaptations help the animals survive in a climate where the temperature is often below freezing. The scaly skin of the desert spiny lizard has the opposite effect. The lizard's skin reflects the Sun's rays and keeps the lizard from becoming too hot. The leathery skin also prevents water loss. An animal's color also can also help it keep warm or cool. The black feathers on a penguin's back allow it to absorb radiant energy. ☑

Thermal Insulators

When you cook, you want the pan to conduct heat easily from the stove to your food. But you don't want the heat to move easily to the handle of the pan. Most pans have handles that are made from insulators. A **thermal insulator** is a material in which heat flows slowly. Materials that are insulators include wood, some plastics, fiberglass, and air. An insulator is the opposite of a conductor. Remember, conductors are materials through which heat flows quickly. Metals like silver, copper, and aluminum are excellent conductors. Materials that are good insulators are poor conductors of heat. That is why many pan handles are made of wood or plastic.

Air as an Insulator Gases, such as air, are usually better insulators than solids or liquids. Some types of insulators contain many pockets of trapped air. These air pockets conduct heat poorly and keep convection currents from forming. When you wear a fleece jacket, the fibers in the fleece trap air between the jacket and your body. This air slows down the flow of your body heat to the colder air outside the jacket. Your body heat warms the air trapped by the fleece and you are wrapped in a blanket of warm air.

☑ **Reading Check**

4. **Describe** How do animals' colors help them control their temperature?

_____

_____

💡 **Think it Over**

5. **Draw Conclusions** Some of the best cooking pans are made of copper. Why do you think this is?

_____

_____

_____

How are buildings insulated?

Materials that are insulators help keep warm air from flowing out of buildings in cold weather. They also keep warm air from flowing into buildings in warm weather. These materials are called insulation. Building insulation is usually made of materials, such as fiberglass or foam, which contain pockets of trapped air. Insulation is placed between a building's outer walls and inner walls and between the ceiling and the attic. It reduces the flow of heat between the building and the surrounding air. ☑

Insulation also helps furnaces and air conditioners work more effectively. Since the heat doesn't escape or enter the building as easily, the furnaces and air conditioners do not have to work as much. This can save a great deal of energy. In the United States, about 50 percent of the energy used in homes is used for heating and cooling.

How does a thermos work?

Have you used a thermos bottle like the one in the figure? A thermos bottle lets you keep soup hot or iced tea cold. It does this by reducing the flow of energy into and out of the liquid in the bottle. The temperature of the liquid hardly changes over a number of hours. A thermos bottle has two glass walls. The air between the two walls is removed so there is a vacuum between the glass layers. The vacuum contains little matter. This vacuum prevents heat transfer by conduction or convection between the liquid and the air outside the thermos bottle.

Outer case

Reflective surface

Vacuum

The inside and outside glass surface of a thermos bottle is coated with aluminum. The coating makes each surface highly reflective. Electromagnetic waves are reflected at each surface. The inner reflective surface prevents radiation from transferring heat out of the liquid. The outer reflective surface prevents radiation from transferring heat into the liquid.

Think about the things you do to stay warm or cool. Sitting in the shade reduces the heat transferred to you by radiation. Opening or closing windows increases or reduces heat transferred by convection. Putting on a jacket reduces the heat transferred from your body by conduction. In what other ways do you control the flow of heat?

✔ Reading Check

6. Identify What does building material insulation contain?

_____

_____

Picture This

7. Explain How does the shiny surface of a thermos bottle prevent heat transfer?

_____

_____

_____

_____

● After You Read

Mini Glossary

conduction: the transfer of thermal energy by collisions between particles in matter

convection: when thermal energy is transferred in a fluid by the movement of warmer and cooler fluid from place to place

radiation: the transfer of energy by electromagnetic waves

thermal insulator: a material in which thermal energy flows slowly

1. Review the terms and their definitions in the Mini Glossary. Write a sentence on the lines below that shows your understanding of the term *thermal insulator*.

_____

_____

2. Complete the table to organize information about how heat is transferred.

How Heat Is Transferred	Definition	Example
Conduction		
Convection		
Radiation		

3. ◀ Mark the Text Think about what you have learned. How did identifying definitions help you as you read the section?

_____

_____

End of Section

Thermal Energy

section ❸ Using Thermal Energy

What You'll Learn

- types of heating systems
- laws of thermodynamics
- how car engines work
- how refrigerators keep food cold

Study Coach

Create a Diagram As you read about internal combustion engines in this section, make a diagram of the steps of the four-stroke cycle. Make sure you label your diagram.

FOLDABLES™

D Build Vocabulary
Make the following Foldable to help you learn the information from this section. You will need to use more than one Foldable.

| Forced Air |
| Radiator |
| Elec. Heating |
| Passive Solar |
| Active Solar |
| Open System |
| Closed System |
| Int. Comb. |

● Before You Read

Cars and buses use heat engines. List three more kinds of devices you know about that use engines.

● Read to Learn

Heating Systems

Most buildings have some kind of heating system. All heating systems need a source of energy. In the simplest and oldest heating system, wood or coal is burned in a stove. The heat that is produced is transferred to the air by conduction, convection, and radiation. One disadvantage of this type of heating system is that the transfer of heat is slow.

What are some common heating systems?

Forced-Air Systems The forced-air system is the most common type of heating system today. In this system, fuel is burned in a furnace. This heats a volume of air. A fan blows the warm air through large pipes, called ducts. The ducts lead to openings, or vents, in each room. Cool air returns through other vents to the furnace where it is reheated.

Radiator Systems In the past, many homes and buildings were heated by radiators. In a radiator system, fuel is burned in a furnace and heats a tank of water. Pipes carry the hot water or steam to radiators in the building. The thermal energy in the water or steam is transferred to the air by conduction. The warm air then moves through the room by convection. After the water cools it flows back to the water tank and is reheated.

Electric Heating Systems Not as common as forced-air systems, electric heating systems consist of electrically heated coils in baseboard heaters or space heaters. The coils heat the surrounding air by conduction and energy is transferred throughout the room by convection.

Solar Heating

The Sun emits an enormous amount of radiant energy that strikes Earth every day. The radiant energy from the Sun can be used to help heat homes and buildings.

What kinds of solar heating systems are there?

Passive Solar Heating In passive solar heating systems, materials inside a building absorb radiant energy from the Sun during the day. The materials heat up. At night, the building begins to cool. Thermal energy absorbed by these materials helps keep the building warm.

Active Solar Heating Active solar heating systems use solar collectors. **Solar collectors** absorb radiant energy from the Sun. The collectors usually are placed on the roof or on the south side of a building. Radiant energy from the Sun heats air or water in the solar collectors. An example of an active solar collector is shown in the figure above. The black metal plate absorbs radiant energy from the Sun. The absorbed energy heats water in pipes. A pump circulates the hot water to radiators in the building. The cooled water is pumped back to the collectors to be reheated.

Glass sheets

Black metal plate

Pipes for liquid

Insulation

Thermodynamics

There is a way to increase the thermal energy of an object without adding heat. Have you ever rubbed your hands together to warm them? No heat is flowing to your hands, but they get warmer. Their thermal energy and temperature increase. You did work on your hands by rubbing them together. The work you did increased the thermal energy of your hands. Thermal energy, heat, and work are related. **Thermodynamics** is the study of the relationship among thermal energy, heat, and work.

How do heat and work increase thermal energy?

You can warm your hands by placing them near a fire. Heat is added to your hands by radiation. You can also rub your hands together as you hold them near a fire. The thermal energy of your hands increases even more. Both the heat from the fire and the work you do increase the thermal energy of your hands. ☑

💡 **Think it Over**

1. **Infer** How do you suppose passive solar heating systems got that name?

_____

_____

_____

Picture This

2. **Conclude** Many parts of a solar collector are black. Why do you think they are black?

_____

_____

_____

☑ **Reading Check**

3. **Name** the two things that can increase thermal energy in a system.

_____

_____

In this example, your hands are a system. A system can be a group of objects, such as a galaxy or a car's engine. It can also be as simple as a ball. A system is anything you can draw a boundary around. The heat transferred to a system is the amount of heat that crosses the boundary. The work done on a system is the work done by something outside the system's boundary.

What is the first law of thermodynamics?

The **first law of thermodynamics** says that the increase in thermal energy of a system equals the work done on the system plus the heat transferred to the system. Doing work on a system is a way of adding energy to a system. So, the temperature of a system can be increased by adding heat to the system, doing work on the system, or both. The first law of thermodynamics is another way of stating the law of conservation of energy. The increase in energy of a system equals the energy added to the system.

What is the difference between an non-isolated system and an isolated system?

A system is an non-isolated system if heat flows across the boundary or if work is done across the boundary. In other words, energy is added to the system. If no heat flows across the boundary and no outside work is done, then the system is isolated. The first law of thermodynamics says that the thermal energy of an isolated system doesn't change. Processes within the system may be changing one form of energy into another. But the total energy of the system stays the same. Energy cannot be created or destroyed. The total energy stays constant in an isolated system.

What is the second law of thermodynamics?

When heat flows from a warm object to a cool object, the thermal energy of the warm object decreases. The thermal energy of the cool object increases. According to the first law of thermodynamics, the increase in thermal energy of the cool object equals the decrease in thermal energy of the warm object. Can heat flow from a cold object to a warm object? This process would not break the first law of thermodynamics. The decrease in thermal energy of the cool object would only have to be equal to the increase in thermal energy of the warm object. However, the natural flow of heat from a cool object to a warm object never happens. It breaks another law.

Think it Over

4. Compare The law of conservation of momentum says that momentum can be transferred from one object to another, but it cannot be created or destroyed. How is this similar to the first law of thermodynamics?

_____

_____

_____

FOLDABLES

E Find Main Ideas Make the following Foldable to take notes about the first and second laws of thermodynamics.

First Law of Thermodynamics

Second Law of Thermodynamics

The **second law of thermodynamics** says that it is impossible for heat to flow from a cool object to a warmer object unless work is done. For example, if you hold an ice cube in your hand, no work is done. So, heat flows only from your hand to the ice, not the other way around.

Converting Thermal Energy to Work

When you rub your hands, friction converts the work you do on your hands to heat. Your hands become warmer. In this example, work was converted completely into heat. Is it possible to convert heat completely into work? This would not break or violate the first law of thermodynamics. But the second law of thermodynamics states it is impossible to build a device that converts heat completely into work.

A **heat engine**, such as in a car, is a device that changes thermal energy into work. The engine converts chemical energy in gasoline into thermal energy. Then it changes some of the thermal energy into work by rotating the car's wheels. About 25 percent of the thermal energy released by burning gasoline is converted into work. The rest is transferred to the engine's surroundings. ☑

How does an internal combustion engine work?

The heat engine in a car is an internal combustion engine. In an **internal combustion engine**, the fuel burns inside the engine in chambers, or cylinders. Car engines usually have four or more cylinders. Each cylinder has a piston inside it. The pistons move up and down. Each up-and-down movement of the piston is called a stroke. The figure shows the four-stroke cycle of a car engine and explains how it works.

✔ **Reading Check**

5. **Identify** What does a car engine convert into thermal energy?

_____

_____

Picture This

6. **Interpret a Diagram** Circle stroke where the spark plug ignites the fuel.

A Intake stroke
The intake valve opens as the piston moves downward, drawing a mixture of gasoline and air into the cylinder.

B Compression stroke
The intake valve closes as the piston moves upward, compressing the fuel-air mixture.

C Power stroke
A spark plug ignites the fuel-air mixture. As the mixture burns, hot gases expand, pushing the piston down.

D Exhaust stroke
As the piston moves up, the exhaust valve opens, and the hot gases are pushed out of the cylinder.

Labels: Intake valve, Fuel-air mixture, Spark plug, Exhaust valve, Cylinder, Piston, Crankshaft, Exhaust gases

Thermal Energy Movers

The second law of thermodynamics says that it is impossible for energy to flow from a cool object to a warmer object unless work is done. So how does the inside of a refrigerator stay cold? A refrigerator is an energy mover. It does work to move energy from a cooler temperature to a warmer temperature. The energy to do the work is provided by an electric outlet.

How does a refrigerator transfer heat?

A refrigerator contains a substance called a coolant. Liquid coolant is pumped through an expansion valve and changed into a gas. When a coolant changes to a gas, it cools. The cold gas is pumped through pipes around the inside of the refrigerator. It absorbs thermal energy, and the inside of the refrigerator cools. The warm gas is then pumped to a compressor. The compressor does work by compressing the gas, making the gas even warmer. Thermal energy flows from the warm gas to the room. As the gas gives off heat, it cools and changes back into a liquid, and the cycle is repeated. ☑

How does an air conditioner work?

An air conditioner is another heat mover. It is similar to a refrigerator, except that the warm air from the room is forced to pass over tubes containing the coolant. The warm air is cooled. The cooled air is forced back into the room. The thermal energy that is absorbed by the coolant is transferred to the air outdoors. Refrigerators and air conditioners are heat engines working in reverse. They use mechanical energy supplied by the compressor motor to move thermal energy from cooler areas to warmer areas.

How does a heat pump work?

A heat pump works like a reversible air conditioner. When it is hot outside, warm air in the house is cooled by absorbing thermal energy in the coolant and releasing the thermal energy outside. When it is cold outside, energy is removed from the outside air and released into the house. As long as temperatures are above $-18°C$ energy can be taken from the outside air.

How do energy transformations produce heat?

Many energy transformations convert one form of energy into a more useful form. Usually when these energy transformations occur, some thermal energy is produced. For example, when the shaft of an electric motor rotates, it creates friction. The friction converts mechanical energy into thermal energy. The thermal energy produced is no longer in a useful form. It is transferred to the surroundings by conduction and convection.

✔ **Reading Check**

7. Identify When the liquid coolant in a refrigerator passes through an expansion valve, it changes into

_____.

💡 **Think it Over**

8. Explain Why are refrigerators and air conditioners called heat engines working in reverse?

_____

_____

_____

● After You Read

Mini Glossary

first law of thermodynamics: the increase in thermal energy of a system equals the work done on the system plus the heat transferred to the system

heat engine: a device that changes heat into work

internal combustion engine: a heat engine in which the fuel burns inside the engine

second law of thermodynamics: heat cannot flow from a cool object to a warmer object unless work is done

solar collector: a device that absorbs radiant energy from the Sun

thermodynamics: the study of the relationship among thermal energy, heat, and work

1. Review the terms and definitions in the Mini Glossary. On the lines below, write a sentence that describes the relationship between a heat engine and an internal combustion engine.

_____

_____

2. Complete the chart below that describes the parts of the four-stroke cycle.

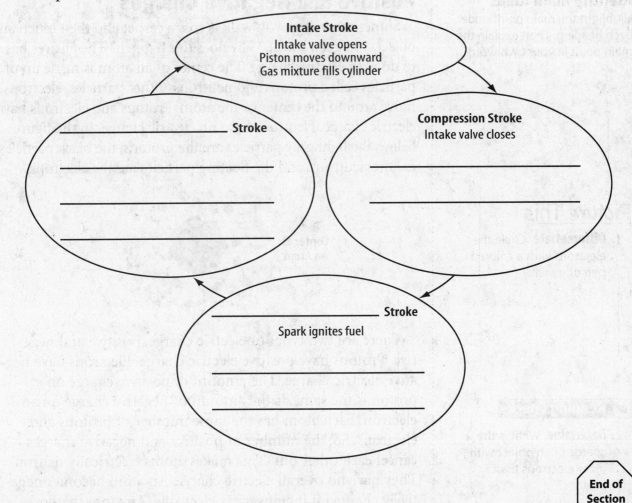

Intake Stroke
Intake valve opens
Piston moves downward
Gas mixture fills cylinder

_____ Stroke

Compression Stroke
Intake valve closes

_____ Stroke
Spark ignites fuel

End of Section

section ❶ Electric Charge

What You'll Learn
- how electric charges exert forces
- about conductors and insulators
- how things become electrically charged

Mark the Text

Identify Main Ideas
Highlight the main point under each heading. Then explain the main point in your own words.

Picture This

1. **Differentiate** Circle the electrons with a colored pen or pencil.

✔ **Reading Check**

2. **Determine** What is the charge of an object with more electrons than protons?

● Before You Read

Think about some electric objects that are plugged into an outlet. But cars, cell phones, and even wristwatches also use electricity. List three things you use every day that use electricity but do not plug in.

_____

● Read to Learn

Positive and Negative Charges

Sometimes, when you walk across a carpet and then touch an object, you get a shock. Why does this happen? The answer has to do with electric charge. The center of an atom is made up of particles called protons and neutrons. Other particles, electrons, move around the center of the atom. Protons and electrons have electric charge. Neutrons have no electric charge. In the figure below, the light gray particles are the protons, the black particles are the neutrons, and the floating particles are the electrons.

Center of
an Atom

There are two types of electric charge, positive and negative. Protons have positive electric charge. Electrons have negative electric charge. The amount of positive charge on a proton is the same as the amount of negative charge on an electron. Each atom has the same number of protons and electrons. So, the number of positive and negative charges cancel each other out. This makes atoms electrically neutral. They have no overall electric charge. An atom becomes negatively charged if it gains extra electrons. An atom that loses electrons becomes positively charged overall. ✔

How is electric charge transferred?

Electrons are bound more tightly to some atoms and molecules than to others. Electrons in the soles of your shoes are bound tightly to the atoms. Electrons in atoms in carpet are not bound as tightly. When you walk on carpet, electrons are transferred from the carpet to the soles of your shoes.

Now, the soles of your shoes have more electrons than protons. They are negatively charged. The carpet has fewer electrons than protons. It is positively charged. The transfer of electrons changed the electric charge of each object. **Static electricity** is the buildup of electric charges on an object. When there is static electricity, electric charges are not balanced.

Is new electrical charge created?

The electrons that moved to your shoe are not new electrons. The **law of conservation of charge** states that charge can be transferred from one object to another, but it cannot be created or destroyed. An object becomes charged when electric charges move from one place to another.

What happens when electrical charges move?

Have you ever taken clothes out of a dryer and had them cling together? Look at the figure below. Opposite electric charges attract each other. They tend to move toward each other. Electric charges that are the same repel each other. They tend to move away from each other.

Opposite charges attract

When clothes tumble in a dryer, the atoms in some clothes lose electrons. Those clothes become positively charged. The atoms in other clothes gain electrons and become negatively charged. The clothes have opposite charges. Objects that have opposite charges attract each other, so the clothes cling together.

Like charges repel

How do charges exert forces?

The electric force between two charged objects depends on how far apart the objects are. The electric force between two charges decreases as the charges move farther apart.

The electric force also depends on the amount of charge on each object. When the amount of charge on one of the objects increases, the electric force increases.

FOLDABLES

A Build Vocabulary As you read this section, make the following vocabulary Foldable. Write the definition for each vocabulary word under its tab.

Static Electricity

Law of Conservation of Charge

Electric Field

Conductor

Insulator

Charging by Contact

Charging by Induction

Electroscope

Picture This

3. **Illustrate** Look at the figures of like and unlike charges. Highlight the negative charges in one color and the positive charges in other color. Notice the charges only attract when the colors are different.

Picture This

4. Identify Look at the figure of electric fields. Why do the arrows point outward from the positive field? Why do they point inward toward the negative field?

_____

_____

_____

_____

 Think it Over

5. Explain in your own words why usually there is very little electric force between two objects.

_____

_____

_____

_____

What are electric fields?

When a charged balloon comes near your hair, your hair will move toward it. Your hair does not have to touch the balloon for an electric force to act on it. So what makes your hair move? It is positively charged. The balloon is negatively charged. They are attracted to each other because they have opposite electric charges.

Move of a Positive Charge in Electric Fields

There is an electric field around every electric charge. The **electric field** exerts a force that attracts or repels other electric charges. The figure shows two electric fields. The arrows show the direction a positive charge would move in each electric field. Your hair has the positive charge and moves to the balloon.

The force of gravity between you and Earth seems very strong. However, electric forces are much stronger. Electric forces between the protons and the electrons hold the particles in atoms together. For example, the electric force between a proton and an electron in a hydrogen atom is a thousand trillion trillion times larger or, 10^{39} times larger than the attractive gravitational forces between the same proton and electron.

Forces Between Atoms Atoms also are held together by electric forces. Electric forces between atoms cause chemical bonds. These electric forces are also much greater than the gravitational forces between the atoms.

Forces Between Objects Many of the forces that act on objects are due to the electric forces between atoms and molecules. All atoms contain electrically charged protons and electrons. When atoms or molecules get close enough, they can exert forces that attract or repel. For example, when you push on a door, the atoms in your hand get close to the atoms in the door. The atoms are close enough to exert forces on each other. The forces between the atoms in your hand and the atoms in the door cause the door to move.

Conductors and Insulators

Remember the example of electrons moving from the carpet to your shoe? If you reach for a metal doorknob after walking on carpet, you might see a spark. Electrons moving from your hand to the doorknob cause the spark. How did those electrons move from your shoe to your hand?

What is a conductor?

Electrons can move more easily in some materials than in others. A **conductor** is a material in which electrons can move easily. Your skin is a better conductor than your shoes. Electrons move from your shoes to your skin, spreading to your hands. The best electric conductors are metals. Atoms in metals have electrons that are able to move easily through the metal. Copper is one of the best conductors.

What is an insulator?

An **insulator** is a material in which electrons cannot move easily. In insulators, electrons are held tightly to atoms. The plastic coating around an electric wire keeps you from getting a dangerous electric shock when you touch the wire coating. Wood, rubber, and glass are other good insulators.

Charging Objects

Just like the clothes in the dryer, when two materials are rubbed together, electrons can be transferred between them. One object will have a negative charge. The other will have the same amount of positive charge. **Charging by contact** is the transfer of charges by touching or rubbing.

How can something be charged at a distance?

Remember, electric forces change when objects move closer together. If a charged object is moved near a neutral object, electrons on the neutral object will move around.

Think about the balloon that was charged by rubbing it on your hair. The charged balloon doesn't need to touch the hair to make the hair move toward it. The same is true if you hold the charged balloon close to a wall. The extra electrons on the balloon repel the electrons in the wall. The electrons in the wall move away from the balloon. Now there is a positively charged area on the wall. The negatively charged balloon is attracted to the positive area of the wall. A charged object rearranging the electrons on a nearby neutral object is **charging by induction**. The wall was charged by induction. The balloon will stick to the wall. An electric force holds it there. ☑

💡 **Think it Over**

6. **Apply** Which is a better conductor, a flagpole or a flag? Why?

_____

_____

_____

_____

💡 **Think it Over**

7. **Explain** why it is hard for electrons to move through an insulator.

_____

_____

_____

_____

✔ **Reading Check**

8. **Compare** What is the difference between charging by contact and charging by induction?

_____

_____

_____

_____

What is lightning?

Have you ever seen lightning hit the ground? Lightning is a large static discharge—a transfer of charge between two objects. It happens if there is a buildup of static electricity.

A large amount of static electricity is formed when air moves around in thunderclouds. Areas of positive and negative charge build up. When enough charge builds up, there is a static discharge between the cloud and the ground. As the electric charges move through the air, they run into atoms and molecules, making the atoms and molecules light up.

What is thunder?

Lightning makes a bright light. It also creates powerful sound waves. Thunder is the sound that lightning makes. The electric energy in a lightning bolt rips electrons off atoms in the air. This causes great amounts of heat. The air temperature around the lightning bolt can be as high as 25,000°C. The heat makes the air around the lightning bolt move faster. This rapid movement of air produces the sound waves that you hear as thunder.

Why is grounding important?

Lightning can cause damage and injury because it releases a great amount of energy. One way to avoid the damage is to make the charges flow to Earth's surface. Earth is a large neutral object that is also a conductor. Because Earth is so large, it can absorb a great amount of excess charge. Grounding provides a path for electric charges to move to Earth. For example, a metal lightning rod on top of a building provides a path to move excess charges to Earth's surface.

Detecting Electric Charge

An **electroscope** can detect when an object has an electric charge. One type of electroscope is a glass beaker with a metal rod inside it, as shown. The metal rod connects to a knob at the top of the beaker. There are two metal branches, or leaves, at the bottom of the metal rod. The metal leaves hang down when there is no charge to the rod. When an object with a negative charge touches the knob, electrons travel down the rod to the leaves. Both leaves gain negative charges. When an object with a positive charge touches the knob, it attracts electrons that move up the rod. The leaves have a positive charge. When the leaves have a charge, they repel each other and spread apart.

Knob

Metal rod

Metal leaves

Think it Over

9. **Analyze** Why do you think a lightning rod is made of metal?

_____

_____

_____

Picture This

10. **Draw** On the figure, draw what the leaves look like if they have a charge.

● After You Read

Mini Glossary

charging by contact: transferring charges by touching or rubbing

charging by induction: when electrons on a neutral object are moved by a charged object

conductor: a material in which electrons can move easily

electric field: a region in which charged particles attract or repel

electroscope: a device that detects electric charge

insulator: a material in which electrons cannot move easily

law of conservation of charge: charge can be transferred from one thing to another, but it cannot be created or destroyed

static electricity: the buildup of electric charges on an object

1. Read the terms and their definitions in the Mini Glossary above. Use the words __insulator__ and __conductor__ in a sentence that shows that you understand them.

_____

_____

2. Column 1 lists some of the concepts you learned about in this section. Column 2 gives a fact about each concept. Write the letter of the fact on the line next to the concept that matches it.

Column 1	Column 2
_____ 1. transferring charge	**a.** static electricity is discharged between a cloud and the ground
_____ 2. conservation of charge	**b.** electrons cannot move easily in some materials
_____ 3. insulator	**c.** electrons can move from one object to another
_____ 4. lightning	**d.** charge cannot be created or destroyed

3. ◄ **Mark the Text** You highlighted the main points to help you understand electric charge. How did you decide what the main points were?

_____

_____

End of Section

What You'll Learn
- what makes current flow
- how batteries work
- what Ohm's law says

Mark the Text

Identify Details Use one color to highlight each question heading. Then use another color to highlight the answer to the question.

✓ Reading Check

1. **Explain** How many electrons must move past a point every second to equal one ampere of electric current?

_____

_____

● Before You Read

Have you noticed that one end of a battery is marked with a plus sign and the other end with a minus sign? What happens if you put the batteries in a flashlight in the wrong direction?

_____

● Read to Learn

Current and Voltage Difference

You have read about the ways electric charges move. One example is the spark that can jump between your hand and a metal doorknob. **Electric current** is the net movement of electric charges in one direction.

To understand net movement, consider the movement of electrons in all materials. In all materials, electrons move in every direction. Since the electrons are not moving in the same direction, there is no electric current. When electric current flows in a wire, the electrons still move in all directions, but they also drift in the direction that the current flows. The drifting of the electrons is the net movement in one direction.

Electric current is measured in units called amperes. Amperes are also called amps. The symbol for amperes is the letter A. Amperes measure the electrons that flow past one point. One ampere is equal to 6,250 million billion electrons moving past a point every second. ✓

What is voltage difference?

Even though the electrons are moving in all directions, an electric force acts on the charges to make them flow in one direction. Voltage is the electric force that makes charges move.

Voltage is also like the force that acts on water in a pipe. Water flows from higher pressure to lower pressure. In the same way, electric charge flows from higher voltage to lower voltage. A **voltage difference** is related to the force that makes electric charges flow. Voltage difference is measured in units called volts. The symbol for volts is V.

Copyright © Glencoe/McGraw-Hill, a division of The McGraw-Hill Companies, Inc.

What is an electric circuit?

Look at the figure. It shows an electric current doing work by lighting a lightbulb. Electric current must have a closed "loop-like" path to follow. If there is no closed path to follow, the current stops. An **electric circuit** is a closed path that electric current follows. If the circuit in the figure is broken by taking away one part, such as the battery or the lightbulb, current will not flow. It will also not flow if a wire is broken or cut. The lightbulb will not light.

Lightbulb

Electron flow

Battery

Batteries

A circuit needs a voltage difference to keep electric current flowing in it. A battery can provide the voltage difference that keeps electric current flowing. Look at the figure of the circuit again. The positive end and the negative end of a battery are called the terminals. When a closed path connects the terminals, the current will flow.

How do dry-cell batteries work?

The batteries used in a flashlight are called dry-cell batteries. Look at the figure below of a dry-cell battery. The battery has two electrodes. One electrode is a carbon rod. The other electrode is a zinc container.

Around the electrodes is a moist paste. The paste is called an electrolyte. The electrolyte contains chemicals that are conductors. The electrolyte lets charges move from one electrode to the other electrode. This kind of battery is called a dry cell because the electrolyte is a paste, not a liquid.

Positive terminal

Plastic insulator

Moist paste

Carbon rod

Zinc container

Negative terminal

Dry cell

Picture This

2. **Describe** Look at the figure of a circuit. Which direction do electrons flow in the circuit, away from the negative terminal or away from the positive terminal?

_____

_____

FOLDABLES

B **Build Vocabulary**
Make a Foldable as shown. Write the definitions under the tabs and add information as you read this section.

Dry-Cell Batteries

Wet-Cell Batteries

Picture This

3. **Locate** What is used as an electrolyte in a dry-cell battery?

_____

4. **Describe** when the chemical reaction occurs in a dry-cell battery.

_____

_____

Picture This

5. **Compare** What two things are shown in both the figure of the dry-cell battery and the figure of the wet-cell battery?

_____

_____

Making Electricity When the two terminals of a dry-cell battery are connected in a circuit, there is a reaction between the zinc and the chemicals in the electrolyte. Electrons move between some of the compounds in this chemical reaction. The carbon rod becomes positive. The positive terminal is marked with a plus sign (+). Electrons build up on the zinc, making it the negative terminal. The negative terminal is marked with a minus sign (−). The voltage difference between the terminals causes current to flow through a closed circuit.

How do wet-cell batteries work?

Another kind of battery is the wet-cell battery shown below. A wet cell has two connected plates made of different metals. The metals are in a conducting solution. Chemical reactions transfer electrons from the lead plates to the lead dioxide plates. This battery is called a wet cell because the conductor is a liquid. A wet-cell battery contains several wet cells that are connected. Together the cells give a larger voltage difference than each of the cells alone.

Negative terminal
Positive terminal
Lead plate
Battery solution
Wet cell
Partition
Lead dioxide plate

What is a lead-acid battery?

Lead-acid batteries are wet-cell batteries. They are usually used in cars. A lead-acid battery has six separate wet cells that are connected. The cells are made of lead and lead dioxide plates. The plates are in a sulfuric acid solution. A chemical reaction gives a voltage difference of about 2 V in each cell. There are six cells, so the total voltage difference is 12 V.

How are electric outlets different from batteries?

Electric outlets, such as wall sockets, also give a voltage difference. This voltage difference usually is much higher than the voltage difference a battery gives. Most wall sockets give a voltage difference of 120 V. Some outlets have a voltage of 240 V that is needed for large appliances, such as electric ovens and clothes dryers.

Resistance

Flashlights use dry-cell batteries to make the current that lights up the lightbulb. What makes a lightbulb glow? Part of the circuit is a thin wire in the bulb. The wire is called a filament. The electrons in the current flow through the filament. As they move, they bump into the metal atoms in the filament.

The electrons bump into the metal atoms, turning some of their electrical energy into thermal energy. The metal filament gets hot enough to glow. The radiant energy lights up the room. ☑

How do materials resist current?

Electric current loses energy when it moves through material because of resistance. **Resistance** is the tendency for a material to oppose, or go against, the flow of electrons. Resistance turns electrical energy into thermal energy and light.

Almost all materials have electrical resistance. Materials that are electrical conductors have less resistance than materials that are electrical insulators. Resistance is measured in units called ohms (Ω).

What can affect resistance?

The temperature, length, and thickness of a material can affect the electric resistance of the material. Usually, the hotter something is, the more resistance it has. The resistance of an object also depends on its length and thickness. The longer the circuit is, the more resistance it has. Resistance also increases as the wire gets thinner.

A lightbulb filament is a thin piece of tungsten wire made into a short coil. The uncoiled wire is about 2 m long and very thin. Tungsten is a good conductor, but since the wire is so long and thin, it has resistance. The resistance makes the filament glow. The more resistance a filament has, the brighter it glows. ☑

The Current in a Simple Circuit

A simple electric circuit has three main parts. First, it has a source of voltage difference, such as a battery. Second, it has a device that has resistance, like a lightbulb. Third, it has conductors, such as wires. The conductors connect the resistance device to the battery terminals. When the wires are connected to the battery terminals, the path is closed and current flows.

Two electric circuits are shown in the figures on the next page. Each circuit is a battery connected to a lightbulb by wires and a rod. The circuit on the right is shorter because the wires are closer together on the rod. That circuit has less resistance.

☑ **Reading Check**

6. **List** the three types of energy needed to make a lightbulb in a flashlight glow.

_____

_____

_____

☑ **Reading Check**

7. **Recognize Cause and Effect** How does the resistance of a filament affect its glow?

_____

_____

_____

Picture This

8. Compare In which circuit will the light be brighter, the one on the left or the one on the right?

_____

The voltage difference, current, and resistance in a circuit are related. If the voltage difference stays the same as the resistance decreases, the current in the circuit increases. If the wire is short, the lightbulb will be brighter. If the resistance doesn't change, increasing the voltage difference increases the current. If you use a larger battery, the lightbulb will be brighter.

What is Ohm's law?

There is a relationship between voltage difference, current, and resistance in a circuit. **Ohm's law** states that the current in a circuit equals the voltage difference divided by the resistance. If I stands for electric current, Ohm's law can be written as:

$$\text{current (in amperes)} = \frac{\text{voltage difference (in volts)}}{\text{resistance (in ohms)}}$$

$$I = \frac{V}{R}$$

Ohm's law can also be used to measure resistance. Change the equation so that resistance, R, is alone on one side. Do this by multiplying both sides of the equation by R. Then divide both sides by I. The new equation is:

$$R = \frac{V}{I}$$

Suppose a current of 0.5 A flows in a 75-W lightbulb. The voltage difference between the ends of the filament is 120 V. Find the resistance of the filament.

$$R = \frac{V}{I}$$

$$= \frac{120}{0.5}$$

$$= 240$$

The resistance is 240 Ω.

Applying Math

9. Solve the equation to show how $I = \frac{V}{R}$ becomes $R = \frac{V}{I}$. Show your work.

_____

_____

_____

_____

● After You Read

Mini Glossary

electric circuit: a closed path that electric current follows

electric current: the rate at which electric charges move in one direction past one point

Ohm's law: the current in a circuit equals the voltage difference divided by the resistance

resistance: the tendency for a material to oppose the flow of electrons, changing electric energy into heat and light

voltage difference: something related to the force that makes electric charges flow

1. Read the terms and their definitions in the Mini Glossary above. Rewrite the definition of *resistance* in your own words on the lines below.

   _____

   _____

2. Complete the table below to describe a simple circuit. The first column lists the parts of a circuit. In the second column, give an example of each part of a circuit. Under the heading *Function*, write a short description of what job each part does in the circuit.

Parts of a Simple Circuit		
Part	**Example**	**Function**
Voltage difference		
Source of resistance		
Conductors		

3. **Mark the Text** As you read this section, you highlighted the question headings and their answers. Why was using two colors helpful?

   _____

   _____

End of Section

section ❸ More Complex Circuits

What You'll Learn

- the difference between series and parallel circuits
- why circuit breakers and fuses are important
- how electric power is calculated

Study Coach

Create a Quiz As you read this section, think of five quiz questions. Write them down. After you read the section, answer the quiz questions you wrote.

FOLDABLES™

❻ Gather Information Use two quarter-sheets of notebook paper to organize information about series circuits and parallel circuits. Include terms and calculations.

Series Circuit   Parallel Circuit

● Before You Read

Do adults ever remind you to turn off the lights when you leave a room? Like many things we use, electricity is not free. How does the electric company know how much electric energy you use?

_____

_____

● Read to Learn

Series and Parallel Circuits

Think of your home. How many things are plugged into electric outlets? You might think of lamps, computers, televisions, and clocks.

As you read in the last section, a circuit includes three parts. The first part is something that provides a voltage difference. It can be a battery or an electric outlet. The second part is something that uses electric energy and provides resistance. Lightbulbs and hair dryers are two examples. The third part is a conductor that connects the other parts. An example of a conductor is a wire. These three parts form a closed path for the electric current to travel on.

Think about using a hair dryer. The dryer needs to be plugged into an electric outlet. A generator at a power plant produces the voltage difference that ends up at the wall outlet. The voltage difference makes electric charges move when the circuit is complete. The dryer and the circuit in your house have conducting wires. The wires carry the current.

Closing the Circuit When you turn the hair dryer switch to the on position, you close the circuit and current flows. The hair dryer turns electrical energy into thermal energy to warm the air, and mechanical energy to run the fan.

Opening the Circuit When you turn the hair dryer off, you open the circuit. This breaks the path of the current. To use electrical energy, you need a complete circuit. There are two kinds of circuits, series circuits and parallel circuits.

What is a series circuit?

One kind of circuit is called a series circuit. In a **series circuit**, the current has only one loop to flow through. Series circuits are used in flashlights. ☑

How does an open circuit affect a series circuit?

Some older strings of holiday lights will not work if just one lightbulb is burned out. The lights are connected in a series circuit. In a series circuit, the parts are wired one after another. The amount of current is the same through every part. When any part of a series circuit is disconnected, no current flows through the circuit. This is called an open circuit. One burned-out bulb makes the string of lights an open circuit.

Where are parallel circuits used?

What would happen if your home were wired in a series circuit? If you turned off one light, the circuit would be open. All the other lights and appliances in your house would go off. This is why houses are wired with parallel circuits. **Parallel circuits** have at least two paths for current to move through.

Look at the parallel circuit in the figure. The parallel circuit divides the current into two paths. This lowers the resistance. Remember Ohm's law from the last section. More current flows through the paths that have lower resistance.

Houses, cars, and most electric systems use parallel circuits. When one path is opened, the current still flows through the other paths. One part can be turned off without turning off the whole circuit.

☑ Reading Check

1. Identify How many loops does a series circuit have?

_____

Picture This

2. Illustrate In the space below, draw a parallel circuit that has three paths. Label the parts of the circuit.

Household Circuits

Many things in your house use electric energy. You don't see all the wires, because they are hidden behind the walls, ceilings, and floors. The wiring is mostly a combination of parallel circuits. The circuits are connected in an organized and logical way.

Look at the figure below showing the wiring in a house. There is a main switch and a circuit breaker or fuse box. These are like the electric headquarters for the house. Parallel circuits branch out from the circuit breaker or fuse box. The circuits run to wall outlets, appliances, and lights.

Picture This

3. **Highlight** In the figure, use a highlighter to trace a path of a circuit from the meter to the wall socket on the far side of the room.

In a house, many appliances use current from the same circuit. If more appliances are plugged in on a circuit, more current will flow through the wires. As more current flows through the wires, more heat is produced in the wires. If the wires get too hot, the insulation can melt. The bare wires can touch, get hot, and cause a fire. To keep the wires from getting too hot, household circuits include either a fuse or a circuit breaker. ☑

What are fuses?

A fuse is a safety device that stops the wires from getting too hot. A fuse is a small glass tube with a piece of metal inside. If the current is too high, the metal melts. When it melts, it breaks the flow in the circuit. The current stops. To get the current to flow again, you need to replace the old fuse with a new one.

Before you replace the fuse, you should turn off or unplug some of the appliances on the circuit. Using too many household appliances at the same time is the main cause for a blown fuse.

✔ Reading Check

4. **Describe** What happens to the current flow in a circuit when more appliances are added to the circuit?

_____

_____

How does a circuit breaker work?

A circuit breaker is another device that keeps a circuit from overheating. A house usually has a metal box, called a breaker box, which contains many circuit breakers. A circuit breaker is a switch, like a light switch, that has a piece of metal inside. If the current in the circuit is too high, the metal warms up and bends. When the metal bends, it flips the switch and opens the circuit. The flow of current stops before the wires get too hot.

You can usually reset the circuit breaker by flipping the switch inside the breaker box back to its original position. But, before you flip the switch, you should turn off or unplug some of the appliances on the circuit. Otherwise, the circuit breaker will flip the switch off again if too many appliances are using the current in the circuit.

Electric Power and Energy

Electrical energy is useful because it is easy to change into other kinds of energy. For example, it can be changed to thermal energy in a hair dryer. It can also be turned into light, or radiant energy, in a lamp or mechanical energy in a fan. **Electric power** is the rate at which electrical energy is changed into another form of energy.

Different appliances use different amounts of electric power. Appliances are usually marked with a power rating. The power rating tells how much power the appliance uses. Appliances that have electric heating elements, such as ovens and hair dryers, usually use the most power.

How is electric power calculated?

The amount of electric power something uses depends on the voltage difference and the current. You can use the following equation to calculate electric power.

$$\text{electric power} = \text{current} \times \text{voltage difference}$$
$$\text{(in watts)} \qquad \text{(in amperes)} \qquad \text{(in volts)}$$
$$P = IV$$

The unit for power is the watt. The abbreviation for watt is W. The watt is a small unit of power. Because of this, electric power usually is measured in kilowatts. *Kilo-* means "thousand." One kilowatt equals 1,000 watts. The abbreviation for kilowatt is kW. You may see this symbol, or something similar, if you look at an electric bill.

Copyright © Glencoe/McGraw-Hill, a division of The McGraw-Hill Companies, Inc.

Think it Over

5. **Apply** Think about the circuits in your house. Why is it a good idea to have circuit breakers and fuses?

_____

_____

_____

_____

Applying Math

6. **Calculate** The current in an electric clothes dryer is 15 A when it is plugged into a 120-V outlet. How much power does the clothes dryer use? Show your work. Show your answer in kW.

_____

_____

_____

7. Infer Why do you think electric companies charge by the amount of electric energy used as opposed to the amount of electric power used?

_____

_____

_____

_____

How is electrical energy calculated?

Using electric power costs money. However, electric companies charge by the amount of electrical energy used, not the amount of electrical power. Electrical energy usually is measured in units of kilowatt hours. The abbreviation for kilowatt hours is kWh. Kilowatt hours can be calculated using this equation:

$$\underset{\text{(in kWh)}}{\text{electrical energy}} = \underset{\text{(in kW)}}{\text{electric power}} \times \underset{\text{(in hours)}}{\text{time}}$$

$$E = Pt$$

How much does it cost to use electric energy?

You can figure out how much it costs to use an appliance. You do this by multiplying the electric energy used by the cost of each kilowatt hour. Suppose you leave a 100-W lightbulb on for 5 h. The amount of electric energy it uses is

$$E = Pt = (0.1 \text{ kW}) (5 \text{ h}) = 0.5 \text{ kWh}$$

If the power company charges $0.10 per kWh, the cost of using the light for 5 h is

$$\text{cost} = (\text{kWh used}) (\text{cost per kWh})$$
$$= (0.5 \text{ kWh}) (\$0.10/\text{kWh}) = \$0.05$$

So, in this example, it costs five cents to use a 100-W lightbulb for 5 h.

The cost of using some household appliances is given in the table. The cost of $0.15 per kWh was used in the calculations.

Cost of Using Home Appliances			
Appliance	Hair Dryer	Stereo	Color Television
Power rating	1,000	100	200
Hours used daily	0.25	2.0	4.0
kWh used monthly	7.5	6.0	24.0
Cost per kWh	$0.15	$0.15	$0.15
Monthly cost	$1.13	$0.90	$3.60

● After You Read

Mini Glossary

electric power: the rate at which electric energy is changed into another form of energy

parallel circuit: a circuit with at least two paths for current to move through

series circuit: a circuit with just one loop for current to move through

1. Read the terms and their definitions in the Mini Glossary above. On the lines below, write a sentence that shows your understanding of the difference between a series circuit and a parallel circuit.

_____

_____

2. Complete the graphic organizer.

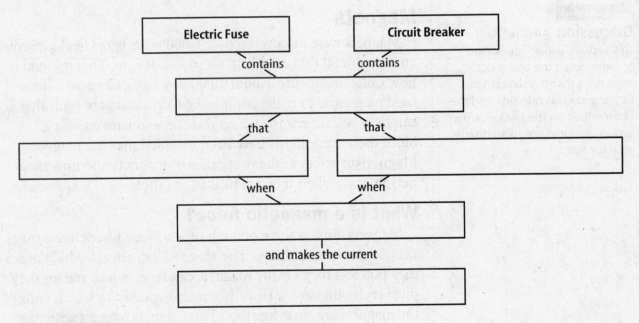

3. Study Coach You used the Create-a-Quiz strategy as you read this section. Look at the quiz questions you wrote. How many of them can you answer correctly? Did this strategy help you understand and remember what you read?

_____

_____

End of Section

chapter 7 Magnetism and Its Uses

section ❶ Magnetism

What You'll Learn

- how a magnet applies force
- how temporary and permanent magnets act
- magnetic materials and magnetic domains

Study Coach

Discussion After reading this section, use an index card to write down the two most important things you learned. Put one idea on one side and the second idea on the back. Form a group of four students to discuss your topics.

✔ Reading Check

1. Infer What causes the ways magnetic forces interact with each other?

● Before You Read

Think about a magnet that you have used. Tell what it looked like and the kinds of materials it attracted.

_____

_____

● Read to Learn

Magnets

Magnets were discovered over 2,000 years ago. Greeks discovered a mineral that could attract pieces of iron. This mineral is now called magnetite. About 1,000 years ago, Chinese sailors used magnetite to make compasses. Compasses are tools that can help you determine which direction you are traveling. Since then many items have been invented that use magnets. **Magnetism** refers to the properties of magnets and how magnets interact when they are near each other.

What is a magnetic force?

Magnets apply a force on each other. These forces make magnets do one of two things. The magnets can attract, which means they pull together. Or the magnets can repel, which means they push each other away. How they react depends on which ends of the magnets are close together. Two magnets interact with each other even before they touch. As the magnets move closer together, the force between them increases. As the magnets move farther apart, the force decreases.

What is a magnetic field?

The way magnetic forces interact with each other is caused by magnetic fields. The **magnetic field** exerts a force on other magnets and objects that are made of magnetic material. The magnetic force is strongest close to the magnet. ✔

Lines can represent a magnetic field. The figures below show what magnetic fields might look like. The arrows show that magnetic fields have direction.

What are magnetic poles?

Magnetic poles are the places on a magnet that exert, or put forth, the strongest magnetic force. All magnets have a north and a south pole. As shown, the north and south poles are at the opposite ends of a bar magnet. The lines that represent the magnetic field are closest together at the poles.

The next figures show the magnetic poles of two magnets with different shapes. A horseshoe-shaped magnet has its north and south poles at its two ends. The magnetic field lines start at the north pole and end at the south pole. Look at the disk magnet and the bar magnet. Like all magnets, the magnetic field lines of the disk magnet and the bar magnet go from north pole to south pole.

Magnetic Fields

How do magnetic poles interact?

Remember, two magnets can either attract or repel each other. This depends on which poles of the magnets are placed close together. Two north poles will repel each other. The same is true for two south poles. However, a north pole and a south pole always attract each other. Like magnetic poles repel each other and unlike poles attract each other.

How do magnets affect compasses?

A compass needle is a small bar magnet. The force exerted by another bar magnet will make the compass needle turn. The needle turns until it lines up with magnetic field lines. The south pole in a compass needle is attracted to the north pole of a magnet.

How does Earth act like a magnet?

Earth is like a giant bar magnet. It is surrounded by a magnetic field and has a north and a south magnetic pole. Earth also has a north and a south geographic pole. The geographic poles are at opposite ends of Earth—one in the north and one in the south. These are different from the magnetic poles. The south magnetic pole is near the geographic north pole.

Picture This

2. **Highlighting** Using a highlighter, trace the lines of the magnetic fields for all three types of magnets shown on this page. At which magnetic pole did you always start ?

_____

FOLDABLES

Ⓐ **Organizing Information** Make the following Foldable to help you organize information about magnetic fields, magnetic poles, and how magnets repel and attract.

Why does a compass point north?

A compass needle lines up with Earth's magnetic field lines. The needle always points toward Earth's geographic north pole. Remember that magnetic poles are attracted only to their opposite. So, even though the compass is pointing at the geographic north pole of Earth, the north pole of the compass is pointing at the south magnetic pole of Earth.

The figure shows the magnetic field lines around Earth. No one is sure what causes Earth's magnetic field. Earth's inner core is made of iron and nickel. One theory suggests that this may produce Earth's magnetic field.

North geographic pole

South geographic pole

Picture This

3. Determine Write an N on the figure to show where Earth's north magnetic pole is. Write an S on the figure to show where Earth's south magnetic pole is.

Magnetic Materials

A magnet will not attract all metal objects. For example, a magnet will not attract aluminum foil. Only a few metals, such as iron, cobalt, and nickel are attracted by magnets. These metals can be made into permanent magnets. Permanent magnets keep their magnetism even after they have been removed from a magnetic field. Think back to what you have learned about electrons. Recall that electrons have magnetic properties. In most elements, the magnetic properties of the electrons cancel out. But in iron, cobalt, and nickel, they don't cancel out. Each atom in these metals acts like a small magnet with its own magnetic field. ☑

Even though the atoms in iron, cobalt, or nickel have magnetic fields, objects made from them do not always act like magnets. For example, a nail is made from iron. If you hold an iron nail close to a refrigerator door and then let it go, it will fall to the floor. However, you can make the nail act like a magnet temporarily and it will stick to the door.

What are magnetic domains?

In magnetic materials, the magnetic field made by each atom exerts a force on other nearby atoms. This causes the atoms to rotate and form a magnetic domain. A **magnetic domain** is a large group of atoms with their magnetic poles lined up in the same direction. Because the atoms are lined up, a domain acts like a magnet. A domain has a north pole and a south pole.

✔ Reading Check

4. Explain Why do the atoms in nickel act like a magnet?

_____

_____

How do domains line up?

An iron nail has a large number of magnetic domains that act like magnets. So why doesn't a nail act like a magnet? The poles of the domains point in different directions. Since the domains do not line up, the magnetic fields cancel each other out. Therefore, the nail does not act like a magnet. The figure below shows the magnetic domains of a nail. ☑

One way to make the domains line up is to touch a bar magnet to the nail. The domains will rotate and point in the same direction because of the magnetic field of the magnet. Now the nail acts like a magnet. The second figure shows domains that have lined up.

If the bar magnet is taken away, the atoms in the nail move around and bump each other. This causes the domains to move out of line. This is why a nail is not always a magnet.

How can you make a permanent magnet?

Permanent magnets can be made by placing a magnetic material, such as iron, in a strong magnetic field. The strong magnetic field causes the magnetic domains to line up and combine to make a strong magnetic field inside the material. This field keeps the atoms from bumping the domains out of line. The material becomes a permanent magnet.

Heating a permanent magnet causes it to lose its magnetism. Heat causes the atoms in the magnet to move faster. This moves the domains out of line. The permanent magnet will then lose its magnetic field.

Can a pole be isolated?

Suppose a magnet is broken into two pieces as shown in the figure. Is one piece a north pole and the other piece a south pole? Remember, each atom in a magnetic material acts like a tiny magnet. So, every magnet is made up of many smaller magnets that are lined up. Both pieces of a broken magnet have their own north and south poles.

5. **Explain** Why doesn't a nail act like a magnet?

_____

_____

Picture This

6. **Observe** Look at the second figure. How are the north and south poles lined up?

_____

_____

Picture This

7. **Infer** If the two pieces of magnet were put back together, what would happen to the north and south poles at the broken edge?

_____

_____

● After You Read

Mini Glossary

magnetic domains: groups of atoms with magnetic poles that line up in the same direction

magnetic field: the lines of force around a magnet

magnetic poles: the north and south pole where the forces of the magnet are strongest

magnetism: the properties of magnets and how magnets interact with each other

1. Review the definitions of the vocabulary words. Write a sentence that explains how Earth is a magnetic field.

   _____

   _____

2. Complete the chart below. List the different things you learned about magnetism that have the word *magnetic* as part of their description.

Magnetism
1. magnetic force
2.
3.
4.
5.

3. **Study Coach** Review the ideas your group wrote on the index cards. Write one idea that you all agreed was important. How did this idea help you to understand magnetism?

   _____

   _____

   _____

End of Section

Magnetism and Its Uses

section ❷ Electricity and Magnetism

● Before You Read

List some machines that use motors to make them work.

_____

_____

● Read to Learn

Electric Current and Magnetism

In 1820, a Danish physics teacher discovered that there is a connection between electricity and magnetism. He was doing a demonstration using an electric current. There was a compass near the electric circuit. He noticed that the compass needle changed direction depending on the flow of the electric current. This made the teacher hypothesize that an electric current produces a magnetic field around the wire. He also hypothesized that the direction of a magnetic field changes with the direction of the electric current. The figure shows that when the current in the wire changes direction, then the direction of the magnetic field lines also change. A strong current in the wire makes a strong magnetic field, and moving away from the wire weakens the magnetic field.

Magnetic field lines · Electron flow

Electron flow · Magnetic field lines

What is electromagnetism?

Moving electric charges are surrounded by magnetic fields. So a magnet close to an electric wire will be affected by the current. Scientists determined that magnetism and electric current must be related. The interaction between magnets and electric charges is called **electromagnetism**. The force that is produced by the interaction of electric force and magnetic force is called the **electromagnetic force**. The electromagnetic force is a fundamental force, like gravity.

What You'll Learn

- how an electric current and electromagnet produce magnetic fields
- about electromagnets and electric motors

Mark the Text

Highlighting As you read this section, use a highlighter to mark the most important ideas in each paragraph.

FOLDABLES

❸ **Note Cards** As you read this section, make note cards out of half sheets of paper to write notes about the three main topics.

Electric Current
Electromagnets
Electric Motors

1. **Describe** a solenoid.

_____

_____

_____

_____

💡 **Think it Over**

2. **Explain** How is a solenoid used to make an electromagnet?

_____

_____

_____

Picture This

3. **Draw** In the second figure, change the drawing of the electromagnet so that it will be stronger than the one shown.

Electromagnets

You can make a magnetic field stronger. This is done with a solenoid. A **solenoid** (SOH luh noyd) is a single wire carrying electric current that is wrapped into a cylinder-shaped coil. It looks like a spring. The figure to the right is a solenoid. ✔

Solenoid

Electron flow

S   N

What does a solenoid do?

A solenoid makes a magnetic field stronger. Suppose there is an electric current flowing through a wire. If a loop is formed out of the wire, the magnetic fields inside the loop combine. This makes the magnetic field inside a loop stronger than the magnetic field around a straight wire. Since a solenoid has many loops, the magnetic field inside it is even stronger.

The solenoid is used to make an electromagnet. An **electromagnet** is a solenoid wrapped around an iron core. The figure to the right is an electromagnet. The solenoid's magnetic field temporarily magnetizes the iron core. The magnetic field inside the electromagnet can be 1,000 times stronger than the field inside a solenoid without an iron core.

Electromagnet

S   N

Electron flow

How do electromagnets behave?

Electromagnets are temporary magnets. The magnetic field is present only when current is flowing in the solenoid. The magnetic field of an electromagnet can be made stronger by adding more coils of wire or by adding more current.

An electromagnet acts like any other magnet. It has a north and a south pole, attracts magnetic materials, and is attracted or repelled by other magnets. If put in a magnetic field, an electromagnet will line itself up along the magnetic field lines.

Electromagnets are useful because you can control how they act by changing the electric current flowing through the solenoid. When the current flows in the electromagnet and it moves toward or away from another magnet, electric energy is changed into mechanical energy. The mechanical energy will do work. Electromagnets make mechanical energy to do the work in many devices, such as stereo speakers and electric motors.

What makes an electromagnet rotate?

A permanent magnet can apply forces on an electromagnet to make it rotate. The figure below shows an electromagnet between the north and south poles of a permanent magnet. The north and south poles of the electromagnet are attracted to the opposite poles of the permanent magnet. This causes a downward force on the left side of the electromagnet in the figure and an upward force on the right side. These forces make the electromagnet rotate until opposite poles are lined up.

Electromagnet

Electromagnet

Permanent magnet

Electron flow

Permanent magnet

Electron flow

The electromagnet continues to rotate until its poles are next to the opposite poles of the permanent magnet. Once the north and south poles of the electromagnet are in opposite directions, the electromagnet stops rotating.

How do electromagnets make sound?

There is an electromagnet in the speaker you use when you listen to a CD. The electromagnet is connected to a speaker cone made from paper, plastic, or metal. A permanent magnet surrounds the electromagnet.

Changing Electric Current Recall that increasing the current passing through a wire increases the strength of a magnetic field. A CD player produces a voltage, a measure of electrical potential energy that can be changed into other forms of energy. As voltage increases, more electrical potential energy is ready to be changed into other forms of energy. The CD's voltage produces an electric current in the electromagnet next to the speaker cone. The CD contains information that changes the amount of electric current and its direction. The changing electric current causes the direction and the strength of the magnetic field around the electromagnet to change. A change in direction causes the electromagnet to attract or repel a permanent magnet. This makes the electromagnet move back and forth causing sound.

The electromagnet changes electrical energy to mechanical energy. The mechanical energy vibrates the speaker cone so it reproduces the sound that was recorded on the CD. ☑

Picture This

4. Identify Which pole of the electromagnet is attracted to the north pole of the permanent magnet?

_____

_____

☑ Reading Check

5. Determine What causes the speaker cone to vibrate?

_____

_____

_____

_____

How does a fuel gauge work?

A <u>galvanometer</u> is a device that uses an electromagnet to measure electric current. Look at the first figure. It is a galvanometer. An electromagnet is located between the poles of a permanent magnet and is connected to a small spring. The electromagnet rotates until the force applied by the spring is balanced with the magnetic forces on the electromagnet. A needle is attached to the electromagnet, so it turns also.

Scale
Needle
Spring
Permanent magnet
S · N
Wires carrying current to coil
Electromagnet

The second figure is a fuel gauge from a car's dashboard. The fuel gauge is a galvanometer. When the amount of gasoline in the car's fuel tank changes, the needle in the gauge moves. This movement is started by a sensor in the car's fuel tank that tells when the fuel level changes. This sensor sends an electric current to the galvanometer. The current changes as the amount of fuel changes. The current change causes the electromagnet to turn. This makes the needle move to different positions on the gauge. The gas gauge is set so that when the fuel tank is full, the needle moves to the full mark on the gauge.

Electric Motors

On hot days, do you use a fan to keep cool? A fan uses an electric motor. An <u>electric motor</u> is a machine that changes electrical energy into mechanical energy. The wires carrying electric current produce a magnetic field. This magnetic field acts in the same way as the magnetic field that a magnet produces. Two wires carrying electric current can attract each other as if they were two magnets, as shown in the figure below.

S
N
Electron flow

S
N
Electron flow

Picture This

6. **Analyze** What would make the needle in the fuel gauge turn to the right?

_____

_____

_____

Picture This

7. **Trace** the arrows between both the magnets and the wires that show the attractions.

Where are electric motors used?

Electric motors are used in many types of machines. You can find these machines in industry, agriculture, and transportation. Objects as large as airplanes and cars use electric motors. Objects as small as personal-sized fans also use motors. You can find machines with electric motors in most rooms of your house. Almost every appliance with a moving part contains an electric motor. Electric motors are used in CD and DVD players, computers, hair dryers, electric fans, and many other appliances.

How does a simple electric motor work?

Any machine that changes electric energy into kinetic energy is a motor. A motor is a permanent magnet and an electromagnet connected to an electric source. The electromagnet is positioned so that it is able to spin in the magnetic field of the permanent magnet. Electricity from the source creates the magnetic field of the electromagnet. The opposite poles of the two magnets attract one another and the electromagnet spins. The wire ends of the electromagnet rub against a circular device that is split so that the current reverses in the electromagnet at the midpoint of its spin. This allows it to spin continuously as its poles change and interact with the poles of the permanent magnet. The figure shows a simple motor set up. ☑

Simple Electric Motors

Electron flow

Electron flow

Battery

✔ **Reading Check**

8. Identify What is a machine that changes electric energy into kinetic energy called?

_____

_____

Picture This

9. Determine In the figure on the right, which direction will the loop spin, clockwise or counterclockwise?

_____

_____

● After You Read

Mini Glossary

electromagnetic force: the attractive or repulsive force between electric charges and magnets

electric motor: a device that changes electrical energy into mechanical energy

electromagnet: a single wire carrying an electric current that is wrapped around an iron core

electromagnetism: the interaection between electric charges and magnets

galvanometer: a device that uses an electromagnet to measure electric current

solenoid: a single wire carrying electric current that is wrapped into a coil shaped like a cylinder

1. Review the vocabulary words and their definitions in the Mini Glossary. Write a sentence that shows how a solenoid and an electromagnet are related.

_____

_____

2. Complete the diagram to list three items that use electromagnets to make them work.

Use electromagnets to make them work	

3. **Mark the Text** Look at the parts of the text that you highlighted. How did this help you to learn about electricity and magnetism?

_____

_____

4. What could you do if something you highlighted didn't make sense to you?

_____

_____

End of Section

Magnetism and Its Uses

section ❸ Producing Electric Current

⬤ Before You Read

Name three things you used today that use electrical energy to make them work. Where do you think the electrical energy came from?

_____

_____

_____

⬤ Read to Learn

From Mechanical to Electrical Energy

In 1831, scientists discovered that if they moved a loop of wire through a magnetic field, it caused an electric current to flow in the wire. They also found that moving a magnet through a loop of wire produced a current in the wire. In both cases, the movement caused an electric flow in the wire. In other words, mechanical energy was changed into electrical energy.

The loop of wire or the magnet has to move to make an electric current. This makes the magnetic field inside the loop change. When the magnetic field changes, it causes an electric current to flow in the wire. The current change in the wire can start a current in a nearby coil. **Electromagnetic induction** is the making of a current by a changing magnetic field.

How do electric generators work?

Generators use electromagnetic induction to change mechanical energy into electrical energy. The figure shows a hand turning the handle of a simple generator. The turning handle provides mechanical energy to rotate the coil between the poles of a permanent magnet producing a current in the coil.

Electron flow

What You'll Learn

- what electromagnetic induction is
- how a generator produces an electric current
- the difference between direct and alternating current
- how to change the voltage of an alternating current

Study Coach

Sticky-Note Discussions
Place sticky notes at parts of the section you find interesting or that you have a question about. Write the question on the sticky note.

FOLDABLES™

❻ **Build Vocabulary** As you read th is section, make a vocabulary Foldable to show that you understand the vocabulary terms.

| Electromagnetic Induction |
| Generator |
| Turbine |
| DC |
| AC |
| Transformer |

1. Explain What provides the original mechanical energy for the simple generator in the figure? (*Hint: It is not the handle.*)

_____

_____

Current Flow A simple generator with a turning wire coil is shown in the figure on the previous page and this page. The coil rotates through the magnetic field of the permanent magnet. This makes a current flow through the coil. Each time the coil makes half of a turn, the ends of the coil move past opposite poles of the permanent magnet. This causes the current to change direction. The current in the coil changes direction twice each time it makes one full turn. You can control how often the current changes direction by controlling how fast the generator rotates. In the United States, generators rotate 60 times per second to produce electric current. This is equal to 3,600 turns per minute.

Electron flow

How are electric generators used?

Generators are used in cars. They are called alternators and provide electrical energy for the car's lights. They also provide electrical energy to the spark plugs in the car's engine. The spark plugs ignite (cause a spark that burns) the fuel in the cylinders of the engine. Once the engine is running, the fuel provides the mechanical energy to turn the coil in the alternator.

How is electricity produced for your home?

Electric power plants produce most of the electrical energy. The huge generators in electric power plants operate in a different way from the alternator. The coil does not rotate. Instead, the permanent magnet rotates. Mechanical energy rotates the magnet, and electrical current is produced in the coil.

The electrical energy you use in your home comes from a power plant with huge generators. These generators have many coils of wire wrapped around huge iron cores. The magnets are connected to a **turbine** (TUR bine), a large windmill-like wheel. The turbine rotates when it is pushed by steam, water, or wind, and the rotating magnets produce the electric current in the wire coils.

Power plants use three different kinds of energy to make electricity. They use thermal, wind, or water energy. For thermal energy, power plants burn fossil fuels such as oil, natural gas, and coal, or use heat made by nuclear reactors. The thermal energy is used to heat water and produce steam. When the steam pushes the turbine blades, the thermal energy is changed into mechanical energy. The generator then changes the mechanical energy into the electrical energy you use.

✔ Reading Check

2. Identify What are three types of energy used by power plants?

_____

_____

Wind Energy Some power plants use the mechanical energy in falling water to turn the turbines. Other power plants use wind energy. Fields of windmills, like those in the figure below, use mechanical energy in wind to turn the generators. The propeller on each windmill is connected to an electric generator. The turning propeller rotates a coil or a permanent magnet.

Picture This

3. **Make Connections** What do the propellers rotate?

_____

_____

Direct and Alternating Currents

Have you ever had a power outage in your house? The electrical devices do not work because electrical energy is not coming into your house. Some electrical devices use batteries as a primary source of energy. A battery-operated radio is an example of this type of device. The current produced by a battery is different from the current from an electric generator.

A battery produces a direct current. A **direct current** (DC) is a current that flows in only one direction through a wire. A CD player or any other appliance that plugs into a wall outlet uses alternating current. An **alternating current** (AC) reverses or changes the direction of the current twice during each rotation of the coil. Electronic devices that use batteries for backup energy, such as a radio, usually need direct current to operate. When the radio is plugged into a wall outlet, electronic parts in the radio change the alternating current to direct current. ☑

Think it Over

4. **Apply** List two devices that use DC only and two devices that use AC only.

_____

_____

_____

_____

Transmitting Electrical Energy

Have you ever seen power lines along a highway? They carry electrical energy from a power plant to buildings. When electrical energy travels through wires, some of the electrical energy changes into heat because of electrical resistance in the wires. As wires get longer, there is more electrical resistance and the heat increases. One way to reduce the heat produced in a power line is to send the electrical energy at high voltage. This voltage in the lines is too high for appliances to use. A transformer is used to decrease the voltage before it enters your home.

Reading Check

5. **Explain** Why can some radios work on both batteries and when plugged into a wall outlet?

_____

_____

_____

6. Identify What causes the magnetic fields in the primary coil to change direction?

_____

_____

Picture This

7. Identify In the two figures, circle the labels that are the same.

Transformers

A **transformer** is used to increase or decrease the voltage of an alternating current. A transformer has a primary coil and a secondary coil. Both coils are wrapped around the same iron core. An input voltage of alternating current passes through the primary coil. This causes the coil's magnetic field to magnetize the iron core. When the current in the primary coil changes direction, this causes the magnetic fields in the primary coil and the iron core to also change directions. This causes an output voltage in the secondary coil. ✔

The figures show two kinds of transformers, a step-up transformer and a step-down transformer.

60 volts AC in — Primary coil 10 turns of wire — Increase 2 times — Secondary coil 20 turns of wire — 120 volts AC out

Step-up Transformer

240 volts AC in — Primary coil 40 turns of wire — Decrease 2 times — Secondary coil 20 turns of wire — 120 volts AC out

Step-down Transformer

Step-Up Transformer A step-up transformer increases voltage so that the output voltage is greater than the input voltage. The secondary coil then has more turns than the primary coil. In the first figure, an input voltage of 60 volts in the primary coil provides an output voltage of 120 volts in the secondary coil. The secondary coil has twice as many turns as the primary coil has. Therefore, the output voltage is twice as large as the input voltage.

Step-Down Transformer A step-down transformer decreases voltage so that the output voltage is less than the input voltage. The secondary coil then has fewer turns than the primary coil. In the second figure, the input voltage of 240 volts in the primary coil is changed to an output voltage of 120 volts in the secondary coil. The secondary coil has half as many turns as the primary coil has. Therefore, the output voltage is one-half of the input voltage.

What path does an alternating current follow?

Power plants usually make alternating current because the voltage can be increased or decreased with transformers. As the electrical energy leaves the power plant, a step-up transformer increases the voltage. This electrical energy is carried along power lines. When the electrical energy leaves the power lines to enter a building, a step-down transformer decreases the voltage. Even though the voltage is changed, the amount of electrical energy is not changed.

● After You Read

Mini Glossary

alternating current (AC): electrical current that changes its direction twice during each rotation of a coil

direct current (DC): electrical current that flows in only one direction through a wire

electromagnetic induction: a changing magnetic field producing an electric current in a wire

generator: a device that uses electromagnetic induction to change mechanical energy into electrical energy

transformer: a device that increases or decreases the voltage of an alternating current

turbine: a large wheel that rotates when it is pushed by water, wind, or steam

1. Review the definitions of the vocabulary words in the Mini Glossary. Choose one of the words and write what it means in your own words.

_____

_____

2. Complete the chart below to organize information you have learned about electric current.

3. **Study Coach** Think about what you have learned in this section. Look at the parts you marked with sticky notes. How did these notes help you learn?

_____

_____

End of Section

Energy Sources and the Environment

section ❶ Fossil Fuels

What You'll Learn
- what fossil fuels are and how they formed
- how fossil fuels are used

Mark the Text

Locate Information
Underline every heading in the section that asks a question. Then, highlight the answers to those questions as you find them.

FOLDABLES

A Describe Make the following Foldable to help describe fossil fuels and how they are converted to energy.

Describe fossil fuels.

Explain how they are converted into energy.

● Before You Read

Write what you think the energy source is for the heating system, water heater, and stove in your home.

_____

● Read to Learn

Using Energy

Energy is used in many ways. Furnaces use thermal energy to heat buildings, and stoves use it to cook food. Air conditioners use electrical energy to remove thermal energy from buildings. Vehicles use mechanical energy to move people and materials.

How is energy used?

Remember, the law of conservation of energy says that energy cannot be created or destroyed. Energy can only transform, or change, from one form to another. When you use energy, you change it to another form. For example, the chemical energy in fuel changes to thermal energy in stoves. The thermal energy heats the food you cook.

Some energy changes are not useful. For example, about 10 percent of the electrical energy that flows through power lines changes to thermal energy in the lines. This thermal energy is the result of friction between the flowing electrons and the atoms in the wire.

How is energy used in the United States?

The United States uses more energy than any other country. Look at the figure at the top of the next page. The first circle graph shows energy use in the United States. Energy is used in homes for heating and cooling, running appliances, lighting buildings, and heating water. Energy used for transportation powers cars, trucks, and airplanes. Businesses use energy to heat, cool, and light stores and offices. Factories and farms use energy to make products and grow crops.

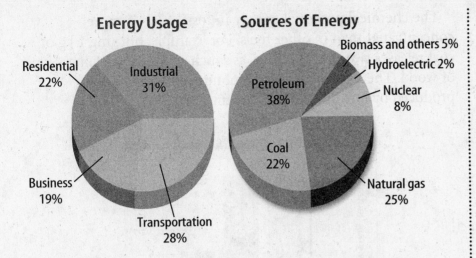

Energy Usage

Residential 22%
Industrial 31%
Business 19%
Transportation 28%

Sources of Energy

Biomass and others 5%
Hydroelectric 2%
Petroleum 38%
Nuclear 8%
Coal 22%
Natural gas 25%

Applying Math

1. **Interpret a Graph** What percentage of energy used in the United States is used by businesses and industries together?

_____

The second circle graph shows the main sources of the energy used in the United States. Almost 85 percent of the energy used comes from burning petroleum, natural gas, and coal. Nuclear power plants provide about eight percent of the energy used in the United States.

Making Fossil Fuels

A car might use several gallons of gasoline, or fuel, in one hour of freeway driving. Did you know that it takes millions of years to make fuels like oil, natural gas, and coal? These fuels are called fossil fuels. **Fossil fuels** are formed from the decayed remains of ancient plants and animals.

Petroleum, or oil, and natural gas form when matter made up of dead plants and animals piles up on the ocean floor. The matter is buried slowly under layers of sediment. Chemical reactions caused by heat and great pressure change the matter into oil and natural gas. The oil and gas may bubble up to the surface or be trapped beneath a layer of rock. Coal is formed from dead plants in ancient swamps in much the same way. ☑

Why are fossil fuels concentrated energy sources?

Fossil fuels are energy sources. They store chemical energy in the bonds between the atoms. When a fossil fuel burns, a chemical reaction takes place. The carbon and hydrogen atoms in the fossil fuel combine with oxygen in the air to form carbon dioxide and water. Heat and light are produced when chemical potential energy is converted in the chemical reaction. Chemical bonds in fossil fuels are the source of the potential energy.

Reading Check

2. **Explain** What is necessary to change matter into fossil fuels?

_____

_____

_____

Reading Essentials **125**

The chemical potential energy in fossil fuels is more concentrated than in other fuels. For example, burning 1 kg of coal releases two to three times as much energy as burning 1 kg of wood. The figure shows the amount of energy that is produced by burning 1 g of different fossil fuels.

Applying Math

3. **Compare** About how much more energy is released by burning 1 g of gasoline than 1 g of wood?

_____

_____

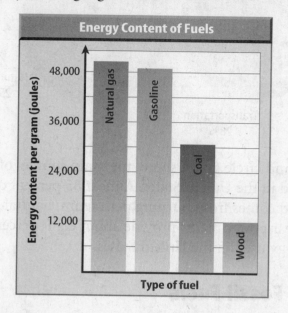

☑ Reading Check

4. **Identify** What two types of atoms do petroleum molecules contain?

_____

_____

Petroleum

Petroleum is a liquid fuel formed by decayed organisms. It is a mixture of thousands of chemical compounds. Most of these compounds are hydrocarbons, which means their molecules contain only carbon atoms and hydrogen atoms. Petroleum, or crude oil, is pumped from wells deep in Earth's crust. Millions of gallons of petroleum are pumped every day. ☑

How are hydrocarbons separated?

Petroleum is a mixture of many different hydrocarbon compounds. Some are thick and heavy, like asphalt. Others are very light, like gasoline. Heavy hydrocarbon compounds have large molecules with many carbon and hydrogen atoms. Light hydrocarbon compounds have small molecules with few carbon and hydrogen atoms.

Petroleum compounds must be separated or refined in a process called fractional distillation. At a refinery, crude oil is heated in the bottom of a tall, slender tank. The compounds boil and turn to vapor. Hydrocarbons with the lowest boiling points rise to the top of the tank. Those with high boiling points remain liquid and are drained from the bottom of the tower.

Most of the hydrocarbons that are separated at a refinery end up as fuels such as gasoline, jet fuel, diesel fuel, and heating oil. The rest are used as ingredients in other chemical products.

What are other uses for petroleum?

Not all of the products we get from petroleum are turned into fuels. In the United States, about 15 percent of petroleum-based substances are used for other products. Plastics and synthetic fabrics are made from the hydrocarbons found in petroleum. Lubricants, such as grease and motor oil, and the asphalt used in roads also are made from petroleum.

Natural Gas

Natural gas and petroleum are formed by the same chemical processes. However, natural gas is a gas and petroleum is a liquid. Underground, natural gas is usually found above petroleum deposits. Natural gas is made up mostly of the hydrocarbon methane, CH_4. It also contains other hydrocarbon gases such as propane, C_3H_8, and butane, C_4H_{10}. Natural gas provides energy for cooking, heating, and manufacturing. About one-fourth of the energy used in the United States comes from natural gas. Natural gas contains more energy per kilogram than petroleum or coal. It also burns more cleanly than other fossil fuels and produces fewer pollutants.

Coal

Coal is a solid fossil fuel. It is found in underground mines. From 1900 to 1950, coal provided more than half of the energy that was used in the United States. Now, almost two-thirds of the energy used comes from petroleum and natural gas. Only about one-fourth comes from coal. Almost 90 percent of all the coal that is used in the United States is burned by power plants to generate electricity. ☑

Where does coal come from?

Coal deposits, the places where coal can be found, were once ancient swamps. Over thousands of years, plants that lived in the swamps died. Coal formed from the dead plants. The amount of coal worldwide is estimated to be 20 to 40 times greater than the supply of petroleum.

Coal is a solid mixture of hydrocarbons and other chemicals, such as sulfur compounds. Because of the chemical compounds coal contains, burning it often produces more pollutants than burning petroleum or natural gas. The sulfur compounds in coal, when burned, produce sulfur dioxide. Sulfur dioxide is a major cause of acid rain.

Think it Over

5. **Identify** List three products that are made from petroleum.

_____

_____

_____

✔ Reading Check

6. **Use Percentages** About what percent of all the coal that is used in the United States is used to produce electricity?

_____

How Electricity is Generated

Water tank · Water · Water · Steam · Turbine · Fuel · Steam · Intake pipe · Cooling water · Generator · Transformer · Power lines

Picture This

7. Observe Circle the part of the figure that shows where the electric current is produced.

Applying Math

8. Calculate What percent of energy is converted to thermal energy during fossil fuel conversion?

_____

_____

_____

Generating Electricity

Burning fossil fuels produces almost 70 percent of the electrical energy used in the United States. The chemical energy contained in fossil fuels is converted or changed to electrical energy in electric power stations. Follow the process in the figure above. First, the fuel is burned and converted to thermal energy. The thermal energy heats water and produces steam under high pressure. Then the steam spins the blades of a turbine. The turbine is connected to an electric generator. The spinning turbine turns magnets inside the generator. This produces electric current. Finally, the electric current is sent to homes, schools, and businesses through power lines.

Efficiency of Power Plants

When fossil fuels are burned, not all the chemical energy in the fuel changes into electrical energy. Some of the energy is converted into thermal energy. Efficiency is the percentage of the total energy that is usable. You can find the overall efficiency of a power plant by multiplying the efficiencies of each stage of the process. In fossil fuel-burning power plants, the overall efficiency is only about 35 percent. Only about 35 percent of the chemical energy contained in fossil fuels is delivered as electrical energy. The rest of the energy is converted to thermal energy that is transferred to the environment.

The Costs of Using Fossil Fuels

Fossil fuels are useful for making electricity. They also provide power for transportation. But they have some harmful side effects. Petroleum products and coal give off smoke when they are burned. The smoke contains small particles called particulates. The particulates can cause breathing problems for some people.

What are the effects of burning fossil fuels?

Burning fossil fuels also releases carbon dioxide. The amount of carbon dioxide in Earth's atmosphere has increased greatly over the past 100 years. The increase is due to burning fossil fuels. An increase of carbon dioxide in the atmosphere could cause Earth's surface temperature to increase. ✔

What are the problems with using coal?

There is more coal than any other fossil fuel. However, burning coal releases more pollutants than oil or natural gas do. Many power plants that burn coal remove some pollutants before they are released into the air. Mining coal also can be dangerous. Some miners suffer from lung diseases that are caused by breathing coal dust for long periods of time.

Nonrenewable Resources

Fossil fuels are nonrenewable resources. <u>**Nonrenewable resources**</u> are resources that cannot be replaced by natural processes as quickly as they are used. Fossil fuel reserves are decreasing. Reserves are the amounts of fossil fuels remaining in the ground. At the same time, the population and industrial needs for fuel are increasing. The graph below shows that over the next 50 years, oil production might decrease from more than 25 billion barrels to about 5 billion barrels per year.

Conserving Fossil Fuels

Since reserves of fossil fuels are decreasing, the price of fossil fuels is increasing. The need for energy keeps increasing because the population of the world is increasing. One way to meet energy needs and to reduce the use of fossil fuels is to get energy from other sources.

✔ Reading Check

9. List two harmful side effects of using fossil fuels.

_____

_____

_____

Picture This

10. Reading Graphs In what year was global oil production the highest?

_____

• After You Read

Mini Glossary

fossil fuel: fuels that are formed from the decayed remains of ancient plants and animals

nonrenewable resource: resources that cannot be replaced by natural processes as quickly as they are used

petroleum: a liquid fuel formed by decayed organisms

1. Review the terms and their definitions in the Mini Glossary. Explain why fossil fuels are called nonrenewable resources.

_____

_____

2. Complete the graphic organizer below to organize the information you learned in this section about fossil fuels.

	Comes From	Mainly Used For	Effects of Burning
Petroleum			
Natural Gas			
Coal			

3. ◄ **Mark the Text** Think about what you have learned. How did underlining the questions and their answers help you as you read the section?

_____

_____

_____

End of
Section

Energy Sources and the Environment

section ❷ Nuclear Energy

● Before You Read

What do you know about nuclear energy? Write what you know about nuclear energy on the lines below.

_____

_____

Copyright © Glencoe/McGraw-Hill, a division of The McGraw-Hill Companies, Inc.

What You'll Learn

■ how a nuclear reactor works
■ advantages and disadvantages of nuclear energy

● Read to Learn

Nuclear Fusion

A different type of nuclear reaction, thermonuclear fusion, is the source of the Sun's energy. In **fusion**, two very small nuclei collide and stick, or fuse, together at very high temperatures. The most common fusion in the Sun fuses two hydrogen ions into a helium nucleus. This reaction makes even more energy than fission. A small amount of mass is converted into energy. Fusion is the most concentrated energy source known.

What are the advantages and disadvantages of fusion?

Fusion has advantages and disadvantages. One advantage is it uses hydrogen as fuel. Hydrogen is abundant on Earth. Another advantage is that a fusion reaction produces helium, which is not radioactive. One disadvantage is that fusion can take place only at extremely high temperatures. A great deal of energy needs to be used to reach these temperatures. The amount of energy used in this process can be more than the energy produced in the reaction. Another problem is how to contain a reaction that happens at such high temperatures. Until solutions to these problems are found, the use of fusion as an energy source is not possible.

Study Coach

Make Flash Cards For each heading in this section, think of a question your teacher might ask on a test. Write the question on one side of a flash card. Then write the answer on the other side. Quiz yourself until you know the answers.

1. **Identify** How is nuclear fission controlled in a nuclear reactor?

_____

_____

_____

Using Nuclear Energy

Electric power plants that do not burn fossil fuels have been developed. Some of these are nuclear power plants. Since fossil fuels are not burned, nuclear power plants do not cause pollution or release carbon dioxide. Nuclear power plants convert nuclear energy to electrical energy through a process called nuclear fission. Nuclear **fission** is the process of breaking apart the nucleus of an atom. This releases energy. In the process, an extremely small amount of matter is converted into a huge amount of energy. Today almost 20 percent of the electricity produced in the United States comes from nuclear power plants. Nuclear power plants produce about nine percent of all the energy used in the United States. In 2003, there were 65 nuclear power plants in the United States. These plants contained 104 nuclear reactors.

Nuclear Reactors

A **nuclear reactor** uses the energy from controlled nuclear reactions to generate electricity. Nuclear reactors can have different designs, but they all have the same major parts. Nuclear reactors contain a fuel that can undergo nuclear fission. They also have control rods that can speed up or slow down the nuclear reactions. Nuclear reactors have a cooling system that keeps the reactor from being damaged by the heat that is produced. The actual fission of the atoms of the fuel happens in a small part of the reactor known as the core. ☑

What is nuclear fuel?

The fuel for a nuclear reactor must be an element whose nucleus can undergo fission, or split. One element that is used is uranium. Uranium dioxide is the fuel that is used most often in nuclear reactors. One isotope of uranium, called U-235, has a nucleus that can split. However, only about 0.7 percent of naturally occurring uranium is the U-235 isotope. Uranium must be enriched to increase the amount of U-235. Uranium used as nuclear fuel is enriched so that it contains three to five percent U-235.

Copyright © Glencoe/McGraw-Hill, a division of The McGraw-Hill Companies, Inc.

What is in the reactor core?

The reactor core contains small, enriched uranium dioxide pellets that are sealed inside metal fuel rods. The pellets look like pencil erasers. They are placed end to end in a tube. The tubes are bundled into what are called fuel rods. Then they are covered with a metal alloy. A typical reactor core contains about 100,000 kg of uranium in hundreds of fuel rods. For every kilogram of uranium that undergoes fission in the core, 1 g of matter is converted into energy. The energy released by this 1 g of matter is equal to the energy released by burning more than 3 million kg of coal.

How does nuclear fission happen?

The figure shows a uranium nuclear fission reaction. When a neutron hits the nucleus of a U-235 atom, the nucleus splits into two smaller nuclei. Two or three neutrons are released at the same time. The smaller nuclei are called fission products. The neutrons from the first reaction hit other U-235 nuclei. Every uranium nucleus that splits apart releases neutrons that cause other uranium atoms to split apart. This process is called a nuclear chain reaction.

What happens when a nucleus splits?

The number of nuclei that split in a nuclear chain reaction can more than double at each step of the reaction. So, a huge number of nuclei can be split after only a small number of reactions. Nuclear chain reactions take place in less than a second. If nuclear chain reactions aren't controlled, the chain reaction could create an explosion.

Applying Math

2. **Calculate** Suppose that over a period of time, 100 kg of nuclear fuel is converted to energy in a nuclear power plant. How many kilograms of coal would have to be burned to release the same amount of energy? Write your answer in both numerals and words.

_____

_____

_____

_____

Picture This

3. **Identify** Circle the fission products.

How is the chain reaction controlled?

To control a nuclear chain reaction, some of the neutrons that are released when U-235 splits apart must be stopped from hitting other U-235 nuclei. Control rods containing boron or cadmium metals are used to absorb some of the neutrons from the reactions. The rods are inserted into the reactor core. Moving the control rods deeper into the reactor makes them absorb more neutrons. This slows down the chain reaction. Eventually, only one of the neutrons released in the fission of each of the U-235 nuclei strikes another U-235 nucleus. When this happens, energy is released at a constant rate.

Nuclear Power Plants

Nuclear fission reactors produce electricity in almost the same way as other power plants. Thermal energy from the reactor is used to boil water and produce steam. The figure below shows a nuclear reactor. Cool water, or coolant, is pumped through the reactor. The coolant absorbs heat from the core and then is pumped to a heat exchanger in the boiler. The thermal energy is transferred from the coolant and boils water to make steam. This steam then drives a turbine. The turbine rotates an electric generator.

The overall efficiency of nuclear power plants is similar to the efficiency of fossil fuel power plants.

Picture This

4. **Identify** In the figure, circle where the coolant absorbs heat, put a box around where the heat is exchanged, circle the turbine, and put a box around the generator.

The Risks of Nuclear Power

Nuclear power plants don't use fossil fuels. They don't release pollutants into the atmosphere. They don't produce carbon dioxide. These are all advantages over fossil fuel power plants.

However, there are disadvantages to producing energy from nuclear fission. The mining of uranium can damage the environment. Also, water that is used as a coolant in the reactor core must be cooled before being released into streams and rivers. If the water is not cooled properly, its excess heat could harm fish and other animals and plants in the water.

What happens when radiation is released?

A serious risk of nuclear power is the escape of harmful radiation from power plants. If released, radiation could damage living organisms. To prevent accidents, nuclear reactors and plants have many safeguard systems, strict safety precautions, and highly trained workers. Yet, accidents have happened.

A reactor overheated in 1986 in Chernobyl, Ukraine. The building was partially destroyed, and radioactive materials escaped. As a result, 50 people died of radiation sickness. Up to 600,000 people may have been exposed to radiation that was carried by winds. ☑

The Disposal of Nuclear Waste

The U-235 in nuclear fuel is used up after about three years. The used rods contain radioactive fission products and the remaining uranium. **Nuclear waste** is any radioactive by-product that is created when radioactive materials are used.

What happens to nuclear waste?

There are two kinds of nuclear waste. One is low-level waste and the other is high-level waste.

Low-Level Waste Low-level nuclear wastes contain a small amount of radioactive material. They usually do not contain radioactive materials that have long half-lives. A half-life is how long it takes for half of the material to decay. Products of some medical and industrial processes are low-level wastes. These include items of clothing used in handling radioactive materials. Air filters from nuclear power plants and discarded smoke detectors are also low-level wastes. Low-level nuclear wastes are usually sealed in containers and buried deep in the ground. When it is diluted enough, low-level waste sometimes is released into the air or water.

High-Level Waste High-level nuclear waste contains larger amounts of radioactive material. High-level waste is made by nuclear power plants and nuclear weapons programs. Some of this material has long half-lives and will remain radioactive for tens of thousands of years. High-level waste is usually stored in deep pools of water or disposed of in extremely strong and long-lasting containers.

☑ **Reading Check**

5. **Explain** What can happen if a person is exposed to too much radiation?

_____

_____

💡 **Think it Over**

6. **Draw Conclusions** What would you say is the main risk from nuclear power plants and nuclear fuel?

_____

_____

_____

_____

● After You Read

Mini Glossary

fission: when a small amount of mass energy is converted into a tremendous amount of thermal energy

fusion: when atomic nuclei combine at very high temperatures resulting in a small amount of mass being converted into a large amount of thermal energy

nuclear reactor: a system that uses the energy from controlled nuclear reactions to generate electricity

nuclear waste: any radioactive by-product that is created when radioactive materials are used

1. Review the terms and their definitions in the Mini Glossary. Explain how nuclear reactors and nuclear waste are connected.

_____

2. Complete the outline to help you organize what you learned about nuclear energy.

> **Nuclear Energy**
>
> I.   How does nuclear fission happen?
>
>   A. _____
>
>   B. The nuclei split, producing energy and more neutrons.
>
>   C. _____
>
> II. How does a reactor work?
>
>   A. Coolant absorbs heat from the core.
>
>   B. _____
>
>   C. Steam drives a turbine that turns a generator.
>
> III. What are advantages of using nuclear energy?
>
>   A. _____
>
>   B. Doesn't cause air pollution or release carbon dioxide.
>
> IV. What are disadvantages of using nuclear energy?
>
>   A. _____
>
>   B. Harmful radiation could escape.

3. **Study Coach** How could you use the flash cards you made to study for a test?

_____

End of Section

Copyright © Glencoe/McGraw-Hill, a division of The McGraw-Hill Companies, Inc.

Energy Sources and the Environment

section ❸ Alternative Energy

● Before You Read

Have you ever heard the term *alternative*? An alternative is a choice or option that is different from the usual one. Write the name of an alternative energy source on the line below.

_____

● Read to Learn

Energy Options

The demand for energy keeps increasing, but supplies of fossil fuels are decreasing. Nuclear reactors produce nuclear waste that has to be disposed of safely. To solve these problems, other sources of energy are being developed. A **renewable resource** is an energy source that is replaced by natural processes as quickly as it is used. Some alternative energy sources are renewable resources.

Energy from the Sun

You can feel energy from the Sun if you step outside on a clear day. The amount of solar energy that falls on the United States in one day is more than the total amount of energy used in the United States in one year. The Sun is expected to produce energy for several billion more years. Because of this, solar energy will not run out. It is considered a renewable resource.

What are solar cells?

Many objects, such as solar-powered calculators, use solar energy for power. They use a **photovoltaic cell** that converts radiant energy from the Sun directly into electrical energy. Photovoltaic cells are also called solar cells.

What are solar cells made of?

Solar cells are made of two layers of semiconductor materials. The top layer has many electrons, but the bottom layer does not. The semiconductors are sandwiched between two layers of conducting metal.

What You'll Learn

- why alternate energy sources are needed
- alternate methods for generating electricity

Mark the Text

Identify Main Points
Highlight each kind of alternative energy source as it is discussed in this section. Circle the main points about each energy source.

FOLDABLES

ⓒ Compare and Contrast Make the following one–sheet Foldable out of notebook paper. Compare and contrast information about alternative energy sources. Label the columns on the right Advantages and Disadvantages.

Alternative Energy Sources

How do solar cells make electricity?

Remember, a flow of electrons is an electric current. In a solar cell, energy from sunlight knocks electrons away from their atoms in the top semiconductor. These electrons flow through the top metal layer, then through a circuit to the bottom metal layer, and then to the bottom semiconductor. An electric current is produced. Only about 7 percent to 11 percent of the radiant energy from the Sun that strikes a solar cell is converted to electrial energy.

How is solar energy used?

It is cheaper to make electricity from fossil fuels than with solar cells. But in some areas, where it is too expensive to run power lines, solar cells are a good source of electrical power.

Another way to generate electricity from solar energy is in a solar-concentrating power plant. In a solar-concentrating power plant, mirrors focus sunlight onto a receiver. Heat from the sunlight boils water to produce steam. The steam turns a turbine, making electricity. The world's largest solar-concentrating power plant is in the Mojave Desert in California. The plant generates over 350 megawatts of power. This is enough electrical power to meet the needs of about 500,000 people.

Energy from Water

Water behind a dam has gravitational potential energy. When water is released from a dam, its potential energy is changed to kinetic energy because the water is in motion. The kinetic energy of water flowing from a dam can be used to spin a turbine and generate electricity. The figure shows how a dam makes electricity.

Applying Math

1. **Use Statistics** How many megawatts of electricity would be needed for a city of 3 million people?

Picture This

2. **Draw Conclusions** What moves the turbine in hydroelectric power plants?

What is hydroelectricity?

Electricity produced from the energy of moving water is called **hydroelectricity**. A dam that produces hydroelectricity is a hydroelectric dam. About 8 percent of the electrical energy used in the United States is produced by hydroelectric power plants. Not all of the kinetic energy of flowing water is changed to electrical energy in a hydroelectric dam. Some of the energy is lost to friction, and some water escapes without moving the turbine. But, because no heat or steam is needed to spin the turbine, hydroelectric dams are almost twice as efficient as fossil fuels or nuclear power plants. ☑

What are advantages and disadvantages of hydroelectricity?

The lakes formed by dams can provide water for drinking and irrigation. These lakes also may be used for boating and swimming. After a hydroelectric dam is paid for, the cost of the electricity it produces is relatively cheap.

However, artificial dams can disturb the balance of natural ecosystems. For example, dams can prevent migratory fish from moving up rivers to lay their eggs. Also, the lakes formed by dams destroy land habitats when the land is submerged. Some rivers that would make great sources for hydroelectric power are often too far from cities that need the power.

Energy from the Tides

Gravity from the Sun and Moon causes bulges in Earth's oceans that move westward as Earth rotates. The movement of the bulges causes the height of the ocean to rise and fall continually. These rises and falls are ocean tides that can generate hydroelectric power. As the tide comes in, the moving water spins a turbine in a hydroelectric dam. The water then is trapped behind the dam. As the tide lowers, the water behind the dam flows back out to the ocean. This spins the turbines again and generates more electrical power. ☑

Hydroelectricity from tides causes almost no pollution. It is about as efficient as hydroelectricity from dams. However, there aren't many places on Earth where tides are great enough for generating hydroelectricity. There is only one power plant in North America that produces electricity from tides. It is in Annapolis Royal, Nova Scotia in Canada.

☑ **Reading Check**

3. **Explain** Why are hydroelectric dams almost twice as efficient as fossil fuels or nuclear power plants?

_____

_____

_____

☑ **Reading Check**

4. **Identify** What causes tides?

_____

_____

5. **Think Critically** Which do you think would work best near your community, a hydroelectric dam or a windmill farm? Explain.

_____

_____

_____

_____

Harnessing the Wind

Windmills have been used for centuries to pump water. Now, windmills are made to generate electricity from wind's energy. A windmill has a propeller that is turned by the wind. The propeller then spins a generator. Windmills often are placed together in large groups called windmill farms in areas that are regularly windy. Some windmill farms contain several hundred windmills.

Because the wind is never constant, electricity from windmills is not always reliable. Windmills are also only 20 percent efficient. Researchers are trying to increase this efficiency. Wind energy has some disadvantages. Windmills change the landscape and they make noise. They can also confuse migrating birds. However, windmills do not use nonrenewable natural resources and they do not pollute the atmosphere or water.

Energy from Inside Earth

The interior of Earth is hot from the decay of radioactive elements. This heat is called geothermal heat. Geothermal heat causes the rock beneath Earth's crust to soften and melt. This hot molten rock is called magma. The thermal energy that is contained in hot magma is called **geothermal energy**.

Where does magma escape from Earth?

In some places, Earth's crust has cracks or thin spots that allow magma to rise near the surface. Active volcanoes are places where magma escapes from Earth's crust. In other areas, magma heats water underground, forming hot springs and geysers, such as Old Faithful in Yellowstone National Park. In some areas, water heated by geothermal energy is pumped to houses to provide heat.

How does a geothermal power plant generate electricity?

Geothermal energy also can be used to make electricity. A geothermal power plant is shown in the figure on the next page. Where magma is close to Earth's surface, the surrounding rocks are also hot. Water is pumped into the hot rocks through a well. The hot rocks boil the water to produce steam. The steam then returns to the surface. It rotates the turbines that spin generators.

Geothermal power plants generate electricity without using fossil fuels. The also create very little pollution. But, their efficiency is only about 16 percent. Also, there aren't many places where magma is close enough to Earth's surface to use the geothermal energy.

Geothermal Power Plant

Cooling tower / Condenser

Electricity

Generator

Turbine

Pump

Steam

Fractures in rock

Magma

Alternative Fuels

The use of fossil fuels could be reduced if cars used other energy sources. Cars have been developed that use electrical energy supplied by batteries. Hybrid cars use both electric motors and gasoline engines. Hydrogen gas is another possible alternative fuel. Hydrogen may one day replace gasoline as a fuel for automobiles. Burning hydrogen instead of gasoline is better for the enviroment. It produces only water vapor when it burns and creates no pollution.

What are biomass fuels?

Could any other materials be used to heat water and produce electricity like fossil fuels and nuclear fission? Biomass can be burned to convert stored chemical energy to thermal energy. **Biomass** is renewable organic matter, such as wood, sugarcane fibers, rice hulls, and animal manure. Burning biomass for energy is probably the oldest use of natural resources to meet human energy needs.

Picture This

7. **Highlight** Use a highlighter to follow the path of the water and steam in the geothermal power plant. Be sure to start at the pump.

Think it Over

8. **Draw a Conclusion** Why might burning biomass as a source of energy be harmful to the environment?

_____

_____

_____

● After You Read

Mini Glossary

biomass: renewable organic matter, such as wood and plant fibers

geothermal energy: thermal energy contained in hot magma

hydroelectricity: electricity produced from the energy of moving water

photovoltaic cell: a device that converts radiant energy from the Sun into electrical energy

renewable resource: an energy source that is replaced almost as quickly as it is used

1. Review the terms and their definitions in the Mini Glossary. Is hydroelectricity a renewable resource? Why or why not?

_____

_____

2. Complete the chart below to summarize what you know about these alternative energy sources.

Energy Sources			
Solar	**Hydroelectricity**	**Tidal**	**Wind**
Advantage:	Advantage:	Advantage:	Advantage:
Disadvantage:	Disadvantage:	Disadvantage:	Disadvantage:

3. ◀ **Mark the Text** You identified main points as you read this section. Would you recommend this strategy to a friend who is reading this chapter? Why or why not?

_____

End of Section

Energy Sources and the Environment

chapter 8

section ❹ Environmental Impacts

● Before You Read

How do you think the total number of people on Earth in 1900 compares to the human population in 2000? How will the population change in the next 100 years?

_____

_____

_____

● Read to Learn

Population and Carrying Capacity

Humans share Earth with many other living things. There are large and small animals. Plants also share Earth. Bears, ants, fish, and trees are examples of populations. A **population** is the total number of individuals of one species occupying the same area. That area can be large or small. A human population can be of one community, such as Los Angeles, or the entire planet.

Is Earth's population increasing?

In 2000, there were 6.1 billion humans on Earth. Each day, approximately 200,000 more humans are born. Earth is experiencing a population explosion. The word *explosion* is used because the rate at which the population is growing is rapidly increasing.

The growth rate of the human population is shown in the graph on the next page. Notice the slow population growth through the 1800s. After the mid-1800s, the population increased much faster. By 2050, the population is predicted to be about 9 billion—one and a half times what it is now.

What You'll Learn

- the causes and effects of pollution on land, water, and air
- methods used to control the types and sources of pollution
- how to protect and preserve natural resources

Study Coach

Create a Quiz As you read the text, write a quiz question for each topic. When you have finished reading, try to answer your questions correctly.

FOLDABLES

Ⓓ Think Critically Make a 3-tab Foldable. Take notes on human impacts to land, water, and air.

What are population limits?

Each person uses space and resources. Population size depends on the amount of resources that are available on Earth and how the population uses those resources. A population without resources will suffer and the population size could decrease.

Humans once thought that Earth had an endless supply of natural resources, such as fossil fuels, clean water, rich soils, and metals. It's now known that Earth's resources are limited. The planet has a carrying capacity. **Carrying capacity** is the largest number of individuals of a particular species that the environment can support. Unless Earth's resources are treated with care, they could disappear and humans could reach their carrying capacity. ✔

People and the Environment

How will you affect the environment over your lifetime? By the time you're 75 years old, you will have produced enough garbage to equal the mass of 11 African elephants—53,000 kg. You will have consumed 18 million liters of water. That's enough water to fill 68,000 bathtubs. If you live in the United States, you will have used several times as much energy as a person living elsewhere in the world.

How do daily activities affect the environment?

Every day, the environment is affected by human activity. The energy for electricity often comes from fossil fuels. Fossil fuels must be mined and then burned to create energy. Each of these steps changes the environment.

Water that is used must be treated to make it as clean as possible before being returned to the environment. You eat food, which needs soil to grow. Many vegetables are grown using chemical substances, such as pesticides to kill insects and herbicides to kill weeds. These chemicals can get into water supplies. If the concentration of chemicals is too great, it can threaten the health of living things. ✔

✔ Reading Check

1. **Define** What is carrying capacity?

_____

_____

✔ Reading Check

2. **Identify** Why are pesticides and herbicides used when growing vegetables?

_____

_____

_____

What are pollutants?

Many of the products humans use are made of plastic and paper. Plastic is made from oil. The process of refining oil can produce **pollutants,** which are substances that contaminate the environment. Several of the steps used in making paper can damage the environment. Trees are cut down. Trucks use oil to transport trees to the paper mill. Water and air pollutants are given off in the papermaking process.

Land Usage

Land is not often thought of as a natural resource. Yet it is as important to humans as clean air, clean water, and oil. Land is used for agriculture, logging, garbage disposal, and building. These activities can affect Earth's land resources.

How does agriculture affect the land?

About 16 million km² of Earth's total land surface is used for farming. To feed a growing population, more food needs to be produced. Farmers use chemical fertilizers to increase the amount of food grown on each km² of land. Herbicides and pesticides are also used to reduce weeds, insects, and other pests that can damage crops.

Organic farming techniques include the use of natural fertilizers, crop rotation, and biological pest controls. These methods help crops grow without using chemicals. However, organic farming cannot currently produce enough food to feed all of the humans on Earth.

When farmland is tilled, soil is exposed. Without plant roots to hold soil in place, the soil can be carried away by running water and wind. Several centimeters of topsoil may be lost in one year. In some places, it takes more than 1,000 years for new topsoil to develop.

How can soil erosion be reduced?

Some farmers till, or plow under, the soil before planting a new crop. Other farmers practice no-till farming and plant seeds between the stubble of last year's vegetation instead of tilling. Another method is contour plowing. Rows are tilled across hills and valleys so that runoff from rain traps soil in the plowed rows, reducing soil erosion. Other techniques include planting vegetation between rows to stop wind erosion. Also a cover crop that is not harvested may be planted in the off season. These can add nutrients back to the soil when plowed under. ☑

✔ Reading Check

3. **Summarize** What is one method used in farming to reduce soil loss?

_____

_____

What are forest resources?

According to the Food and Agriculture Organization of the United Nations, about one-fourth of the land on Earth is covered by forest. About 55 percent of this forest is found in developing countries. The other 45 percent occurs in developed countries.

Deforestation is the clearing of forested land for agriculture, grazing, development, or logging. It is estimated that the amount of forested land in the world decreased by 0.24% (94,000 km²) each year between 1990 and 2000. Most of this deforestation occurred in tropical regions.

Deforestation is a serious concern for many people. Many plants and animals that live in tropical rain forests cannot survive in other places. If their homes are destroyed, they may become extinct. Many of these plants might be important for developing new medicines. Scientists currently are researching how deforestation affects plants and animals.

Cutting trees can affect climate as well. Water from tree leaves evaporates into the atmosphere where it can condense to form rain. If many trees are cut down, less water enters the atmosphere and that region receives less rainfall. This is one way humans can affect the water cycle.

How does development affect land?

Between 1990 and 2000, the number of kilometers of roadways in the United States increased by more than 13 percent. Building highways often leads to more paving as office buildings, stores, and parking lots are constructed.

Paving land prevents water from soaking into the soil. Instead, it runs into sewers and streams. Streams are forced to move a larger volume of water. If enough rainwater flows into streams, it increases stream discharge. Then streams may flood.

Many communities get their drinking water from underground sources. When land is covered with roads, sidewalks, and parking lots, a reduced amount of rainwater soaks into the ground to refill underground water supplies.

Some communities and businesses have begun to preserve areas rather than pave them. Land is being set aside for environmental protection. ☑

Think it Over

4. Explain How can the cutting of trees affect an area's climate?

_____

_____

_____

✔ Reading Check

5. Identify Where do many communities get their drinking water?

_____

_____

_____

What are hazardous wastes?

Land also is used to dispose of garbage. About 60 percent of our trash is put into sanitary landfills. Some of the wastes thrown away are dangerous to organisms. Wastes that are poisonous, cause cancer, or can catch fire are called **hazardous wastes**. Hazardous wastes used to be put in landfills with regular household garbage. In the 1980s new laws made it illegal for industries to put hazardous wastes into sanitary landfills. New technologies have since been developed to help recycle hazardous wastes. Now there is less need to dispose of them.

Is any land being preserved?

Not all land on Earth is being used to store waste or produce something humans can use. Some land remains mostly uninhabited by humans. National forestlands, grasslands, and national parks in the United States are protected from development. The map below shows some areas of the country that are used for national parks and preserves. Many other countries also set aside land for natural preserves. As the world population continues to grow, the strain on the environment may worsen. Preserving some land in its natural state will benefit future generations.

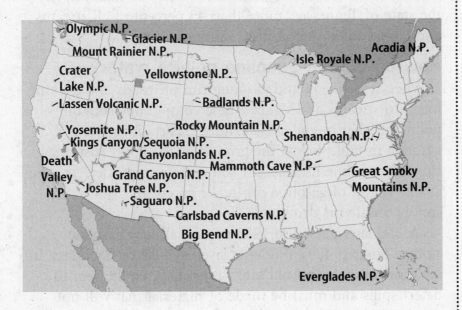

Think it Over

6. **Predict** What do you think could happen to the wildlife in national parks if the parks were not protected from development?

_____

_____

_____

Importance of Clean Water

All organisms need water. Plants need water to make food from sunlight. Some animals, such as fish and whales, live in water. Humans and other land animals cannot live without drinking water.

What happens if water isn't clean? Polluted water contains chemicals and organisms that can cause disease and death in many living things. Water also can be polluted with sediments, such as silt and clay.

Sediment makes water cloudy, blocking the sunlight that underwater plants need to make food. Sediment covers the eggs of organisms that live in the water, preventing them from getting the oxygen they need to develop.

💡 **Think it Over**

7. **Describe** two ways that metals enter bodies of water.

_____

_____

Can metals cause water pollution?

Many metals, such as mercury, lead, and nickel, can be poisonous. Lead and mercury in drinking water can harm the nervous system. Yet these metals are part of many useful items, such as paints, stereos, and batteries. Today, environmental laws control the amounts of metals released with wastewater from factories. Because metals remain in the environment for a long time, metals released many years ago still are polluting bodies of water today.

Mining also releases metals into water. For example, in the state of Tennessee, more than 43 percent of all streams and lakes contain metal from mining activities.

How do oil and gasoline pollute water?

When it rains, oil and gasoline run off roads and parking lots into streams and rivers. These compounds contain pollutants that might cause cancer. Gas stations store gasoline in tanks below the ground. If the tanks rust, they can leak gasoline into nearby soil and groundwater. As little as one gallon of gasoline can make an entire city's water supply unsafe for drinking.

Federal laws passed in 1988 require all new gasoline storage tanks to have a double layer of steel or fiberglass. In addition, all new and old tanks must have equipment to detect spills and must be made of material that will not develop holes. These laws help protect soil and groundwater.

What is sewage?

When you flush a toilet or take a shower, the water goes down a drain. This wastewater, called sewage, contains human waste, household detergents, and soaps. Human waste contains harmful organisms that can make people sick. In most cities, sewage treatment facilities purify sewage before it enters a body of water.

Causes of Air Pollution

Cities can be exciting places. They are centers of business, culture, and entertainment. However, cities also have many cars, buses, and trucks that burn fuel for energy. The brown haze you sometimes see over a city forms from the exhaust of these vehicles. Air pollution also comes from burning fuel in factories, generating electricity, and burning trash. Dust from plowed fields, construction sites, and mines also contributes to air pollution.

What is smog?

One type of air pollution found in and around cities is called smog. Major sources of smog include cars, factories, and power plants. A combination of pollutants and natural conditions lead to the formation of smog.

How does smog form?

A hazy blanket of smog that forms with the help of sunlight is called **photochemical smog**. When gasoline is burned, nitrogen and carbon compounds are released and pollute the air. The figure below shows how these compounds produce smog.

Smog

D Ozone and other compounds form smog.

B In the presence of sunlight, nitrogen compounds release oxygen atoms.

C Oxygen molecules (O_2) in air combine with oxygen atoms to form ozone (O_3).

A Car exhaust contains nitrogen and carbon compounds.

Picture This

8. Identify Circle the step where photochemical smug is formed.

Reducing Air Pollution

Air pollutants cross the borders between states and countries, going wherever the wind carries them. This makes them difficult to control. For example, burning coal in midwestern states causes acid rain in the northeastern United States and Canada. ✓

What is acid rain?

Acids form when sulfur oxides from coal-burning power plants and nitrogen oxides from cars combine with moisture in the air. When acidic moisture falls to Earth as rain or snow, it is called **acid rain**.

How has the United States reduced air pollution?

The United States Congress has passed several laws to protect the air. One was the Clean Air Act of 1990. It helps to reduce air pollution from cars, power plants, and other industries. The Clean Air Act is summarized in the table below.

Clean Air Regulations	
Urban air pollution	All cars manufactured since 1996 must reduce nitrogen oxide emissions by 60 percent and hydrocarbons by 35 percent from their 1990 levels.
Acid rain	Sulfur dioxide emissions had to be reduced by 14 million tons from 1990 levels by the year 2000.
Airborne toxins	Industries must limit the emission of 200 compounds that cause cancer and birth defects.
Ozone-depleting chemicals	Industries were required to immediately cease production of many ozone-depleting substances in 1996.

Can you help reduce air pollution?

The Clean Air Act can work only if we all cooperate. Cleaning the air takes money, time, and effort. How might you help reduce air pollution? You might change your lifestyle by walking, riding a bike, or using public transportation to get to a friend's house instead of asking for a car ride. You also can set the thermostat in your home lower in the winter and higher in the summer. ✓

✓ **Reading Check**

9. **Explain** Why do states and countries find air pollution difficult to control?

_____

_____

_____

💡 **Think it Over**

10. **Predict** Look at the chart on this page. What might people do to reduce air pollutants even further?

_____

_____

_____

✓ **Reading Check**

11. **Identify** two ways you can reduce air pollution.

_____

_____

● After You Read

Mini Glossary

acid rain: acidic moisture, with a pH below 5.6, that falls to Earth as rain or snow and can damage forests, harm organisms, and corrode structures

carrying capacity: the largest number of individuals of a given species that the environment can support.

hazardous waste: wastes that are poisonous, cause cancer, or can catch fire

photochemical smog: hazy, yellow-brown blanket of smog found over cities that is formed with the help of sunlight, contains ozone near Earth's surface, and can damage lungs and plants

pollutant: substance that contaminates the environment

population: total number of individuals of one species occupying the same area

1. Review the terms and their definitions in the Mini Glossary above. Then write a sentence using one of the terms to show how humans can or should take care of the environment.

_____

_____

2. Complete the concept map to show the types of air pollution.

3. ◀ **Study Coach** Review the questions you wrote as you read this section. What resources could you use to find answers to your questions? Did the questions you wrote help you understand the information in this section?

_____

_____

_____

End of Section

Introduction to Waves

chapter 9

section ❶ The Nature of Waves

What You'll Learn
- how waves transfer energy but not matter
- about mechanical, transverse, and longitudinal waves

● Before You Read

Write what you think a wave is and list two types of waves you have seen or heard about.

_____

_____

● Read to Learn

What's in a wave?

Imagine that you are watching a surfing championship on television. As you look at the surfers riding the giant waves, you heat up some leftover pizza in the microwave. You call a friend to tell her about the surfing. Your friend tells you to turn on your radio to listen to a song you both like. You have just experienced three different types of waves. You saw the waves in the ocean. You cooked using microwaves. Sound waves were produced by the television, your friend's voice, and the radio.

A **wave** is a repeating disturbance or movement that transfers energy through matter or space. For example, ocean waves disturb the water and transfer energy through it. In an earthquake, powerful waves transfer energy through Earth. Light is a type of wave that can travel through empty space to transfer energy from one place to another, such as from the Sun to Earth.

Waves and Energy

Have you ever watched a pebble fall into a pool of water and seen ripples form? The pebble causes a disturbance in the water. Some of the kinetic energy from the falling pebble transfers to the water molecules that are close by. The water molecules pass the energy along to other water molecules that are next to them. In this way, the energy passes from molecule to molecule until it is far from where the pebble first fell. A wave is formed. It carries the energy through the water. ☑

Reading Check

1. **Identify** What do waves transfer with them?

_____

Do waves carry matter?

Suppose you are in a boat on a lake. The waves in the water bump against your boat. You notice that the boat moves up and down and maybe even back and forth a little. After the waves have moved past, your boat has not moved to a different place. The waves don't carry the water along with them. They only carry energy as they move. This is true for all waves. They carry energy without moving matter from place to place.

How can you make a wave?

A wave will travel as long as there is both energy and a medium to carry it. Think of the pebble and the pool of water. The ripples eventually stop and the water is smooth again. There is no more energy to carry.

The figure shows a hand making a wave with a rope. As the hand moves up and down, a wave begins to travel along the rope. The wave moves along the rope until it reaches the end. Once the wave reaches the end, the rope is still because there is no more energy to transfer.

Producing a Wave

The hand in the figure is moving up and down. Anything that moves up and down or back and forth repeatedly is vibrating. The hand moving up and down is a vibrating movement. Vibrations cause all waves.

Mechanical Waves

Sound waves travel through the air to reach your ears. Ocean waves travel through water to reach the shore. Both kinds of waves move their energy through a medium. A **medium** is the matter through which a wave travels. A medium can be a solid, a liquid, or a gas. It can also be a combination of these forms of matter. The medium for sound waves is air. The medium for ocean waves is water. Not all waves need a medium in order to travel. Light waves and radio waves are examples of waves that do not need a medium. They can travel through space. Waves that can travel only through matter are called mechanical waves. The two types of mechanical waves are transverse waves and longitudinal waves. ✔

Picture This

2. **Mark the Figure** Use a pen or a pencil to circle the parts of the rope that show the location of the wave.

✔ Reading Check

3. **Identify** What must a mechanical wave travel through?

_____

_____

Picture This

4. Explain Why are the coils closer together at the middle of the spring in the figure?

_____

_____

_____

Think it Over

5. Analyze When a wave travles through a spring, why does it look like the whole spring is moving?

_____

_____

_____

How does matter move in a transverse wave?

In a <u>transverse wave</u>, the matter in the medium moves at a right angle to the direction the wave travels. An ocean wave moves across, horizontally, while the water it passes through moves up and down, vertically. The wave and the matter in the medium move at right angles to each other.

How does matter move in a longitudinal wave?

In a <u>longitudinal wave</u>, matter in the medium moves back and forth in the same direction that the wave travels. The figure below shows how a longitudinal wave moves along a coiled spring. Suppose you hold onto both ends of a spring. You squeeze some coils together at one end of the spring, then let go of them. A wave travels along the spring.

If you tie a piece of yarn to one of the coils of the spring, you can see the back-and-forth movement in the spring. As the wave passes through the coil with the yarn, the yarn moves in the direction of the wave. After the wave passes the yarn, it moves back to its original position. However, the wave continues to move to the end of the spring. The yarn moves back and forth as the wave passes. The wave carries energy with it, but it doesn't carry matter. Another name for longitudinal waves is compressional waves.

Loungitudinal wave

How do sound waves move?

Sound waves are longitudinal waves. Some sound waves travel through air. If you pick a string on a guitar, the string vibrates. The vibration pushes nearby air molecules close together. The air molecules squeeze together like the coils in the spring. Then the compressions travel through the air to make a wave. Sound waves also can travel through mediums such as water and wood. The particles in those mediums also squeeze together and move apart when sound waves travel through them.

When a sound wave reaches your ear, it causes your eardrum to vibrate. Your inner ear sends signals to your brain. Your brain understands these signals as sound.

How do water waves move?

Water waves look like transverse waves. They are actually a combination of transverse and longitudinal waves. The figures below show the movement of water in a wave. The small arrows show the direction of the wave and the large arrows show the movement of the water. As a wave goes by, the water moves up and down. The water also moves back and forth for a short distance in the same direction that the wave moves. Waves have both high and low points. Water pushes forward or backward toward the high part of the wave. This causes the low part of the wave to form as the first figure shows. Then as the wave passes, the water that was pushed forward or backward moves back to where it was.

Water is pushed aside.

Water returns to where it was.

Picture This

6. Observe Use your highlighter to trace the arrows that show the movement of water when it is pushed forward or backward toward the high part of the wave.

The up-and-down and back-and-forth motion causes the water to move in circles. An object that floats on the surface of the water takes in some of the energy from the waves. This causes it to bob in a circular motion.

How are ocean waves formed?

Wind blowing across the ocean surface causes most ocean waves to form. The changing speed of the wind acts like a vibration on the water. The size of the waves depends on the speed of the wind, the length of time the wind blows, and how far it travels over the water. ☑

✔ Reading Check

7. Identify What usually starts the movement of a wave in an ocean?

_____

_____

What are seismic waves?

Forces in Earth's crust can cause parts of the crust to shift, bend, or even break. When this happens, Earth's crust vibrates and releases energy. This creates seismic (SIZE mihk) waves that carry energy outward and cause an earthquake.

Seismic waves are a combination of transverse and longitudinal waves. They travel through Earth and along Earth's surface. Objects on Earth's surface move and shake when they take in some of the energy from the seismic waves. The more Earth's crust moves during an earthquake, the more energy is released.

Copyright © Glencoe/McGraw-Hill, a division of The McGraw-Hill Companies, Inc.

● After You Read

Mini Glossary

longitudinal wave: a wave in which the matter in the medium moves back and forth in the same direction that the wave travels

medium: the matter through which a wave travels

mechanical wave: waves that travel only through matter

transverse wave: a wave in which matter in the medium moves back and forth at a right angle to the direction the wave travels

wave: a repeating disturbance or movement that transfers energy through matter or space

1. Review, the terms and their definitions in the Mini Glossary above. What is the difference between a transverse wave and a compressional wave?

_____

_____

2. Complete the diagram below to list what you learned about mechanical waves.

End of Section

Introduction to Waves

chapter 9

section ❷ Wave Properties

● Before You Read

Have you ever sat high in the stadium at a baseball game and heard the sound of a ball being hit by a bat? Did you notice that you heard the sound after you saw the ball being hit? Explain why you think this happened.

_____

_____

_____

● Read to Learn

The Parts of a Wave

How are sound waves, water waves, and seismic waves different? Some waves have more energy than others. Some travel faster than others. There are other ways that waves are different.

Remember that transverse and longitudinal waves act differently as they travel through a medium. The first figure below shows a transverse wave. Notice that the medium (the rope) has alternating high points and low points. A **crest** is the high point of a transverse wave. A **trough** is the low point of a transverse wave. The second figure shows a longitudinal wave in a coiled spring. It does not have crests or troughs. Instead, the wave creates areas where the coils are close together. This area is called a **compression**. In other areas, the coils are spread apart. A **rarefaction** (RAYR uh fak shun) is the area in a longitudinal wave where the medium is most spread out.

Crest

Trough   Rest position

Rarefaction

Compression

What You'll Learn

- what wavelength, frequency, period, and amplitude are
- how frequency and wavelength are related
- how a wave's energy and amplitude are related
- how to calculate wave speed

Mark the Text

Underline As you read this section, look for important information about the properties of waves. Underline all of the ideas that you think are important to understand and remember.

Picture This

1. **Describe** the pattern shown in the medium by the transverse wave.

_____

_____

Copyright © Glencoe/McGraw-Hill, a division of The McGraw-Hill Companies, Inc.

Picture This

2. Analyze In the figure showing the transverse wave, how do the sizes of the wavelengths compare?

_____

_____

_____

✔ **Reading Check**

3. Explain How is the frequency of a wave measured

_____

_____

_____

Wavelength

Another way to describe a wave is by its wavelength. A **wavelength** is the distance between one point on a wave and the nearest point just like it. The first figure below shows a transverse wave. Commonly the wavelength is measured from the top of one crest to the top of the next crest. You could also measure the wavelength from the bottom of one trough to the bottom of the next trough.

The second figure below shows a compressional wave. The wavelength is the distance from the center of one compression to the center of the next compression. You could also measure the wavelength from the center of one rarefaction to the center of the next rarefaction.

Humans cannot hear all sounds. You can hear sounds that have wavelengths between a few centimeters and about 15 m. The highest-pitched sounds you hear have wavelengths that are smaller than a few centimeters. The deepest sounds you hear have wavelengths that are about 15 m.

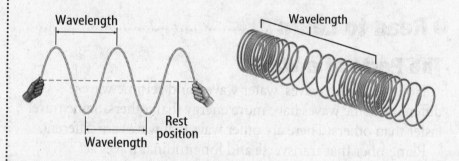

Wavelength

Wavelength

Rest position

Wavelength

Frequency and Period

Have you ever heard someone use the word *frequency*? It usually refers to how often something happens in a given period of time. Frequency has a similar meaning when referring to waves. The **frequency** of a wave is the number of wavelengths that pass a point each second. To find the frequency of a transverse wave, count the number of crests or troughs that pass by a point each second. In the same way, to find the frequency of a compressional wave, count the number of compressions or rarefactions that pass by a point each second.

When you tune a radio to a certain station, you are choosing radio waves of a certain frequency. Frequency is expressed in hertz (Hz). A frequency of 1 Hz means that one wavelength passes per second. In SI units, 1 Hz is the same as 1/s.

Wave Periods The <u>period</u> of a wave is the amount of time it takes one wavelength to pass a point. Periods are measured in units of seconds. The period of an ocean wave is two seconds if it takes that long for two consecutive crests to pass a point.

How are frequency and wavelength related?

There is a relationship between frequency and wavelength. If you make transverse waves with a rope, you can increase the frequency by moving the rope up and down faster. The frequency increases because more crests or troughs pass by a point in one second. Moving the rope faster also makes the wavelength shorter. The distance from crest to crest or trough to trough is shorter. This relationship is always true: as frequency increases, wavelength decreases.

Look at the two figures below. The waves in the ropes show that as the frequency of waves increases, their wavelengths decrease. The first figure shows one wavelength passing by in one second. The frequency of the first wave is 1 Hz. The second figure shows two wavelengths passing by in one second. The frequency of the second wave is 2 Hz. As more wavelengths pass by in one second, the wavelengths get shorter. In the two figures, the frequency increases from 1 Hz to 2 Hz, so the wavelengths decrease. If you move a rope up and down five times in 1 s, the frequency of the wave is 5 Hz. However, the wavelengths would be shorter than for the rope with a frequency of 2 Hz.

Distance traveled in 1 s

Rest position

One wavelength

Distance traveled in 1 s

One wavelength   One wavelength

Wave Speed

Look back at the question at the beginning of this section. It asks if you have ever sat up high in the stadium at a baseball game, heard the sound of a ball being hit by a bat, and realized that you heard the sound after you saw the ball being hit. You saw the baseball being hit before you heard it because light waves travel through gases much faster than sound waves. Air is a gas. The light waves reflected from the ball reached your eyes before the sound waves created by the bat hitting the ball reached your ears.

Picture This

4. **Identify** What kind of waves are shown in the figure?

_____

_____

Think it Over

5. **Explain** if you are watching fireworks from a distance, why do you see fireworks explode before you hear the noise?

_____

_____

_____

What determines wave speed?

The speed of a wave depends on the medium it is traveling through. Light waves travel faster than sound waves through gases. Sound waves usually travel faster than light waves through liquids and solids. Sound waves are longitudinal waves. Longitudinal waves travel faster in liquids and solids than they do in gases. Also, sound waves usually travel faster in a material when the temperature of the material is increased. For example, sound waves travel faster in air at 20°C than in air at 0°C. ☑

How do you calculate wave speed?

You can calculate the speed of any wave by multiplying its frequency times its wavelength. The Greek letter *lambda* (λ) represents the wavelength, f represents the frequency, and v represents the speed of the wave. The wave speed equation is:

$$\text{speed (in m/s)} = \text{frequency (in Hz)} \times \text{wavelength (in m)}$$
$$v = f\lambda$$

Why does multiplying the frequency unit (Hz) by the distance unit (m) result in the unit for speed (m/s)? Recall that the SI unit Hz is the same as 1/s. So multiplying m × Hz equals m × 1/s, which equals m/s.

Using the equation, you can calculate the speed of a wave traveling in water. If the wave has a frequency of 500 Hz and a wavelength of 3 m, what is the speed of the wave?

Step 1   What do you know?
$f = 500.0$ Hz
$\lambda = 3.0$ m

Step 2   Write the equation and put in the known numbers.
$v = f\lambda$
$v = 500.0 \text{ Hz} \times 3.0 \text{ m}$

Step 3   Solve the equation.
$v = 1{,}500.0$ m/s
The speed of the wave is 1,500 m/s.

Amplitude and Energy

Why do some earthquakes cause terrible damage, while others hardly are felt? This is because waves can carry different amounts of energy. **Amplitude** is a measure of the energy that a wave carries. The greater a wave's amplitude, the more energy the wave carries. Amplitude is measured differently for longitudinal and transverse waves.

6. **Determine** Do light waves or sound waves travel faster through gases?

_____

7. **Calculate** What is the speed of a sound wave that has a frequency of 150 Hz and a wavelength of 0.00002 mm? Show your work.

_____

_____

_____

How is amplitude measured for compressional waves?

The amplitude of a longitudinal wave depends on how tightly the medium is pushed together at the compressions. The tighter the medium is pushed together at the compressions, the greater its amplitude. The greater the amplitude is, the more energy the wave carries. Think of a coiled-spring toy. It takes more energy to push the coils tightly together than it does to barely move them.

Compare the compression in the figure on the left with the compressions in the figure on the right. The coils in the compressions of the figure on the right are closer together. The second wave has greater amplitude and more energy compared to the first wave.

Look at the rarefactions in the two springs. The closer the coils are in the compression, the farther apart they are in the rarefaction. So the less dense a medium is at the rarefactions, the more energy the wave carries.

How is amplitude measured for transverse waves?

Have you ever been knocked over by an ocean wave? If so, you know that higher waves carry more energy than shorter waves do. Remember that the greater a wave's amplitude, the more energy the wave carries. So a tall ocean wave has a greater amplitude than a short ocean wave. The amplitude of a transverse wave is measured differently than the amplitude of a compressional wave. The figure below shows that the amplitude of a transverse wave is the distance from the crest or trough of the wave to the rest position. The greater this distance is, the greater the amplitude is.

Picture This

8. **Identify** In the figure on the right, are the coils in the rarefactions closer together or farther apart than in the figure on the left?

_____

_____

Picture This

9. **Draw** Using the same rest position, draw a wave with a greater amplitude than the one shown. Then draw a wave with a smaller amplitude.

● After You Read

Mini Glossary

amplitude: the measure of the energy that a wave carries

crest: the high point of a transverse wave

compression: the more dense region of a longitudinal wave

frequency: the number of wavelengths that pass a fixed point in 1 s

period: the amount of time it takes one wavelength to pass a fixed point

rarefaction: the section in a compressional wave where the material is less crowded and more spread out

trough: the low point of a transverse wave

wavelength: the distance between one point on a wave and the nearest point just like it

1. Review the terms and their definitions in the Mini Glossary above. Explain how a transverse wave's wavelength is different from its amplitude.

_____

_____

_____

2. Complete the concept map below to list the properties of waves that you learned about in this section.

Properties of Waves

3. **Mark the Text** Look back at the important ideas you underlined about wave properties. How did underlining these ideas help you learn the information?

_____

_____

End of Section

Introduction to Waves

section ⊜ The Behavior of Waves

● Before You Read

Think about a time when you walked down an empty hallway and heard the echo of your footsteps. Write what you think caused the echo.

_____

_____

● Read to Learn

Reflection

Suppose you and a friend are the last students to leave your school building. You shout to your friend at the other end of the hallway. Your voice echoes throughout the hallway. You also notice your reflection in one of the glass windows. These are both examples of wave reflection. Wave reflection causes the echo you hear and the image you see of yourself. Wave reflection happens when a wave strikes an object or surface and bounces off it. All types of waves—including sound, water, and light waves—can be reflected.

How do light waves reflect?

What happens when you see your face in a mirror? First, light waves strike your face and bounce off. Then, the reflected light strikes the mirror and reflects back to your eyes.

What are echoes?

Echoes are reflected sound waves. When you called to your friend in the school building, your voice echoed around the hall. Sound waves formed when you shouted. The waves traveled through the air to your ears and to other objects. The waves reflected off the walls, floor, and ceiling and then came back to your ears. You could hear your voice again, a few seconds after you first heard your voice. This caused the echo. Dolphins and bats use echoes to determine where objects are. Dolphins make clicking noises and listen to the echoes.

What You'll Learn

- the law of reflection
- how waves change direction
- what are refraction and diffraction
- how waves interfere with each other

Study Coach

Create a Quiz Write questions on index cards as you read this section. After you read, form a group of three students. Take turns asking each other your questions and answering them.

FOLDABLES

C Build Vocabulary
Make the following Foldable. It will help you understand the content of this section by defining the terms used.

Reflection
Refraction
Diffraction
Interference
Standing Waves
Resonance

What is the law of reflection?

The figure below shows a flashlight shining on a mirror. The light beam that strikes the mirror is called the incident beam. The light beam that bounces off the mirror is called the reflected beam. The line that is at a right angle to the mirror is called the normal. The angle made by the incident beam and the normal is the angle of incidence. The angle made by the reflected beam and the normal is the angle of reflection. The law of reflection says that the angle of incidence is equal to the angle of reflection.

Normal

Reflection   **Refraction**

Picture This

1. **Label** On the figure of the flashlight and the mirror, label the incident beam as *i* and the reflected beam as *r*.

Refraction

Suppose you put a pencil in a glass of water and then look at it from the side of the glass as shown above. The pencil looks like it is broken in two at the water line. But if you pull the pencil out of the water, you see that it is not broken. What makes the pencil appear broken?

What is refraction?

Remember that a wave's speed depends on the medium it is moving through. When a wave moves from one medium to another, such as from water to air, it changes speed. If this wave travels at an angle when it passes from one medium to another, it changes direction, or bends, as it changes speed. __Refraction__ is the bending of a wave caused by a change in its speed as it moves from one medium to another. Two things happen to the light waves as they move from the air to the water in the glass. The light waves change speed and direction. They are refracted so the pencil appears to be broken.

When does light bend?

When a light wave passes into a material in which it slows down, it bends toward the normal. When a light wave passes into a material in which it speeds up, the wave bends away from the normal. ☑

Reading Check

2. **Identify** What can change the speed of a wave?

_____

_____

Copyright © Glencoe/McGraw-Hill, a division of The McGraw-Hill Companies, Inc.

Refractions The figures show what refractions look like. The figure on the left shows light waves moving from air to water. The light waves slow down when they enter the water. This makes them change direction and bend toward the normal. The figure on the right shows light waves moving from water to air. The waves speed up and bend away from the normal.

Air to water

Water to air

Picture This

3. **Highlight** Look at the figure on the right. Use your highlighter to trace the angle that shows the light bending away from the normal.

How does light bend in water?

Have you ever noticed that the feet of someone standing in a swimming pool look closer to the surface than they really are? Your brain wants to think that the light waves travel in a straight line. But refraction causes the light waves from the feet to bend away from the normal as they pass from the water to the air.

This is similar to the pencil in the glass of water. The pencil looks broken at the surface of the water. The light waves coming from the part of the pencil above the water are not bent. The light waves that move from the air to the water in the glass change speed and bend. This makes the part of the pencil that is under water look like it has shifted.

Diffraction

Suppose you are in a classroom and you hear music coming from another room. The sound waves bend around corners and travel from the room down the hall to where you are. Refraction does not cause sound waves to bend. Instead, they bend because of diffraction. **Diffraction** takes place when an object causes a wave to change direction and bend around it. ☑

Light waves can diffract, too. Light waves do not diffract as much as sound waves do. Suppose you walk toward the room where you hear the music. As you walk toward the open door, you can see light coming out of the room. Light waves bend around the edges of the open door. But the amount of light that bends is not enough for you to be able to see around the corner and into the room. Yet, you can hear the music that is being played in the room.

Reading Check

4. **Identify** Circle the term that describes the bending of a wave around an object.
   a. reflection
   b. refraction
   c. diffraction
   d. rarefaction

Picture This

5. Trace Using a pencil, trace along the waves in the figure that have passed through the opening. What happens to the waves after they pass through the opening?

_____

_____

_____

Think it Over

6. Apply Why is it hard to see around a corner?

_____

_____

_____

When do water waves diffract?

Ocean waves refract when they strike an island. The waves change direction and bend around the island. Diffraction and refraction both cause waves to bend, but there is a difference. Waves refract when they pass through an object. They diffract when they pass around an object.

Waves also can diffract when they pass through a narrow opening. The figure shows water waves passing through a small opening in a barrier. They diffract and spread out after they pass through the opening. The waves are bending around the corners of the opening.

How much will a wave bend?

To find out how much a wave will bend when it strikes an object, compare the size of the object to the wavelength. When an object is smaller than the wavelength, the waves bend around it. Suppose you shine a large spotlight on a very thin tree. The object is smaller than the wavelength, so the light waves bend around the tree. The shadow is narrower than the tree. A large amount of diffraction takes place. Suppose you shine a small flashlight on a very large tree. The object is larger than the wavelength, so the light waves will not bend around the tree easily. The shadow is wider than the tree. Almost no diffraction takes place.

How do sound waves bend around corners?

Think back to the example of the music coming from another room in your school. You can hear the sounds before you reach the door. The wavelengths of sound waves are about the same size as the door opening. The sound waves diffract around the doorway and spread out into the hallway.

How is this different from the light waves in the room? You aren't able to see into the room because light waves have a much shorter wavelength than sound waves. This is why the light waves are not diffracted by the doorway as much as the sound waves are.

How do radio waves diffract?

AM radio waves have longer wavelengths than FM radio waves. So, AM radio waves can diffract around big objects much more than FM radio waves. You can get more AM than FM radio stations when you are near tall buildings and hills.

Interference

Suppose you throw two pebbles into a still pond. Each pebble causes ripples to form around it. The waves of ripples travel toward each other. When the two waves meet, they pass right through each other and continue moving. **Interference** is the point where two waves meet each other and overlap to form a new wave. The new wave lasts only as long as the two waves continue to overlap. There are two kinds of interference, constructive and destructive.

What is constructive interference?

The left-hand figure below shows constructive interference. In constructive interference, the waves meet at the same point and add together. This happens when the crests of transverse waves overlap each other. The troughs of the waves also overlap. The amplitudes of the two waves add together to make a larger wave. So, the new wave has a higher crest and a lower trough. The amplitude of the new wave is the sum of the amplitudes of the other two waves. Constructive interference also happens when two compressional waves overlap. When the compressions of one wave overlap the compressions of another, the compressions add to make a combined wave with a larger amplitude. If they are sound waves, the new wave is louder.

Constructive Interference   Destructive Interference

What is destructive interference?

In destructive interference, the overlapping waves subtract from each other. This happens when the crest of one transverse wave overlaps with the trough of another transverse wave. The amplitudes of the two waves combine to make a new wave with a smaller amplitude. The figure above on the right shows the new wave that forms during destructive interference.

The same is true for longitudinal waves. When the compression of one wave overlaps the rarefaction of the other wave, destructive interference occurs. The compressions and rarefactions combine to form a smaller wave. In sound waves, destructive interference causes a decrease in loudness.

💡 Think it Over

7. **Describe** a time when you heard or used the word *interference* or *interfere*.

_____

_____

_____

Picture This

8. **Use Models** In both figures, use a highlighter to trace over the new wave that forms during either constructive or destructive interference.

Standing Waves

Suppose you and a friend are holding the ends of the same rope. You both shake the rope and make waves the same size that travel toward each other. Interference happens when the waves from one end overlap waves from the other end. A new wave forms when a crest of one of your waves meets a crest of one of your friend's waves. The new wave has a larger amplitude.

When a crest of one wave meets a trough of another wave, the waves cancel each other out. Then there is no movement. A **standing wave** is a wave pattern that forms when two equal-sized waves travel in opposite directions and continuously interfere with each other. The interference of these two waves makes the rope vibrate and creates a pattern of crests and troughs. This makes it look like the rope is standing still. Nodes are the places where the two waves cancel each other. The nodes always stay in the same place on the rope. The wave pattern vibrates between the nodes. ✔

What are some standing waves?

When you use a bow to play a violin, the string vibrates and creates standing waves. The standing waves in the string make a rich, musical tone. Other instruments also use standing waves to create music. A flute creates standing waves in a column of air. The material in a drumhead vibrates to make standing waves.

Resonance

Bells of different sizes and shapes make different sounds. When you strike a bell, it vibrates at its own natural frequencies. All objects have their own natural frequencies of vibration. The frequencies depend on the size and shape of the object. Frequency also depends on the kind of material that the object is made of.

There is another way to make an object vibrate at its natural frequencies. Suppose you have a tuning fork with a single natural frequency. Imagine that a sound wave of the same frequency strikes the tuning fork. Because the sound wave has the same frequency, the tuning fork will vibrate. **Resonance** is when an object vibrates by absorbing energy from another object that is vibrating at the same natural frequencies.

Resonance can cause an object to absorb a large amount of energy. An object vibrates more strongly as it keeps absorbing more energy at its natural frequencies. If enough energy is absorbed, the object can vibrate too much and break apart.

✔ Reading Check

9. **Describe** What happens when a crest of one wave meets a trough of another wave of the same amplitude?

_____

_____

💡 Think it Over

10. **Summarize** How can resonance cause an object to break?

_____

_____

_____

● After You Read

Mini Glossary

diffraction: the bending of a wave around a barrier

interference: the point where two waves meet each other and overlap to form a new wave

node: location where interfering standing waves cancel

refraction: the bending of a wave caused by a change in its speed as it moves from one medium to another

resonance: the vibration of an object by absorbing energy from another object that is vibrating at its natural frequencies

standing wave: a wave pattern that forms when two equal-sized waves travel in opposite directions and interfere with each other

1. Review the key terms and their definitions in the Mini Glossary. How are refraction and diffraction the same? How are they different?

   _____

   _____

   _____

2. List the main ideas you learned about the behavior of waves. For help, use the main headings in the section.

   The Behavior of Waves

   1. _____

   2. _____

   3. _____

   4. _____

   5. _____

3. ◄ **Study Coach** Look at the questions you wrote on index cards during this section. How did writing these questions help you learn?

   _____

   _____

End of Section

section ❶ The Nature of Sound

What You'll Learn
- how sound travels
- what changes the speed of sound
- how your ears allow you to hear

Mark the Text

Identify the Main Point
Highlight the main idea of each paragraph. Then review what you highlighted after you finish reading the section.

● Before You Read

Place your hand on your throat and hum a tune. How is the movement that you feel related to the sound you made?

_____

_____

● Read to Learn

What causes sound?

An amusement park is a noisy place. Music is playing, videogames are beeping, and people on rides are screaming. It can be hard to hear your friends talking. These sounds are all different, but they all have something in common. Each sound is made by an object that vibrates.

Remember that vibration is a quick, rhythmic back-and-forth movement. The sounds of your friends' voices and people screaming are made by the vibrations of their vocal cords. Music and noises from videogames are made by vibrating speakers. All sounds are made by something that vibrates.

Sound Waves

Think of a radio speaker. When the speaker vibrates, it bumps into nearby molecules in the air. Some of the speaker's energy is transferred to air molecules. These air molecules bump into other air molecules. They pass the energy on to the new molecules. The energy that came from the vibrating speaker keeps moving from one molecule to another. The collisions and the energy they transmit make a sound wave.

Sound waves are longitudinal waves. Recall that there are two areas in longitudinal waves. One area is the compression where the molecules of the medium are very close together. The other area is the rarefaction where the molecules of the medium are more spread out. A longitudinal wave has alternating areas of compressions and rarefactions.

In each figure, the radio speaker is vibrating. In the first figure, the speaker is vibrating outward. When this happens, molecules in the air are pushed together to form a compression. The second figure shows the speaker moving inward. Air molecules have room to spread out. A rarefaction forms. The back-and-forth vibrations form compressions and rarefactions.

Picture This

1. **Communicate** Look at the figures to the left. Describe how the particles in a compression are different from the particles in a rarefaction.

_____

_____

_____

_____

Compression

When the speaker vibrates outward, molecules in the air next to it are pushed together to form a compression.

Rarefaction

When the speaker vibrates inward, the molecules spread apart to form a rarefaction.

How does a sound wave travel?

The air molecules around the speaker bump into other molecules. The speaker keeps vibrating, so a series of compressions and rarefactions forms a sound wave. This sound wave travels out from the speaker for your ear to hear.

Moving Through Materials

Did you ever swim underwater and hear sounds? Most sounds that you hear travel through air. Sound can also travel through solids, liquids, and other gases. Sound waves travel in any material, or medium. But sound waves cannot travel through empty space. There are no molecules to transmit a sound wave.

How does the speed of sound change in different materials?

How quickly a sound wave moves through a material depends on the material and whether it is solid, liquid, or gas. The table shows that sound travels slowest through gases and fastest through solids. Molecules in liquids and solids are closer than molecules in gases. When molecules are close, they transmit energy more quickly. Loud and soft sounds travel through a material at the same speed.

Speed of Sound in Different Mediums	
Materials	**Speed of Sound (m/s)**
Air	347
Cork	500
Water	1,498
Brick	3,650
Aluminum	4,877

Applying Math

2. **Use a Table** Suppose lightning strikes two kilometers away from you. Find how long it would take the sound of the thunder to reach your ears by dividing the distance by the speed of sound through air. Round your answer to the nearest tenth of a second.

_____

Why do close molecules transmit sound more quickly?

Imagine a large group of people standing in a line. They are passing a bucket of water from person to person. If everyone stands far apart, each person has to take the time to walk the bucket to the next person. But if everyone stands close together, each person quickly can hand the bucket to the next person.

The molecules in solids and liquids are like the people standing close together in the line. The particles in gases are farther apart, like the people standing far apart in line. The closer the molecules are to each other, the faster they can transfer energy. ☑

How does temperature affect the speed of sound?

The speed of sound waves also depends on the temperature of a medium. As temperature increases, molecules move faster. It is easier for molecules to bump into each other if they are moving quickly. If the particles in a medium are bumping into each other more often, more energy can be transferred faster. So the higher the temperature, the faster sound waves move. For example, when the air temperature is 0°C, sound travels through the air at 331 m/s. But at a temperature of 20°C, sound travels at 343 m/s.

Human Hearing

Our vocal cords and mouths produce many different kinds of longitudinal waves. But how does your brain make sense of sound waves? Your ears and your brain work together. They turn longitudinal waves into something that you can understand. Making sense of sound waves involves four steps. First, your ears gather the longitudinal waves. Next, your ears amplify the waves, or make them stronger. The amplified waves are changed to nerve impulses that travel to the brain. Finally, the brain makes sense of the nerve impulses.

How does the outer ear gather sound waves?

When you think of your ear, you probably picture the part that is on the outside. But the human ear has three parts: the outer ear, the middle ear, and the inner ear. Look at the figure on the next page. It shows the three parts of the human ear. The outside part of the ear together with the ear canal and the eardrum make up the outer ear. The outer ear is where sound waves are gathered. Like a funnel, it helps catch the sound waves, then sends them into the ear canal.

✔ **Reading Check**

3. **Explain** Why does sound travel faster in solids and liquids than it does in gases?

_____

_____

_____

💡 **Think it Over**

4. **Explain** Hearing aids are placed in a person's outer ear. This makes it easy for the person to put them in and take them out. What do hearing aids do when they are placed in a person's outer ear?

_____

_____

_____

_____

The sound waves travel along the ear canal to the eardrum. The **eardrum** is a tough membrane, or tissue, about 0.1 mm thick. The eardrum is stretched over the end of the ear canal. When sound waves reach the eardrum, they transfer their energy to the eardrum and it vibrates.

Outer Ear
Gathers sound waves

Middle Ear
Amplifies waves

Inner Ear
Converts waves to nerve impulses

Hammer
Anvil
Cochlea
Stirrup
Ear canal
Eardrum

What does the middle ear do?

When the eardrum vibrates, it transfers the vibrations into the middle ear. There are three tiny bones in the middle ear that amplify the vibrations. The bones are called the hammer, anvil, and stirrup. They are a lever system that makes the force and pressure of the sound waves stronger. The stirrup is connected to a membrane in the oval window. When the stirrup vibrates, the membrane vibrates too.

What does the inner ear do?

When the membrane in the oval window vibrates, the sound vibrations are sent into the inner ear. The inner ear contains the cochlea. The **cochlea** (KOH klee uh) is a spiral-shaped structure that is filled with liquid. It also contains tiny hair cells. When the hair cells in the cochlea begin to vibrate, nerve impulses are sent to the brain. When the nerve impulses reach the brain, they are interpreted as sounds. It is the cochlea that changes sound waves to nerve impulses.

If the hair cells in the cochlea are damaged or destroyed, a person can lose some hearing ability. This damage can happen when someone is exposed to loud sounds. Scientists are finding that the hair cells may be able to repair themselves.

Picture This

5. Use a Scientific Illustration Using a highlighter, draw a circle around the structures that make up the outer ear. Then use two different colors of highlighters to draw circles around the structures of the middle ear and of the inner ear.

FOLDABLES

A Drawing and Identifying On the inside of a 3-tab book Foldable, draw a human ear. Identify the three sections of the ear and label all the parts found in each section.

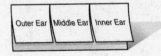

Outer Ear | Middle Ear | Inner Ear

Think it Over

6. Think Critically Write two things you can do to help protect your hearing.

_____

_____

_____

_____

● After You Read

Mini Glossary

cochlea: a spiral-shaped structure in the inner ear that turns vibrations into nerve impulses

eardrum: a tough membrane, or tissue, at the end of the ear canal that transmits vibrations to the middle ear

1. Review the terms and their definitions in the Mini Glossary. Choose one term and use it in a sentence that demonstrates your understanding of how it makes hearing possible.

_____

_____

2. Complete this graphic organizer to outline how vibrations travel through the air to be interpreted as sounds by the brain.

Object creates vibrations

which are transferred through the air by

compressions   and

to the outer ear, then transmitted through the

ear canal

to the

where they are made stronger by the

hammer

and transmitted into the

where the

change them into brain waves.

End of Section

section ② Properties of Sound

● Before You Read

Do your parents ever ask you to turn down the volume on the television? Why do you think some sounds might seem too loud to one person and just right to another?

_____

_____

● Read to Learn

Intensity and Loudness

Suppose you turn down the volume on your radio. The notes sound the same, but the sound is not as loud. What happens to the sound waves? The quieter sound waves do not carry as much energy as louder sound waves do.

The amount of energy a wave carries depends on its amplitude. For a longitudinal wave, amplitude is related to the density of the molecules. Density means how close together molecules are.

Compare the figures. Molecules vibrate with a lot of energy. Sound waves with dense compressions vibrate with a lot of energy. The second sound wave shown has a high-amplitude. The compressions are much denser in a high-amplitude sound wave. Molecules act in the opposite way in the rarefactions. The rarefactions are less dense in the high-amplitude sound wave.

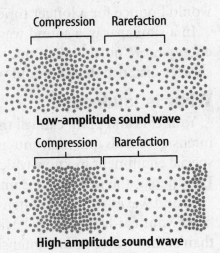

Low-amplitude sound wave

High-amplitude sound wave

Remember that sound waves are longitudinal waves. The wave carries energy, but it doesn't carry matter. Matter compresses and expands as a sound wave passes through the matter.

What You'll Learn
- how amplitude, intensity, and loudness are related
- how sound is measured
- the relationship between frequency and pitch
- the Doppler effect

Mark the Text

Identify Definitions As you read the section, highlight the headings that are questions. Highlight the answers in a different color.

FOLDABLES

ⓑ Build Vocabulary
Make the following vocabulary book Foldable to define the vocabulary terms of this section.

Intensity
Loudness
Decibel
Pitch
Ultrasound
Doppler Effect

1. **Apply** How can you increase the intensity of the sound coming from a radio?

_____

_____

What is intensity?

Imagine sound waves moving from a radio, through a small window, and then to your ear. The amount of energy that passes through the window in 1 s is a measure of intensity. **Intensity** is the amount of energy that flows through a certain area in a specific amount of time. When you turn down the volume of your radio, you reduce the energy carried by the sound waves. So you are also reducing their intensity.

Intensity affects how far away a sound can be heard. Think about whispering with a friend. The sound waves you make have low intensity. They do not travel far. You have to sit close together to hear each other. Now think about shouting to someone. You can be much farther apart. When you shout, the sound waves have high intensity. They can travel farther.

How are intensity and distance related?

Some of a wave's energy is converted into other forms of energy when it is passed from particle to particle. So intensity affects how far a wave will travel.

What happens when you drop a basketball? The ball held above the ground has potential energy. When the ball falls, the potential energy is changed into energy of motion. When it hits the ground and bounces up, some of the ball's energy is transferred to the ground. The ball doesn't have enough energy to bounce back as high as it was. The ball transfers a small amount of energy with each bounce. Finally, the ball has no more energy and it stops bouncing. If you held the ball higher above the ground, it would have more energy and would bounce for a longer time before it stopped.

In a similar way, a sound wave with low intensity loses its energy more quickly than one with high intensity. It travels a shorter distance than a sound wave of higher intensity.

What is loudness?

Your ears and brain can tell the difference between low-intensity sounds and high-intensity sounds. You do not need special equipment to know which sounds have greater intensity. **Loudness** is the way humans understand sound intensity. ☑

Sound waves with high intensity carry more energy. They make your eardrums move back and forth a greater distance than sound waves of low intensity do. The bones of the middle ear change the increased movement of the eardrum into increased movement of the hair cells in the inner ear. This makes you hear a loud sound instead of a quiet one. As the intensity of a sound waves increases, the loudness of the sound you hear increases.

How is loudness measured?

It is hard to say how loud too loud is. Two people might not agree on whether a noise is too loud. A sound that seems fine to you may seem much too loud to your teacher. However, the intensity of sound can be measured. Each unit on the scale for sound intensity is called a **decibel** (DE suh bel). The abbreviation for decibels is dB.

The quietest sound that most people can hear is 0 dB. The intensity of the noises in an average home is 50 dB. Sounds that are louder than 120 dB can cause pain and even permanent hearing loss. During some rock concerts, sound intensity reaches this damaging level. Factories, construction sites, and other workplaces also can have noise levels that might damage hearing. Wearing ear protection, such as earplugs, around loud sounds can help protect against hearing loss. The figure below shows the intensity levels of some sounds in decibels.

Loudness in Decibels

Picture This

3. **Interpret a Graph** How many decibels louder is a jet plane taking off than a vacuum cleaner?

_____

Pitch

Have you ever sung the musical scale do, re, mi, fa, so, la, ti, do? When you sing this scale, your voice starts low and becomes higher with each note. What you hear is a change in pitch. **Pitch** is how high or low a sound seems to be. The pitch of a sound is related to the frequency of the sound waves.

How are frequency and pitch related?

Frequency, being the number of wavelengths that pass a fixed point in one second, is measured in hertz (Hz). One hertz means one wavelength passes by in 1 second. The frequency of a compressional wave is the number of compressions or the number of rarefactions passing by a fixed point each second.

When a sound wave with high frequency hits your ear, many compressions hit your eardrum each second. The vibrations are transmitted to your brain. You understand them as a sound with a high pitch. As the frequency of a sound wave decreases, the pitch becomes lower. A whistle with a frequency of 1,000 Hz has a high pitch. Low-pitched thunder has a frequency of less than 50 Hz.

Think it Over

4. **Explain** Why can the frequency of a compressional wave be counted as either the number of compressions or the number of rarefactions that passes by in one second?

_____

_____

_____

The figure shows different notes and their frequencies. A healthy human ear can hear sound waves with frequencies from about 20 Hz to about 20,000 Hz. Humans can best hear sounds that are between 440 Hz and 7,000 Hz. Most people can hear much softer sounds at this range than at higher or lower frequencies.

Note:	C	D	E	F	G	A	B	C
Sound:	do	re	mi	fa	so	la	ti	do
Frequency:	262 Hz	294 Hz	330 Hz	349 Hz	393 Hz	440 Hz	494 Hz	524 Hz

Are there sounds humans can't hear?

There are some sounds with frequencies that are too high or too low for people to hear. Ultrasonic waves have sound frequencies above 20,000 Hz. People cannot hear sound frequencies that are this high. Dogs can hear sounds with frequencies up to about 35,000 Hz. Bats can hear sounds with frequencies of more than 100,000 Hz. Even though humans can't hear ultrasonic waves, they use them for many things. Doctors use ultrasonic waves to diagnose and treat illnesses. Ultrasonic waves also are used to estimate the size, shape, and depth of underwater objects. ☑

Sound waves with frequencies below 20 Hz are called infrasonic waves. They are also called subsonic waves. These frequencies are too low for most people to hear. Infrasonic waves are produced by sources that vibrate very slowly. Wind, heavy machinery, and earthquakes are some things that produce infrasonic waves. You probably can't hear infrasonic waves. But sometimes you can feel them as a rumble inside your body.

The Doppler Effect

Suppose that you are standing beside a racetrack. Race cars are zooming past. As a car moves toward you, the pitch of its engine becomes higher. As it moves away from you, the pitch becomes lower. The **Doppler effect** is a change in pitch or wave frequency because a wave source is moving.

Applying Math

5. Apply Notice that the musical note C is both the first and the last note of the scale. What do you notice about the frequencies of these two notes?

_____

_____

_____

☑ **Reading Check**

6. Identify What are two ways that people use ultrasonic waves?

_____

_____

_____

What happens when a sound source moves?

As a race car moves, it sends out sound waves. The sound waves are made up of compressions and rarefactions. Look at the first figure. The race car creates a compression, labeled compression A. Compression A moves through the air toward the flagger. Now look at the second figure. By the time compression B leaves the race car, the car has moved forward. Compressions A and B are closer together than they would be if the car had stayed still. Because the compressions are closer together, more compressions pass by the flagger each second than if the car were at rest. So the flagger hears a higher pitch. Look at the second figure again. You can see that the compressions behind the moving car are farther apart. This means the sound of the car has a lower frequency and a lower pitch as the car moves away from the flagger.

What happens when a listener moves?

You also can notice the Doppler effect when you are moving past a sound source that is standing still. Suppose you pass a building with a ringing bell. The pitch of the bell sounds higher as you get closer to the building. The bell sounds lower as you get farther from the building. The Doppler effect happens any time position changes. The change can be the position of the source of a sound or of the listener. The faster the change in position, the greater the change in frequency and pitch. ☑

How is the Doppler effect used?

The Doppler effect also occurs with other kinds of waves, such as electromagnetic waves. Radar guns with electromagnetic waves are used to measure the speed of cars. The radar gun sends radar waves toward a moving car. The waves reflect from the car. The frequency of the waves changes, depending on the speed and direction the car is moving. The radar gun measures the change and finds the speed of the car. Weather radar also uses the Doppler effect to show the movement of winds in storms.

Picture This

7. Draw and Label In the space below, draw and label a diagram that explains the Doppler effect.

✔ **Reading Check**

8. Explain When does the Doppler effect happen?

_____

_____

● After You Read

Mini Glossary

decibel: the unit for measuring sound intensity

Doppler effect: when pitch or wave frequency changes because a wave source is moving

intensity: the amount of energy that flows through a certain area in a specific amount of time

loudness: how humans understand sound intensity

pitch: how high or low a sound is

1. Review the vocabulary terms and their definitions in the Mini Glossary above. What is the difference between intensity and loudness?

   _____

   _____

2. Fill in the blanks below with the following words to make correct statements about the material you read in this section: *intensity, high-amplitude, decibel, subsonic, frequency.*

   As the _____ of a sound wave increases, the pitch increases.

   In a _____ sound wave, the compressions are dense.

   The _____ is the unit for measuring how intense a sound is.

   As the _____ of a sound wave increases, the loudness increases.

3. ◀ **Mark the Text** As you read this section, you highlighted the headings that are questions. Will this strategy help you remember what you read in the section? Why or why not?

   _____

   _____

   _____

End of Section

● Before You Read

Ask a parent or teacher what kind of music he or she liked at your age. Have you ever heard that kind of music? If not, ask the person to sing part of a favorite song. How is it similar to or different from the kind of music you like?

_____

_____

● Read to Learn

What is music?

Music and noise are both caused by vibrations. But there are some important differences between them. Noise has random patterns and pitches. **Music** is made up of sounds that are carefully chosen and have regular patterns.

What are natural frequencies?

Every material when hit, struck, or disturbed in some way will vibrate. The set of frequencies at which the material will vibrate is called its natural frequencies. No matter how you pluck a guitar string, you hear the same pitch because the string always vibrates at its natural frequencies. The guitar string's natural frequencies depend on its thickness and length. It also depends on how tightly the string is stretched. Each string on a guitar is tuned to different natural frequencies. This lets you play different notes on the guitar and make music. Every musical instrument contains something that vibrates at its natural frequencies to create a pitch. It could be strings, a membrane, or a column of air.

What is resonance?

In wind instruments, such as an oboe or a flute, the column of air inside the instrument vibrates. The air vibrates because of resonance. Resonance is the ability of a medium to vibrate by absorbing energy at its own natural frequencies.

What You'll Learn

- the difference between noise and music
- why different kinds of instruments sound differently
- how instruments make music

Mark the Text

Vocabulary Highlight all the vocabulary terms and their definitions in this section.

FOLDABLES

❻ Identify Make the Foldable shown below to hold quarter sheets of paper that contain information on sound quality and musical instruments.

Music

Copyright © Glencoe/McGraw-Hill, a division of The McGraw-Hill Companies, Inc.

A musician plays a brass instrument by vibrating the lips while blowing air into the mouthpiece. The air in the instrument absorbs some of this energy and starts to vibrate. The resonance of the air makes the sound louder. Resonant frequencies amplify the sound of many musical instruments.

Sound Quality

Suppose someone played a note on a flute and then played the same note on a piano. You could tell the difference between the two instruments, even if you had your eyes closed. They would not sound the same. **Sound quality** describes the differences among sounds of the same pitch and loudness. Each instrument has its own sound quality.

Objects can vibrate at other frequencies besides their natural frequencies. This produces sound waves with more than one frequency. Musical instruments get their quality of sound from the combination of frequencies that each kind of instrument has. ☑

What are overtones?

Even though an instrument can vibrate at many different frequencies at the same time, you hear just one note. All of the frequencies do not vibrate at the same intensity. The main tone that is played and heard is the fundamental frequency. On a guitar, the fundamental frequency is made by the entire string vibrating back and forth. Look at the first example in the figure. It shows how the guitar string vibrates at the fundamental frequency. But the string also vibrates to produce overtones.

Guitar Overtones

Fundamental (250 Hz)

First overtone (500 Hz)

Second overtone (750 Hz)

An **overtone** is a vibration with a frequency that is a multiple of the fundamental frequency. Suppose a guitar string vibrates at a fundamental frequency of 250 Hz. Multiples of 250 are 500, 750, 1000, and so on. The guitar string will vibrate at its fundamental frequency of 250 Hz. It can also vibrate at overtones of 500 Hz, 750 Hz, 1000 Hz, and so on. The first two guitar-string overtones are shown in the figure. These overtones create the rich sounds of a guitar. Every instrument has a different number and intensity of overtones. These overtones produce each instrument's own sound quality.

Copyright © Glencoe/McGraw-Hill, a division of The McGraw-Hill Companies, Inc.

✔ **Reading Check**

1. **Explain** Why do different instruments have different sound qualities?

_____

_____

_____

Picture This

2. **Predict** In the box below, draw what you think the vibration that produces the third overtone would look like.

Musical Instruments

A musical instrument is anything that is used to make a musical sound. Violins, cellos, oboes, bassoons, horns, and kettledrums are some of the musical instruments in an orchestra. But people throughout the world play many different kinds of instruments. Australian Aborigines play a woodwind instrument called the didgeridoo (DIH juh ree dew). Caribbean musicians use rubber-tipped mallets to play drums made of steel. An instrument called the nay is played throughout the Arab world. It is similar to a flute.

How do string instruments produce sound?

Violins, electric guitars, and harps are all string instruments with tightly stretched strings. Musicians produce sound by plucking, striking, or moving a bow across the strings. The sound of a vibrating string is soft. A **resonator** (RE suh nay tur) is a hollow space filled with air that makes sound louder when the air inside of it vibrates. As shown in the figure, string instruments usually have a resonator. Imagine a violin string that is stretched tightly between two nails on a board. If you pluck the string, the sound will be very quiet. However, the violin frame and the air inside the body absorb energy from the vibrating string. They begin to vibrate, too. The vibration of the violin body and the air inside makes the sound of the string louder. It also changes the quality of the sound.

Sound waves

Think it Over

3. **Experiment** Close your mouth and gently thump your cheek. Then, open your mouth and thump your cheek again. What causes the difference in sound?

_____

Picture This

4. **Identify** Circle the area where the sound enters the resonator.

What about brass and woodwind instruments?

Vibrating air helps brass and woodwind instruments, like horns, oboes, and flutes, make music. They use different methods to make the air inside vibrate. Brass instruments have cone-shaped mouthpieces, as shown. A mouthpiece is inserted into a metal tube, the resonator. As the musician blows into the instrument, his or her lips vibrate against the mouthpiece. Then air in the resonator starts to vibrate. The resonator produces a pitch.

In brass and wind instruments, the length of the vibrating tube of air determines the pitch of the sound. For example, in flutes and trumpets, the musician changes the length of the resonator by opening and closing holes in the tube. In trombones, the musician slides the vibrating tube of air in and out to make the tube shorter or longer.

How do percussion instruments make sound?

Percussion means hitting or striking. They are hit, shaken, or rubbed to produce sound. People have used drums and other percussion instruments for thousands of years.

Some drums have a cover that is a membrane stretched over one or both of the ends. The area inside the drum is the resonator. When the drummer hits the membrane, it vibrates. The membrane makes the air in the resonator vibrate. The resonator amplifies the sound. Some drums have only one pitch. Others have a pitch that can be changed by making the membrane tighter or looser. ☑

Caribbean steel drums were developed in the 1940s in Trinidad. They are made from 55-gallon oil barrels. The end of a barrel is hammered to make up to 32 different striking surfaces. Each striking surface has its own pitch.

The xylophone is another percussion instrument. It has a series of wooden bars. Each bar has its own resonator that is shaped like a tube. The musician hits the bars with mallets. Different kinds of mallets change the sound quality. Other types of percussion instruments are cymbals, rattles, and bells.

☑ **Reading Check**

5. **Describe** Tell how the pitch of a drum can be changed.

_____

_____

_____

After You Read

Mini Glossary

music: sounds that are carefully chosen and have regular patterns

overtone: a vibration whose frequency is a multiple of the fundamental frequency

resonator: a hollow space filled with air that makes sound louder when the air inside of it vibrates

sound quality: the differences among sounds of the same pitch and loudness

1. Review the vocabulary terms and their definitions in the Mini Glossary above. On the lines below, give examples of a resonator for a string instrument, a wind instrument, and a percussion instrument.

_____

_____

2. Complete the graphic organizer by writing the names of some of the instruments mentioned in this section in the correct category.

3. **Mark the Text** You highlighted the vocabulary terms and their definitions. How did these terms help you understand more about music?

_____

_____

End of Section

chapter 10 Sound

section 4 Using Sound

What You'll Learn

- what affects sound in a concert hall
- how some animals use sound waves
- how sonar is used
- about ultrasound

Study Coach

Outline the Text As you read the section, make an outline of the information in it. Use the headings as the categories of your outline. Fill in the outline with details.

FOLDABLES™

D **Compare** Make a 4-door Foldable like the one shown to organize and compare ideas about sound.

● Before You Read

You use sound when your alarm clock goes off and wakes you up. Write two ways that you have used sound today.

_____

_____

● Read to Learn

Acoustics

Have you ever been in a noisy gymnasium where it is hard to hear someone talking? Sound waves bounce or reflect off the hard surfaces of the floor, ceiling, and walls. A sound and its reflections reach your ears at different times and you hear echoes. The echoing produced by many reflections of sound is called reverberation (rih vur buh RAY shun).

Reverberation is a problem in music performances. It can ruin the sound of the music. Scientists and engineers who design concert halls try to prevent reverberation. They need to understand how the size, shape, and furnishings of the room affect the reflection of sound waves. These scientists and engineers specialize in acoustics. **Acoustics** (uh KEW stihks) is the study of sound. They might recommend that the walls of concert halls be covered with carpets and draperies. Soft materials like these can reduce reverberation.

Echolocation

At night, bats swoop around in darkness without bumping into anything. They even are able to catch insects in the dark. Many species of bats depend on echolocation. **Echolocation** is the process of locating objects by making sounds and interpreting the sound waves that reflect, or bounce back. Bats can get information about an insect, such as how far away it is and how big it is, from the sound waves that reflect from it. Dolphins also use echolocation to locate objects.

Sonar

More than 140 years ago, a ship named the *Central America* disappeared in a hurricane. It held 21 tons of gold that would be worth many millions of dollars today. When the ship was wrecked, there was no way to search for it. The *Central America* and its gold lay at the bottom of the ocean under 2,400 m of water. In 1988, crews used sonar to find the wreck. **Sonar** is a system that uses reflections of underwater sound waves to detect objects.

The figure shows how sonar works. A sound wave is sent toward the ocean floor from a boat on the surface of the water. When the sound wave hits something solid, it is reflected. The sound wave bounces back to the surface. An underwater microphone, called a hydrophone, is attached to the bottom of the boat. The hydrophone picks up the reflected sound. The speed of sound in water is known. So the distance from the boat to the object can be found by measuring how much time it takes the sound to bounce back to the hydrophone.

Hydrophone

Reflected signal

Sonar signal

Many uses have been developed for sonar. Navy ships use sonar to locate submarines. Fishing crews use sonar to find schools of fish. Scientists use sonar to map the ocean floor. Most sonar uses ultrasonic or high-frequency waves. Sound with frequencies above 20,000 Hz is called **ultrasound**.

Other Uses for Ultrasonic Waves

Ultrasonic waves are used in other ways besides echolocation and sonar. They also are used to clean delicate items, such as jewelry. Ultrasonic waves are aimed at a dirt buildup. The ultrasonic waves cause the buildup to vibrate quickly until the dirt breaks up into small pieces. These small pieces can be removed easily.

Picture This

1. **Explain** A fishing crew is using sonar to find schools of fish. The crew's boat is over a rocky area of the ocean floor. What problems might the crew have when using sonar in this area?

_____

_____

_____

_____

Applying Math

2. **Calculate** Sound travels at 1,500 m/s in water. The *Central America* was 2,400 m under the surface of the ocean. How long would it take a sound wave from a ship to bounce off the wrecked ship and return to the surface?

_____

How is ultrasound used in medicine?

One of the most important uses of ultrasonic waves is in medicine. A process called ultrasound uses ultrasonic waves to gather images of the inside of a person's body. In the past, doctors had to perform surgery to do this. Special instruments are used to send ultrasonic waves into a part of a patient's body. The reflected waves create an image that allows doctors to study many conditions such as pregnancy, heart disease, and cancer.

How does ultrasound imaging work?

Like X rays, ultrasound can be used to make images of body parts inside the body. A technician aims the ultrasound waves toward the target area. The sound waves reflect off the organs or tissues. A computer turns the reflected waves into video pictures, or images. The images are called sonograms. Doctors can use sonograms to find medical problems. ☑

Ultrasound is used to examine many parts of the body. These include the heart, liver, gallbladder, pancreas, spleen, kidney, breast, and eye. Doctors use ultrasound to keep track of a developing fetus. Ultrasound is better than X rays for examining soft structures of the body. It is not as useful for examining bones and lungs. Hard tissues and air absorb the ultrasonic waves instead of reflecting them.

How do doctors use ultrasound to treat medical problems?

Doctors use ultrasound to treat some medical problems. Small, hard deposits of calcium can form in kidneys. The deposits are called kidney stones. In the past, doctors had to operate to remove kidney stones. Now ultrasonic treatments are used to break up the stones. As shown in the figure, many ultrasonic waves create vibrations that break the stones into small pieces. The pieces then pass out of the body with urine. Gallstones can be treated in the same way. Patients who are treated with ultrasound recover more quickly than people who have surgery.

Ultrasonic waves

Kidney

3. Define What is a sonogram?

_____

_____

_____

Picture This

4. Infer Kidney stones sometimes leave the kidney and begin to move down behind the pelvic bone. Kidney stones cannot be treated with ultrasound if they are behind the pelvic bone. Why do you think this is so?

_____

_____

_____

● After You Read

Mini Glossary

acoustics: the study of sound

echolocation: the process of sensing things by emitting sounds and interpreting the sound waves that are reflected back

sonar: a way to use the reflection of underwater sound waves to find objects

ultrasound: sound with frequency above 20,000 Hz

1. Read the vocabulary terms in the Mini Glossary above. How are echolocation and sonar similar and different?

_____

_____

_____

2. Complete the diagram with ways that people and animals use sound.

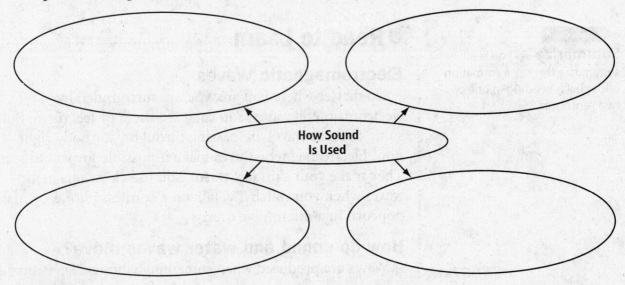

How Sound Is Used

3. Study Coach You made an outline of the information in this section as you read it. How did this strategy help you learn the material in the section? Would you recommend that a friend use this strategy?

_____

_____

End of Section

Electromagnetic Waves

section ❶ What are electromagnetic waves?

What You'll Learn

- how electromagnetic waves are formed
- how electric charges produce electromagnetic waves
- properties of electromagnetic waves

● Before You Read

Light is transmitted by electromagnetic waves. Without light, you would not be able to see. On the lines below, write three things you could not do without light.

_____

_____

_____

Study Coach

Summarize As you read, summarize the key information under each heading in one or two sentences.

● Read to Learn

Electromagnetic Waves

No matter where you are, you are surrounded by electromagnetic waves. You can't see them or feel them. But some of these waves are passing through your body right now. Electromagnetic waves make it possible for you to see. They make your skin feel warm. You use electromagnetic waves when you watch TV, talk on a cordless phone, or make popcorn in a microwave oven.

How do sound and water waves move?

Waves are produced when something vibrates. They carry energy from one place to another. Both sound waves and water waves move through matter. Sound waves move through matter that is solid, liquid, or gas. Water waves move through water, a liquid. Sound waves and water waves travel because energy is transferred from one particle to another particle.

How are electromagnetic waves made?

Electromagnetic waves do not need matter to transfer energy. **Electromagnetic waves** are made by vibrating electric charges and can travel through space where there is no matter. Electromagnetic waves do not transfer energy from particle to particle. Instead, they travel by transferring energy between vibrating electric and magnetic fields. ☑

✔ Reading Check

1. Apply What vibrates when electromagnetic waves transfer energy?

_____

_____

Electric and Magnetic Fields

What happens if you move a magnet close to a metal paper clip? The paper clip moves toward the magnet and sticks to it. The magnet moves the paper clip without touching it because every magnet is surrounded by a magnetic field. There is a magnetic field around magnets even if the space around the magnets contains no matter.

Electric charges are surrounded by electric fields in the same way. An electric field surrounds an electric charge even if the space around the charge has no matter. An electric field allows electric charges to exert forces on each other even when they are far apart.

How do moving charges create magnetic fields?

An electric charge is surrounded by an electric field. Electric charges also can be surrounded by magnetic fields. An electric current in a wire is the flow of electrons in one direction. The movement of these electrons creates a magnetic field around the wire as shown in the figure. So, any moving electric charge is surrounded by an electric field and a magnetic field.

Magnetic field lines

What happens when electric and magnetic fields change?

A changing magnetic field creates a changing electric field. One example of this relationship can be seen in a transformer. A transformer transfers electric energy from one circuit to another circuit. In the main coil of a transformer, changing electric current produces a changing magnetic field. This changing magnetic field then creates a changing electric field in another coil. This electric field produces an electric current in the coil. The reverse is also true. A changing electric field creates a changing magnetic field.

Think it Over

2. **Draw Conclusions** How could you test an object to see if it has a magnetic field?

_____

_____

_____

_____

Picture This

3. **Label** the electrons moving through the wire. What is the flow of electrons in one direction in a wire called?

_____

4. Identify What is produced when an electric charge vibrates?

_____

_____

_____

Picture This

5. Trace Use a highlighter to trace the magnetic field. Use a pencil to trace the electric field.

Making Electromagnetic Waves

Waves are made when something vibrates. **Electromagnetic waves** are made when an electric charge vibrates. When an electric charge vibrates, the electric field around it changes. Remember, a changing electric field creates a changing magnetic field. Then the changing magnetic field creates a changing electric field. How do the changing fields become a wave? Look at the figure below. This process of changing electric and magnetic fields continues. The magnetic and electric fields create each other again and again. ✔

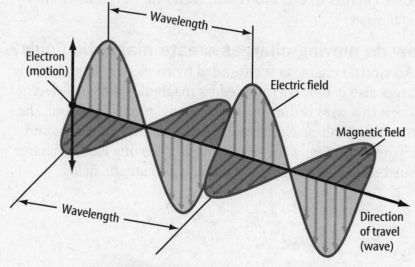

Electromagnetic Wave

Properties of Electromagnetic Waves

An electromagnetic wave travels in all directions. The figure shows a wave traveling in only one direction. The electric and magnetic fields vibrate at right angles to the direction the wave travels. Remember, in a transverse wave, the matter in the medium moves at a right angle to the direction the wave travels. So, an electromagnetic wave is a transverse wave.

All matter contains charged particles that are always moving. Therefore, all objects emit electromagnetic waves. The wavelengths of the waves become shorter as the temperature of the material increases. As an electromagnetic wave moves, its electric and magnetic fields meet objects. These vibrating fields can exert forces on charged particles and magnetic materials. The forces make the charged particles and magnetic materials move. For example, electromagnetic waves from the Sun cause the electrons in your skin to vibrate and gain energy. The energy carried by an electromagnetic wave is radiant energy. Radiant energy makes a fire feel warm. Radiant energy also allows you to see.

What is the speed of electromagnetic waves?

All electromagnetic waves travel at 300,000 km/s in a vacuum. Light is an electromagnetic wave. So, the speed of electromagnetic waves in a vacuum is usually called the speed of light. The speed of light is nature's speed limit. Nothing in the universe can travel faster than the speed of light. The speed of electromagnetic waves through matter depends on what material the waves travel through. Electromagnetic waves usually travel slowest in solids and fastest in gases. The table shows the speed of visible light in some materials.

Speed of Visible Light	
Material	**Speed (km/s)**
Vacuum	300,000
Air	slightly less than 300,000
Water	226,000
Glass	200,000
Diamond	124,000

What are the wavelength and frequency of an electromagnetic wave?

Electromagnetic waves can be described by their wavelengths and frequencies. Look at the figure on the previous page. The wavelength is the distance from the top of one wave to the top of the next wave.

The frequency of any wave is the number of wavelengths that pass a point in 1 s. The frequency of an electromagnetic wave is the same as the frequency of the vibrating charge that makes the wave. This frequency is the number of vibrations of the charge in one second. As the frequency of an electromagnetic wave increases, the wavelength becomes smaller.

Waves and Particles

The difference between a wave and a particle might seem obvious. A wave is a disturbance that carries energy. A particle is a piece of matter. But the difference is really not so clear.

Applying Math

6. **Calculate** Use the table at left. The average distance from the Sun to Earth is about 150,000,000 km. About how long does it take light from the Sun to reach Earth? Round your answer to the nearest minute.

_____

Picture This

7. **Draw and Label** On your own paper, draw two electromagnetic waves. The wavelength of the second wave should be half the wavelength of the first wave. Label the wavelength of each wave.

A **Find Main Ideas** Make two quarter-sheets of paper into note cards to organize information about waves and particles.

Waves as Particles:

Particles as Waves:

Picture This

8. **Describe** Look at the second part of the figure showing electrons being sprayed at two slits. Describe the pattern that they form.

_____

_____

_____

_____

Can a wave be a particle?

In 1887, a scientist named Heinrich Hertz discovered that when light is shined on a metal, electrons are ejected from the metal. Hertz found that whether or not electrons were ejected depended on the frequency of the light and not the amplitude. This result seemed mysterious because the energy carried by a sound wave depends on its amplitude and not on its frequency. Years later, Albert Einstein explained Hertz's discovery. Electromagnetic waves can behave as a particle, called a **photon**, whose energy depends on the frequency of the waves.

Can a particle be a wave?

The discovery that electromagnetic waves could behave as a particle led to other questions. Scientists wondered whether matter could behave as a wave.

Look at the figure below. The first part shows paint particles being sprayed at two narrow openings, or slits. The paint particles cover only the area behind the slits.

The second part shows a beam of electrons being sprayed at two slits. You might expect the electrons to strike only the area behind the slits. But scientists found that electrons form an interference pattern.

Paint particles sprayed at two slits coat only the area behind the slits.

Electrons fired at two slits form an interference pattern, similar to patterns made by waves.

An interference pattern is formed by waves when they pass through two slits and interfere with each other. The pattern is like the one made by water waves in the diagram below. The experiment with two slits shows that electrons can behave like waves. In fact, all particles, not only electrons, can behave like waves.

Water waves make an interference pattern after they pass through two slits.

After You Read

Mini Glossary

electromagnetic waves: waves made by vibrating electric charges that can travel through space where there is no matter

photon: an electromagnetic wave that behaves like a particle and whose energy depends on the frequency of the waves

1. Review the terms and their definitions in the Mini Glossary. Choose one term and use it in a sentence that shows your understanding of the term.

_____

_____

_____

2. Write a fact about electromagnetic waves on the lines under each heading.

Electromagnetic Waves

| How they are formed | Kind of wave | Sometimes behave as |

3. **Study Coach** As you read this section, you summarized the important information under each heading. How did that strategy help you learn the material in this section?

_____

_____

_____

End of Section

Electromagnetic Waves

section ② The Electromagnetic Spectrum

What You'll Learn

- the different kinds of electromagnetic waves
- the properties of electromagnetic waves
- how electromagnetic waves are used

● Before You Read

Radio waves, microwaves, and X rays are examples of electromagnetic waves. On the lines below, write a sentence about how you or someone you know has used electromagnetic waves.

_____

_____

_____

Mark the Text

Identify Definitions As you read this section, highlight each vocabulary term and its definition.

● Read to Learn

A Range of Frequencies

Electromagnetic waves have a wide range of frequencies. Some vibrate once a second. Others vibrate trillions of times a second. The whole range of electromagnetic wave frequencies is called the electromagnetic spectrum. The electromagnetic spectrum is shown below.

Different parts of the electromagnetic spectrum interact with matter in different ways. Because of this, the parts of the spectrum have different names. **Visible light**, the range of electromagnetic waves that humans can see, is a small part of the whole electromagnetic spectrum. But a number of devices have been created to detect the other frequencies. For example, the antenna of a radio detects radio waves.

Picture This

1. **Identify** Which electromagnetic waves have the lowest frequency?

_____

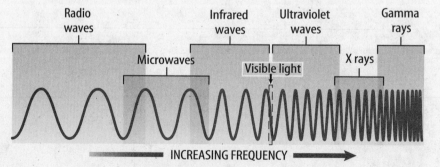

Radio Waves

Even though you can't see them, radio waves are moving everywhere. <u>Radio waves</u> are low-frequency electromagnetic waves with wavelengths longer than about 1 mm. Radio stations make use of these waves. Microphones are used to change sound waves from voices and music into radio waves. The radio waves carry signals that can be picked up by radios. Radios then change the signals back into sound waves. Remember, you hear sounds when compressions and rarefactions from a sound wave reach your ears. A radio wave does not produce compressions and rarefactions, so you cannot hear an actual radio wave. It needs to be turned into a sound wave by a radio before you can hear it.

What are microwaves?

<u>Microwaves</u> are radio waves with wavelengths less than 30 cm. They have a higher frequency and shorter wavelength than the waves used by radios. Microwaves with wavelengths between about 20 cm and 1 cm are used for communication. Cell phones and satellites use microwaves of this wavelength.

How do microwave ovens work?

In microwave ovens, microwaves interact with water molecules in food. The figure shows how microwave ovens work. Each water molecule is positively charged on one side and negatively charged on the other side. There is a vibrating electric field inside a microwave oven. It causes the water molecules in food to rotate billions of times a second. The rotation causes friction. Friction creates thermal energy. Thermal energy made by the interactions between the water molecules heats food. ☑

Normally water molecules are randomly arranged.

The microwaves cause the water molecules to rapidly flip back and forth.

Water molecules

Electromagnetic wave

FOLDABLES

B **Find Main Ideas** Make a layered book Foldable using four sheets of paper to describe each part of the electromagnetic spectrum. Write spectrum on the first tab and label the other tabs.

☑ **Reading Check**

2. **Explain** Why is it impossible to heat items that do not contain water in a microwave oven?

_____

_____

_____

Picture This

3. **Analyze a Diagram** Look at the figure. List two differences between the left part and right part of the figure.

_____

_____

_____

_____

How does radar work?

Radar stands for **RA**dio **D**etecting **A**nd **R**anging. Radar is used to find the position and speed of objects. Radio waves are sent toward an object. The waves bounce off the object and return. The time this takes gives the object's position. Radar can show the position of airplanes, boats, and cars. It also can measure the speed of moving vehicles. ☑

What is magnetic resonance imaging?

Researchers developed a technique called magnetic resonance imaging, or MRI, in the 1980s. MRI uses radio waves to help diagnose illnesses. The patient lies inside a large tube surrounded by a strong magnet, a radio-wave emitter, and a radio-wave detector.

Protons in the hydrogen atoms in bones and soft tissue act like magnets. They line up with the strong magnetic field of the MRI. Energy from radio waves makes some of the protons flip. When the protons flip, they release energy. Different tissues release different amounts of energy. A radio receiver detects the released energy and makes a map of the body's tissues. A picture of the inside of the patient's body is made.

Infrared Waves

When you stand in front of a fireplace, you feel warmth from the fire. Why do you feel heat? The warmth is thermal energy transmitted by infrared waves. **Infrared waves** are electromagnetic waves with wavelengths between about one millimeter and about 750 billionths of a meter.

You use infrared waves every day. A remote control sends out infrared waves to a television. A computer uses infrared waves to read CD-ROMs. In fact, every object gives off infrared waves. Hotter objects give off more infrared waves than cooler objects. Infrared detectors can make pictures of objects from the infrared waves they give off, or emit. Infrared sensors on satellites can identify where plants are growing in a region.

Visible Light

Visible light is the range of electromagnetic waves that people can see. Visible light has wavelengths of about 750 billionths to 400 billionths of a meter. The electromagnetic waves you can see have different wavelengths. You see the different wavelengths as different colors. Blue light has the shortest wavelength. Red light has a longest wavelength. The light looks white if all the colors are present.

💡 Think it Over

5. Predict If your body heat were to show up on an infrared detector, which parts of your body would be giving off the least amount of infrared waves?

Applying Math

6. Write Decimals You can write 10 billionths as a decimal by moving the decimal point in 1.0 ten places to the left. Write 10 billionths as a decimal.

_____

Ultraviolet Waves

Ultraviolet waves are electromagnetic waves with wavelengths from about 400 billionths to 10 billionths of a meter. Ultraviolet (UV) waves have enough energy to enter skin cells. Being exposed to too many UV rays can cause skin damage and cancer. Sunlight contains ultraviolet waves.

Most of the ultraviolet radiation that reaches Earth's surface is longer-wavelength rays. They are called UVA rays. Shorter-wavelength rays are called UVB rays. UVB rays are the rays that cause sunburn. Both UVA and UVB rays can damage the skin and cause skin cancer. ☑

Can UV radiation be useful?

Some UV rays are useful. A few minutes of UVs from the Sun each day helps your body make vitamin D. Vitamin D is needed for healthy bones and teeth.

UVs are also used to sterilize objects such as medical supplies and hospital equipment. When ultraviolet light enters a cell, it damages protein and DNA molecules. This can kill some single-celled organisms such as bacteria.

Ultraviolet waves make some materials light up, or fluoresce. Fluorescent materials absorb ultraviolet waves. Then they emit the energy as visible light. Police detectives sometimes use fluorescent powder to find fingerprints when solving crimes.

What is the ozone layer?

The ozone layer is an area in Earth's upper atmosphere. It is between 20 km to 50 km above Earth's surface. Ozone is a molecule made up of three oxygen atoms. The ozone layer is necessary to life on Earth because it absorbs most of the Sun's harmful ultraviolet waves. Over the past few decades, the amount of ozone in the ozone layer has decreased.

What chemicals harm the ozone layer? Ozone has decreased because of the presence of certain chemicals in Earth's atmosphere. The chemicals are called chlorofluorocarbons, or CFCs. CFCs are used in air conditioners, refrigerators, and cleaning fluids. CFC molecules react with ozone molecules. One chlorine atom from a CFC molecule can break apart thousands of ozone molecules. Many countries are using fewer CFCs and other chemicals that destroy ozone.

If too much ozone is destroyed, the ozone layer will be damaged or lost altogether. Without the ozone layer, everything on the surface of Earth would become exposed to a much higher level of damaging ultraviolet waves.

☑ **Reading Check**

7. Describe How can ultraviolet waves harm you?

_____

_____

_____

💡 **Think it Over**

8. Think Critically What could happen to humans if the ozone layer were destroyed?

_____

_____

_____

_____

9. **Infer** Why might MRIs cause less harm to the body than X rays?

_____

_____

_____

X Rays and Gamma Rays

X rays and gamma rays are the electromagnetic waves with the shortest wavelengths. They also have the highest frequencies. Both X rays and gamma rays are high-energy electromagnetic waves. <u>X rays</u> have wavelengths between 10 billionths and 10 trillionths of a meter. They have enough energy to go through skin and muscle. Doctors and dentists use low levels of X rays to take pictures of internal organs, bones, and teeth. High levels of X rays are dangerous and can cause cancer. X rays are projected only at very specific areas of the body. Lead aprons or shielding are used to protect other areas from exposure. X rays cannot travel through lead.

X Ray of Bone

✓ **Reading Check**

10. **Identify** Which statement is *not* true. (Circle your answer.)

   a. Gamma rays are low-frequency waves.

   b. X rays are high-energy waves.

   c. Gamma rays are used to treat diseases.

<u>Gamma rays</u> are electromagnetic waves with wavelengths shorter than 10 trillionths of a meter. They are the highest-energy electromagnetic waves. They can travel through several centimeters of lead. Gamma rays are produced in the nuclei of atoms. Both X rays and gamma rays are used in radiation therapy. Radiation therapy is used to kill diseased cells in the human body. X rays and gamma rays can kill both healthy and diseased cells. Doctors carefully control the amount of X-ray or gamma-ray radiation the diseased area receives. This reduces the damage to healthy cells. ✓

● After You Read

Mini Glossary

gamma rays: electromagnetic waves with wavelengths shorter than 10 trillionths of a meter

infrared waves: electromagnetic waves with wavelengths between about 1 mm and about 750 billionths of a meter

microwaves: radio waves with wavelengths of less than 30 cm

radio waves: low-frequency electromagnetic waves with wavelengths longer than about 1 mm

ultraviolet waves: electromagnetic waves that have wavelengths from about 400 billionths to 10 billionths of a meter

visible light: the range of electromagnetic waves that people can see

X rays: electromagnetic waves with a wavelength of between 10 billionths of a meter and 10 trillionths of a meter

1. Draw a line to match the name of each ray in the first column with its wavelength in the second column.

Column 1	Column 2
Visible light	less than 30 cm
X ray	400 billionths to 10 billionths of a meter
Microwave	longer than 1 mm
Gamma ray	1 mm to 750 billionths of a meter
Radio wave	10 billionths to 10 trillionths of a meter
Ultraviolet wave	750 billionths to 400 billionths of a meter
Infrared wave	shorter than 10 trillionths of a meter

2. Write the names of the waves listed in question 2 in the electromagnetic spectrum pyramid.

a. _____ b. _____ c. _____

d. _____

e. _____

f. _____ g. _____

INCREASING FREQUENCY →

End of Section

What You'll Learn

- how radio waves transmit information
- the difference between amplitude modulation and frequency modulation
- about radio wave communication

Create a Quiz After you have read this section, create a quiz based on what you have learned. After you have written the quiz questions, be sure to answer them.

FOLDABLES

❻ Summarize Fold one sheet of paper into sixths to summarize the information in this section on radio communication.

Radio Communication
Radio
Television
Telephones
Satellites
GPS

● Before You Read

You learned that all electromagnetic waves have frequencies. Every radio station broadcasts at a certain frequency. What is the frequency of your favorite radio station? What do you think this number means?

_____

_____

● Read to Learn

Radio Transmission

When you listen to the radio, you hear music and words that were created far away. Radio stations need to send sounds over great distances. They do this by changing sound waves to electromagnetic waves. Your radio then changes the electromagnetic waves back to sound waves again.

How does a radio receive different stations?

Each radio station broadcasts at a certain frequency. The tuning knob on a radio lets you choose the frequency, or station, you want to listen to. The frequency of the electromagnetic wave that a radio station uses is the **carrier wave**.

At the radio station, music and voices make air molecules vibrate. This vibration creates sound waves. Microphones convert the sound waves to a changing electric current. The changing electric current is an electronic signal. The signal contains the words and music that the station is broadcasting. The signal is added to the station's carrier wave. The carrier wave has been changed, so it is a modified carrier wave. The modified carrier wave vibrates electrons in the station's antenna. These vibrating electrons create a radio wave. The radio wave travels out in all directions at the speed of light. The radio wave makes electrons in your radio's antenna vibrate.

Modulation The station must broadcast the signal at its assigned frequency, the carrier wave. However, it also has to send information about the sounds your radio will receive. The sounds produced at the radio station are converted into the signal wave, which is added to the carrier wave. The process of adding the signal wave to the carrier wave is called **modulation**. The radio turns the signal wave into sound waves-music and words.

What is AM radio?

The figure below shows the two ways the carrier wave can be modified. The first way is amplitude modulation, or AM. AM radio stations broadcast information by changing the amplitude of the carrier wave. Look at the wave labeled *amplitude modulation*. Notice that the amplitude changes, but the frequency does not. The original sound is changed into an electrical signal that changes the amplitude of the carrier wave. You tune an AM radio to the frequency of the carrier wave. AM frequencies range from 540,000 Hz to 1,600,000 Hz. AM radio stations give their frequencies in kilohertz. A station that gives its frequency as 810 AM means that it is broadcasting at a frequency of 810,000 Hz.

What is FM radio?

The second way a carrier wave can be modified is called frequency modulation, or FM. FM radio stations broadcast information by changing the frequency of the carrier wave. Look at the wave labeled *frequency modulation*. Notice that the frequency of the wave changes, but not the amplitude. The strength of the FM waves is always the same. Because of this, FM signals are usually clearer than AM signals. FM frequencies range from 88 million to 108 million Hz. These are much higher than AM frequencies. FM radio stations give their frequencies in megahertz. *Mega-* means "million," so a station with a frequency of 89.9 FM is broadcasting at a frequency of 89,900,000 Hz.

Carrier wave

Signal wave

Amplitude modulation

Frequency modulation

Copyright © Glencoe/McGraw-Hill, a division of The McGraw-Hill Companies, Inc.

1. **Use Decimals** Suppose a radio station broadcasts at 1250 AM. What is the frequency of the station, in hertz?

_____

_____

Picture This

2. **Interpret a Scientific Illustration** Look at the figure that shows frequency modulation. How does the wavelength of the wave change as its frequency increases?

_____

_____

3. **Calculate** A camera's VGA display is rated at 3.0 megapixels. A pixel is the smallest bit of information in a digital picture. If there are 640 pixels from side to side and 480 from top to bottom, what is the actual number of pixels on the display?

_____

_____

4. **Compare and Contrast** analog and digital signals.

_____

_____

_____

5. **Think Critically** What would happen if you moved into an area where there was no cell phone tower while you were talking on a cell phone?

_____

_____

_____

The Digital TV Revolution

Television might seem like magic, until you know how it works. Television and radio are similar in that they both use radio waves. At a television station, sounds and images are converted into electronic signals. The signals are broadcast by carrier waves.

Until early this century the signals were sent using AM and FM waves. The audio, or sound, information was sent using FM signals and the visual, or picture, information was sent using AM signals. These information signals were analog signals. **Analog signals** are electric signals whose amplitudes change smoothly over time.

In 2009, full-power television stations in the United States began broadcasting only digital signals. A **digital signal** is an electric signal where there are only two possible values: ON and OFF. This is similar to a light switch where the light can be either on or off. ☑

How can digital TV be modulated?

There are many ways to modulate radio waves using this on and off information. In some ways, the signals resemble traditional AM and FM and are called Amplitude-Shift Keying (ASK) and Frequency-Shift Keying (FSK). More complex ways of digital modulation allow more information to be carried by a single wave. In the United States, television stations use multiple amplitude modulations to send data on the carrier wave.

Telephones

Just 20 years ago, you never would have seen someone walking down the street talking on a telephone. Today, cell phones are seen everywhere. When you talk into a telephone, you create sound waves. A microphone turns the sound waves into an electrical signal. In cell phones, this signal is used to create radio waves. The radio waves are transmitted to and from a microwave tower. A cell phone uses one radio signal for sending data to a tower. It uses another signal for receiving data from the tower. At the receiving end, the radio wave is turned back into an electric signal. A speaker in the earpiece changes the electric signal into a sound wave.

Cell phone towers are many kilometers apart. Each tower covers an area called a cell. If you move from one cell to another cell, a control station moves your signal to the new cell.

How do cordless telephones work?

Cell phones and cordless telephones are transceivers. A **transceiver** is a device that transmits one radio signal and receives another radio signal from a base unit. You can talk and listen at the same time using a transceiver because the two signals are at different frequencies.

Cordless telephones, like the one in the figure below, work much like cell phones. With a cordless phone, you must be close to the base unit. Many kinds of cordless phones use the same frequency. Sometimes, when someone nearby is using a cordless telephone, you can hear the conversation on your phone. Many cordless phones have a channel button. You can use the channel button to switch your call to another frequency. ☑

Cordless phone

Radio signals

Base unit

How do pagers work?

Pagers are another way to send signals. A pager is a small radio receiver. Each pager has its own phone number. A caller can dial the pager number and leave a message at a central terminal. The caller enters a call-back number using a telephone keypad or by entering a text message from a computer. At the terminal, the message is changed into an electronic signal. The signal is transmitted by radio waves.

Many kinds of pagers use the same transmission frequencies. Each pager in the area receives all the messages that are transmitted in its frequency. However, each pager has a special identification number. The identification number is sent along with each message. A pager only picks up messages with its identification number. Newer kinds of pagers can send messages as well as receive them.

✔ **Reading Check**

6. **Explain** Why do cordless phones use two different frequencies?

_____

_____

_____

Picture This

7. **Identify** What do the circles in the figure represent?

_____

Communications Satellites

Radio waves cannot be sent through Earth. Instead, radio signals are sent to satellites. The satellites can communicate with other satellites or with stations on the ground.

Thousands of satellites have been launched and orbit Earth. Many of these satellites are used for communication. A radio or television station sends a high-frequency microwave signal to the satellite. The satellite amplifies the signal and sends it back to a different place on Earth. The satellite avoids interference by using a different frequency to receive and to send messages.

What is a satellite telephone system?

If you have a mobile telephone, you can make a phone call from the middle of the ocean using a satellite telephone system. A mobile telephone transmits radio waves directly to a satellite. The satellite sends the signal to a ground station. Then the call is passed on to the regular telephone network. Satellites work well for one-way transmissions. But two-way communications can be difficult. There can be a delay between the sender and the receiver because of the time used to travel to and from the satellite. ☑

Why are satellites used for television signals?

You sometimes see satellite-reception dishes attached to houses. These dishes are receivers for television satellite signals. Satellite television is sometimes used instead of systems that send signals close to the ground. Television satellites use shorter microwaves rather than the longer-wavelength radio waves used for normal television broadcasts. Microwaves travel more easily through the atmosphere.

The Global Positioning System

People sometimes get lost while hiking. If they are carrying a Global Positioning System receiver, this is much less likely to happen. The **Global Positioning System (GPS)** is a system of satellites, ground monitoring stations, and portable receivers that determine a person's or object's exact location on Earth. The system consists of 24 satellites. To determine the location of an object, a GPS receiver measures the time it takes for radio waves to travel from four satellites to the receiver. It gives information around the world 24 hours a day. GPS satellites are operated by the U.S. Department of Defense, but the microwaves they send out can be used by anyone. GPS receivers are used in airplanes, ships, cars, and by hikers.

☑ **Reading Check**

8. **Explain** Why is there sometimes a delay when talking on a satellite phone?

_____

_____

_____

💡 **Think it Over**

9. **Apply** Some cars have GPS systems installed in them. When might a GPS system be helpful?

_____

_____

_____

● After You Read

Mini Glossary

analog signal: electric signals whose amplitudes change smoothly over time

carrier wave: the frequency of the electromagnetic wave that a radio station uses

digital signal: electric signals that have only two values— ON and OFF

Global Positioning System (GPS): a system of satellites, ground monitoring stations, and portable receivers that determines an exact location on Earth

modulation: the process of adding the signal wave to the carrier wave

transceiver: a device that transmits one radio signal and receives another radio signal from a base unit

1. Read the terms and their definitions in the Mini Glossary. In the space below, name some users of the GPS.

_____

2. Complete the list to organize the information from the section about radio transmission.

1.	Sounds make air molecules vibrate and create sound waves.
2.	
3.	
4.	The carrier wave vibrates electrons in the station's antenna and creates a radio wave.
5.	
6.	
7.	The vibrating electrons produce a changing electric current.
8.	
9.	
10.	The sound waves travel to your ears and your brain interprets them as music and voices.

End of Section

section ❶ The Behavior of Light

What You'll Learn

- how light waves interact with matter
- regular and diffuse reflection
- index of refraction
- why prisms bend light

Create a Quiz After you read each paragraph, write a question that you think your teacher might ask on a quiz. When you finish reading the section, try to answer all of your quiz questions.

✔ Reading Check

1. **Determine** Which of the following materials is opaque?

   a. plastic food wrap

   b. poster board

   c. waxed paper

   d. window glass

● Before You Read

Rainbows can appear when the Sun is shining during a rain shower. What do you think causes rainbows in the sky?

_____

_____

● Read to Learn

Light and Matter

Imagine looking around a dark room. Your eyes slowly get used to the darkness. You begin to see things that look familiar. The objects look different than they do with the light on. Everything is a shade of gray or even black. When you turn on the light, the objects are colorful again. What you see depends on the amount of light in the room. It also depends on the color of the objects in the room. For you to see an object, it must reflect light to your eyes.

What are opaque, transparent, and translucent objects?

Objects can absorb light, reflect light, or transmit light. Transmitting light means that light can pass through the object. Whether light is absorbed, reflected, or transmitted depends on what the object is made of. Think of a coffee mug. Most likely, you can't see through it. This is because no light passes through it. It only absorbs or reflects light. A material that only absorbs or reflects light is **opaque** (oh PAYK). ✔

Have you ever seen a drinking glass made of dark-colored glass? You can see light through the glass, but you can't really see what's behind it. **Translucent** (trans-LEW-sunt) materials allow some light to pass through. You can't see clearly through translucent materials.

Matter that is clear, such as glass or water, is transparent. **Transparent** materials allow almost all the light to pass through. You can see clearly through transparent materials.

Reflection of Light

Did you look in a mirror this morning before you left for school? When you saw your reflection, you actually saw light that reflected off of your body. The light then reflected off of the mirror and traveled to your eye. Reflection happens when light waves bounce off objects.

Mirror

i

Normal

r

i = angle of incidence

r = angle of reflection

Applying Math

2. **Observe** Suppose the light from the flashlight hits the mirror at a 60° angle. What will be the angle of the reflected beam of light?

_____

What is the law of reflection?

Have you ever shone a flashlight at a mirror in a dark room? The beam of light does not always shine straight back at you. Depending on how you hold the flashlight, the beam might hit the ceiling or the floor. When light is reflected, it follows the law of reflection. According to the law of reflection, the angle at which a light wave strikes a surface is the same as the angle at which it is reflected. The figure above shows the law of reflection. Light follows this law when it is reflected from any surface, whether it is a mirror or a piece of paper.

How is regular reflection different from diffuse reflection?

You can see your reflection in a window. But you cannot see your reflection in a brick wall. Why not? How well you can see your reflection depends on the smoothness of the surfaces. A glass window has a very smooth surface and a brick wall has a very rough surface. Smooth surfaces like the glass window usually reflect light in an even pattern. They produce sharp images. Reflection of light waves from a smooth surface is called regular reflection. Rough or uneven surfaces like the brick wall reflect light in many directions. Reflection of light from a rough surface is called diffuse reflection.

FOLDABLES

A **Find Main Ideas** Make a Foldable like the one shown to write down the main ideas in this section about reflection of light and refraction of light.

3. Explain Car side mirrors do not work when they are covered with frost. What does frost do to the surface of a mirror to cause a diffuse reflection?

_____

_____

_____

Picture This

4. Identify How do you know that light waves are being refracted when they pass from the air to the water in the glass?

_____

_____

_____

How smooth must a surface be for regular reflection?

Some surfaces seem smooth, but they still cause diffuse reflection. For example, the surface of a metal pot might look smooth. But if you looked at it under a microscope, you would see that it is not smooth. The rough surface causes diffuse reflection.

Refraction of Light

Have you ever seen a spoon sitting in a glass of water, like the one in the figure? If you look at the glass from the side, the spoon looks bent or broken. What you see is caused by the refraction, or bending, of light.

The speed of light depends on the substance it travels through. Light travels at different speeds in air, glass, and water. Refraction is caused by a change in the speed of a light ray. When light passes from one substance to another, it is refracted if it is traveling at an angle.

What is the index of refraction?

The amount that light rays bend when they are refracted depends on the speed that light travels in each material. The greater the difference in speeds, the more the light will be bent. Every material has an index of refraction. The **index of refraction** is a property of a material that indicates how much light slows down when it travels in the material.

The greater the index of refraction, the more light slows down in the material. For example, glass has a greater index of refraction than air. This means light moves more slowly in glass than in air. Glass lenses in eyeglasses, microscopes, and cameras use refraction to focus light.

What are prisms?

A prism is a triangular piece of glass that refracts sunlight into colorful patterns. How does bending light create colors? White light, such as sunlight, is composed of several wavelengths of energy which are detected by the eye as colors. The longest wavelength we call red, and the shortest we call violet. As light enters the glass prism, it slows down and changes its direction (bends). As it leaves the prism it bends a second time as it enters air. The prism's shape and the glass medium separates white light into its various colors which we see in a rainbow pattern.

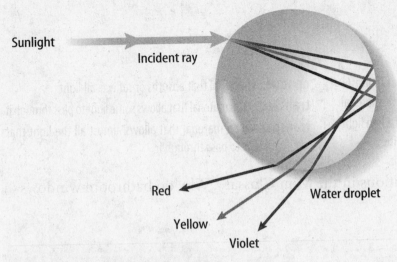

Sunlight

Incident ray

Red

Yellow

Violet

Water droplet

Picture This

5. **Observe** Circle the two areas in the figure that show where sunlight is refracted by the water droplet.

How do rainbows form?

The refracted light from a prism should remind you of a rainbow. Like prisms, raindrops also can refract light. Look at the figure above. Sunlight enters water droplets in the atmosphere. The droplets separate the light into individual colors, depending on wavelength.

Humans can usually see only seven colors in a rainbow or in the refracted light of a prism. The colors are red, orange, yellow, green, blue, indigo, and violet, in order from longest wavelength to shortest wavelength.

What causes mirages?

Have you ever been riding in a car on a hot, sunny day? When you looked ahead, you may have seen what looked like a pool of water on the road. As you got closer, the pool of water seemed to disappear. What you saw was a mirage. A **mirage** is an image of a faraway object. The image is made when light is refracted through air layers of different densities.

Mirages result when the air at ground level is much warmer or much cooler than the air above ground level. As the temperature of air changes, the density of air changes also. These changes in density cause light waves to refract as they pass through air layers with different temperatures. ☑

☑ **Reading Check**

6. **Explain** What happens to light waves as they pass through air layers with different temperatures?

● After You Read

Mini Glossary

index of refraction: a property of a material that indicates how much light slows down when it travels in the material

mirage: an image of a far away object that is made when light is refracted through air layers of different densities

opaque: a material that absorbs or reflects all light

translucent: a material that allows some light to pass through it

transparent: a material that allows almost all the light that strikes it to pass through it

1. Review the terms and their definitions in the Mini Glossary. Why do bathroom windows often have translucent glass?

   _____

   _____

2. Complete the graphic organizer to organize the information you learned in this section about the behavior of light.

<table>
<tr><td>

Reflection

- Happens when light waves _____ off objects.

- The angle at which light strikes a surface and the angle at which it is reflected are the _____.

Regular reflection

- _____ surfaces

- _____ images

Diffuse reflection

- _____ surfaces

- _____ light

</td><td>

Refraction

- The _____ of light.

- A property of a material that indicates how much light slows down when traveling in the material is the

  _____

Prisms

- _____ light

- _____ wavelengths

Rainbows

- formed by light refracting in

  _____

Mirages

- formed when light is refracted through _____

</td></tr>
</table>

End of Section

Light

section ② Light and Color

● Before You Read

Colors are important in everyday life. Name a time when seeing color is important.

_____

_____

Copyright © Glencoe/McGraw-Hill, a division of The McGraw-Hill Companies, Inc.

● Read to Learn

Colors

Why do some apples appear red, while others look green or yellow? You read that white light is made up of light of different wavelengths. Each wavelength is a different color. The color of an object, like an apple, depends on which wavelength(s) of light it reflects. For example, when white light hits a red apple, only the red wavelengths are reflected. This is why the apple appears to be red. The apple absorbs all other wavelengths of light.

Some objects reflect all wavelengths of visible light. Visible light is light that we can detect with our eyes. Objects that reflect all wavelengths of visible light appear to be white. What about black objects? Black is not a color of visible light. Black objects absorb all wavelengths of light. Since almost no light is reflected from these objects, they appear to be black.

How can colors be filtered?

Have you ever worn tinted glasses? Maybe you noticed that tinted glasses change the color of almost everything you see. Yellow glasses make everything look yellow. Red glasses make the world look red. If you put a see-through green plastic sheet over this page, the paper would look green. Tinted glasses and see-through plastic sheets are filters. A <u>filter</u> is a transparent material that lets one or more colors pass through, but absorbs the rest. The color of a filter is the color of light that it lets through.

What You'll Learn

- how you see color
- the difference between light color and pigment color
- what happens when colors are mixed

Mark the Text

Identify the Main Idea
After you read each paragraph, highlight the topic sentence, or the main idea, of the paragraph.

What happens when you look through a colored filter?

When you look at a baseball field, you see green grass. But what would happen if you looked at it through a green filter? Since a green filter lets green light pass, the grass would still look green. Suppose you looked at the same field with a red filter. A red filter only allows red light to pass through. Green grass absorbs all colors, including red, except green. Since the grass absorbs the red and the filter blocks the green, the grass appears black. ✔

Seeing Color

As you approach an intersection the traffic light changes from green to yellow to red. What could happen at the intersection if you couldn't see the color changes? Your safety could depend on seeing colors. How do you see colors?

How does your eye detect light?

Look at the figure below. Light enters your eye through the lens. The lens focuses light onto the retina. The retina is an area on the inside of your eyeball. The retina has two types of cells that absorb light. When retina cells absorb light energy, chemical reactions happen inside the cells. The chemical reactions change light energy into nerve signals that go to your brain.

The two types of retina cells are cones and rods.

What are cones and rods?

Your retina has three types of cones. Cones are sensitive to color and bright light. Each type detects different wavelengths, or colors of light. Red cones detect mostly red and yellow light. Green cones detect mostly yellow and green light. Blue cones detect mostly blue and violet light.

The second type of retina cell is a rod. Rods are sensitive to dim light and aid night vision.

✔ Reading Check

1. Infer What color would green grass appear to be if you looked at it through a red filter?

_____

_____

💡 **Think it Over**

2. Infer Nocturnal animals are awake mostly at night. Which type of retina cells would nocturnal animals probably have the most of?

_____

_____

Picture This

3. Use a Scientific Illustration In the figure, underline the name of the kind of cell that allows you to see colors. Circle the name of the kind of cell that allows you to see in dim light.

How do you interpret color?

When you see a stop sign, how do you know it is red? Red light reflected by the stop sign enters your eye and is focused on the retina. Red cone cells send a signal to your brain. Your brain then understands the signal to mean "red."

What is color blindness?

If one or more sets of cones in your retinas did not work correctly, you would not be able to detect certain colors. This condition is called color blindness. About eight percent of men and one-half percent of women have a form of color blindness. Most people who have color blindness are not truly blind to colors. They have trouble telling the difference between some colors, usually between red and green. People learn to deal with color blindness in many ways. For example, the color of a traffic light can be identified by its position.

Mixing Colors

Have you ever noticed the hundreds of different paint colors customers can choose in a hardware store? You may even have mixed paints to make new colors in art class. It is possible to create different paint colors by mixing pigments. A **pigment** is a colored material that is used to change the color of other substances. A pigment's color depends on the wavelengths of light it reflects.

What happens when you mix colored light?

All the colors you see are made by mixtures of three colors of light—red, green, and blue, known as the primary colors of light. They correspond to the three different types of cones in the retina of your eye. Mixing them in different proportions produces all visible colors. As the figure shows, if red, green, and blue light are mixed equally, the result is white light.

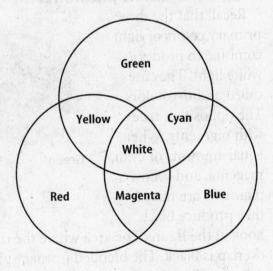

Picture This

5. **Observe** What color appears when red and blue light are mixed?

_____

_____

B **Classify** Make the following Foldable to help you organize information about light color and pigment color.

Light Color:

Pigment Color:

Think it Over

6. **Explain** A hardware store clerk uses green pigment to make green paint. Is green pigment a primary pigment, or is it made from a mixture of primary pigments? Why or why not?

_____

_____

_____

_____

Picture This

7. **Observe** What color appears when magenta and cyan pigments are mixed?

_____

What are paint pigments?

Paints are made with pigments. Paint pigments are made from chemical compounds. Titanium dioxide is a bright white paint pigment. Lead chromate is a yellow pigment that is used to paint the yellow stripes on highways. If you mixed equal amounts of red, green, and blue paint, would you get white paint? No, because mixing paint is different from mixing colored light.

How is mixing pigments different from mixing light?

There are three primary colors of pigments, just as there are three primary colors of light. But the primary pigment colors are different. They are cyan, magenta, and yellow. Cyan is a greenish blue. Magenta is a bluish red. You may have seen these color names if you have ever put ink cartridges in a color printer. You can make any pigment color by mixing different amounts of the three primary pigment colors.

A primary pigment's color depends on the color of light it reflects. Pigments both absorb and reflect many colors. Your eyes and brain, however, see the range of light wavelengths as one color. For example, a yellow pigment appears yellow in white light because it reflects red, orange, yellow, and green light. It absorbs blue and violet light. So, the color of a mixture of primary pigments depends on the primary colors of light that the pigment reflects.

Is there a black pigment?

Recall that the three primary colors of light combine to produce white light. They are called additive colors. The opposite is true with pigments. When equal amounts of cyan, magenta, and yellow pigments are mixed, they produce black.

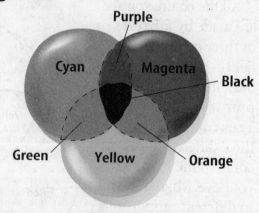

Look at the figure. The area where the three primary pigments overlap is black. The blended primary pigments absorb all colors of light. Because black is the result of no light being reflected, primary pigments are called subtractive colors.

● After You Read

Mini Glossary

filter: a transparent material that lets one or more colors pass, but absorbs the rest

pigment: a colored material that is used to change the color of other substances

1. Review the term and its definition in the Mini Glossary. Write a sentence describing a product you have used that contains pigments.

_____

_____

2. Complete the Venn diagram to organize the information you learned in this section about color.

Colored Light

- The primary colors are

_____ ·

- Mixing all three primary colors produces

_____

All other colors can be produced by mixing the primary colors.

Color Pigments

- The primary colors are

_____

- Mixing all three primary colors produces

_____ ·

3. **Mark the Text** Think about what you have learned. How did highlighting the main idea of each paragraph help you as you read the section?

_____

_____

_____

End of Section

section ⊜ Producing Light

What You'll Learn

- how incandescent and fluorescent lightbulbs work
- about different lighting devices
- about coherent light
- how lasers are used

Study Coach

Outline As you read the section, create an outline to help you remember the different types of lights and how they work.

FOLDABLES

C Find Main Ideas Make the following Foldable to help you find the main ideas about producing light.

| Incandescent |
| Fluorescent |
| Neon |
| Sodium Vapor |
| Tungsten Halogen |
| Laser |

● Before You Read

Lasers have many uses. You have probably seen light that was produced by a laser. Describe what this light looked like.

_____

_____

● Read to Learn

Incandescent Lights

Incandescent light is produced by heating a piece of metal until it glows. Inside a clear lightbulb you can see a small wire called a filament. Filaments in lightbulbs are usually made of tungsten metal. When an electric current flows through the filament, resistance makes the metal hot. The filament gets so hot that it glows brightly.

Fluorescent Lights

You may have fluorescent (floo-RE-sunt) lights in your home or school. A **fluorescent light** uses phosphors to change ultraviolet (UV) radiation to visible light. Phosphors are substances that give off light when they absorb ultraviolet radiation.

A fluorescent bulb, as in the figure, is filled with gas at low pressure. There are electrodes at each end of the tube. When an electric current passes through them, they give off electrons. The electrons then collide with the gas molecules in the tube. The collisions make the gas molecules give off UV radiation. The tube is coated on the inside with phosphors. The phosphors give off visible light when they absorb the UV radiation.

Electrode   Phosphorescent coating

Gas   Bulb

Which type of lighting is more efficient?

Fluorescent lights are more efficient than incandescent lights. A fluorescent bulb uses as little as one-fifth the energy that an incandescent bulb uses to make the same amount of light. Fluorescent bulbs also last much longer than incandescent bulbs. Because fluorescent bulbs are more efficient, they cost less to use over the life of the bulb. They also help reduce energy usage. By reducing energy usage, they could reduce the amount of fossil fuels needed to generate electricity. This would reduce the amount of carbon dioxide and pollutants that are released into Earth's atmosphere.

Neon Lights

Neon lights are glass tubes filled with gas. They work much the same way as fluorescent bulbs. An electric current flows through the tube. The electrons of the electric current collide with gas molecules, as shown in the figure. The collisions produce visible light. If the tube contains only neon gas, the light is bright red. Different colors are produced by adding other gases to the tube.

Sodium-Vapor Lights

Sodium-vapor lights are often used for outdoor lighting. A sodium vapor lamp has a tube filled with neon gas, a small amount of argon gas, and a small amount of sodium metal. When the lamp is turned on, the gas mixture in the tube becomes hot. The sodium metal turns to vapor. The vapor gives off a yellow-orange glow.

Tungsten-Halogen Lights

Tungsten-halogen lights give off bright light. These lights have a tungsten filament, like incandescent bulbs. The filament is inside a quartz tube or bulb. The tube is filled with a halogen gas, such as fluorine or chlorine. The gas allows the filament to get hotter than the filament in an incandescent lightbulb. As a result, the light produced is much brighter. The bulb also lasts longer.

1. **Calculate** Fluorescent lights use one-fifth the energy of incandescent lights. What percent is one-fifth?

_____

Think it Over

2. **Compare** How do the collisions in a fluorescent tube differ from the collisions in a neon tube?

_____

_____

_____

Lasers

Lasers produce light waves that have the same wavelength. A laser's light begins when a number of light waves are given off at the same time. To produce these light waves, a number of atoms are given the same amount of energy. The atoms then release their energy and send off identical light waves. These light waves bounce off mirrors at opposite ends of the laser. One of the mirrors is only partly reflective. It allows some light to escape. This escaped light forms the beam that you see. Some of the light waves do not escape. These waves continue to bounce between the mirrors. This causes other atoms in the laser to produce more identical waves. As this process continues, a steady stream of laser light is produced. ☑

Lasers can be made with gases, liquids, or solids. One kind of laser uses a mixture of helium and neon gases to produce red light. The gases are sealed in a tube with a mirror at each end. The gas atoms get their energy from a flashtube. They lose their energy by giving off light waves.

What is coherent light?

Lasers produce narrow beams of light. Waves of laser light are coherent—they do not spread out. **Coherent light** is light of only one electromagnetic wavelength that travels with its crests and troughs aligned. This means the crests and troughs are always the same distance from each other. The first figure shows coherent light. Coherent light does not spread out because all the waves travel in the same direction. Because the beam does not spread out, the light energy stays concentrated in a small area.

What is incoherent light?

Incoherent light can contain more than one wavelength. The electromagnetic waves are not aligned. The second figure shows incoherent light. Notice that the waves have different wavelengths and they do not travel with their crests and troughs aligned. Light from a lightbulb is incoherent. Beams of incoherent light spread out. The energy of the light waves spreads out with the beam. This makes the intensity of incoherent light much less than the intensity of a laser beam. Incoherent light is better for lighting a room.

✔ **Reading Check**

3. **Describe** Why is one mirror in a laser only partly reflective?

_____

_____

Picture This

4. **Identify** Circle the incoherent light and put a box around coherent light.

Using Lasers

Lasers have properties that make them important tools. They are used in CD players and even in surgery. A laser beam is narrow and does not spread out as it travels. So, lasers can carry energy to small areas. In industry, powerful lasers are used for cutting and welding. Surveyors and builders use lasers for measuring. Scientists use laser light reflected from mirrors on the Moon to measure the Moon's orbit with great accuracy. Information can be sent in pulses of light from lasers. This makes lasers useful in communication. In some telephone systems, pulses of laser light send conversations through long glass fibers called optical fibers.

How do compact discs work?

Compact discs (CDs) are plastic discs with reflective surfaces. They are used to store sound, images, and text in digital form. When a CD is made, information is burned into the surface of the disc with a laser. The laser creates millions of tiny pits, where information is stored.

A CD player also uses a laser to read the disc. Look at the figure below. The figure on the right shows the pits on the bottom surface of a CD. The laser shines on the spinning disc. As the beam hits the pits, different amounts of light are reflected to a light sensor. The reflected light is then converted to an electric signal. The signal creates sound in the speakers.

Spinning CD

Laser

Light sensor

How are lasers used in medicine?

Lasers are often used in eye surgery. They can be used to remove cataracts, reshape the cornea, and repair the retina. Surgeons can use lasers in place of scalpels to cut body tissues. Laser energy seals off blood vessels as it cuts. This reduces bleeding during surgery. Most lasers do not cut deeply through the skin. This makes them useful in removing tumors or birthmarks on the surface of the skin.

Think it Over

5. **Explain** Why is a laser a good tool to make sure something is level?

_____

_____

_____

Picture This

6. **Identify** Circle the part of the figure that shows where information is stored on a CD.

● After You Read

Mini Glossary

coherent light: light of only one wavelength that travels with its crests and troughs aligned

fluorescent light: light that uses phosphors to convert ultraviolet light to visible light

incandescent light: light that is produced by heating a piece of metal until it glows

incoherent light: light that can contain more than one wavelength and travels in many directions with its electromagnetic waves not aligned

1. Review the terms and their definitions in the Mini Glossary. Write a sentence using a term for a device that produces incoherent light.

   _____

   _____

2. Complete the Venn diagram to organize the information you learned in this section about producing light.

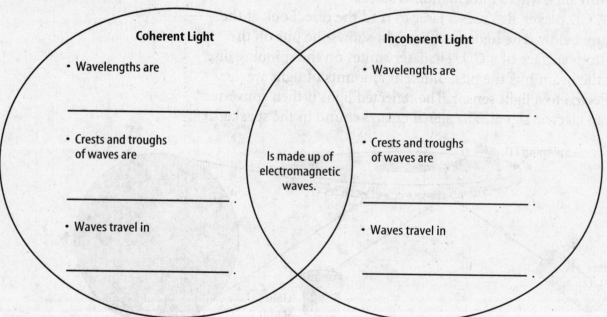

Coherent Light

- Wavelengths are

  _____

- Crests and troughs of waves are

  _____

- Waves travel in

  _____

Is made up of electromagnetic waves.

Incoherent Light

- Wavelengths are

  _____

- Crests and troughs of waves are

  _____

- Waves travel in

  _____

3. **Study Coach** Think about what you have learned. How did outlining as you read help you learn about the different type of lights and lighting devices?

   _____

   _____

End of Section

Light

section ❹ Using Light

● Before You Read

Have you ever worn sunglasses on a very bright day? Describe why you think sunglasses allow you to see more clearly.

_____

_____

Copyright © Glencoe/McGraw-Hill, a division of The McGraw-Hill Companies, Inc.

● Read to Learn

Polarized Light

Do you know what makes "polarized" sunglasses different from other sunglasses? The difference is in the vibrations of the light waves that pass through the lenses. Transverse waves have repeating high points, or crests, and repeating low points, or troughs. If you hold a rope at one can end and shake it, you create a transverse wave in any direction. Light is a transverse wave that can vibrate in any direction. In **linearly polarized light,** the waves vibrate in only one direction.

What are polarizing filters?

Recall that a red filter blocks all colors of light except red. A polarizing filter blocks all of the light, except the light whose magnetic field vibrates in the same direction. A polarizing filter acts like a group of parallel slits. It is similar to transverse waves on a rope passing through a picket fence, as in the figure.

Wave motion transmitted

Wave motion blocked

What You'll Learn

- difference between polarized and unpolarized light
- how a hologram is made
- when total internal reflection occurs
- about optical fibers

Study Coach

Make Flash Cards Each time you read a heading on the page, make a flash card with that heading. Then write the main idea of the information under the heading on the card.

FOLDABLES

❿ Organize Information Make the following Foldable to help you organize information about polarized light, holography, optical fibers, and optical scanners.

Polarized Light	Holography
Optical Fibers	Optical Scanners

Light passes through one polarizing filter. Then it hits another polarizing filter with slits at right angles to the first filter. The first filter only lets light through that was polarized in one way. None of this light is polarized in the same direction as the slits on the second filter. No light gets through the second filter.

Polarized lenses are useful for reducing glare. They do this without affecting your ability to see. Some of the light reflected from a flat surface, such as a lake or car hood, becomes horizontally polarized. The lenses of polarized sunglasses have vertical polarized filters. They block out the reflected light that is polarized horizontally.

Holography

Have you ever seen a three-dimensional image that seems to float in the air? If you walk around the image, you can see it from different angles. It is as if you are looking at a real object. Such images are produced by holography. **Holography** is a method that produces a hologram, a complete, three-dimensional photographic image of an object. ☑

How is a hologram made?

It takes a laser to make a hologram. First, a laser is shined onto an object. The laser light reflects from the object onto photo film. At the same time, a second beam split from the laser is shined directly on the film. The light from the two beams creates a pattern on the film. When the film is developed, the pattern looks nothing like the original object. But when laser light shines onto the film, it produces a holographic image.

How do holograms contain more information than ordinary photos?

A photograph records only the brightness of light that is reflected by an object. A hologram records both the brightness of the light and the direction in which the light travels. A hologram gives more information to your eyes than an ordinary photograph. Holograms are also difficult to copy. This is why some credit cards, identification cards, and product labels have holograms on them. Scientists can use X-ray lasers to produce holograms of microscopic objects. Some day it may be possible to create three-dimensional views of biological cells.

Reading Check

1. **Analyze** What kind of photographic image is a hologram?

_____

_____

Think it Over

2. **Draw Conclusions** Why is a holographic image of a microscopic object better than a two-dimensional image of the object?

_____

_____

_____

Optical Fibers

Optical fibers are transparent glass fibers that carry light from one place to another. They can transmit light because of a process called total internal reflection.

What is total internal reflection?

Remember that as light passes from one substance to another, it is refracted. Refraction happens because the speed of the light is different in the two substances, such as water and air. Light rays also can be reflected.

Critical Angle Look at the figure. The line that is perpendicular to a surface is called the normal. When light travels from water to air, the direction of the light ray is bent away from the normal. If the underwater light ray makes a larger angle, the light ray in the air bends closer to the surface of the water. At a certain angle, the refracted ray is traveling along the surface of the water. This is called the critical angle. For a light ray traveling from water into air, this is about 49°.

Total Internal Reflection If the angle of the original beam is increased past 49°, refraction stops. Instead of being refracted, the light ray is reflected off the surface boundary of the water. It is reflected into the water, as if there were a mirror at the surface. No light escapes the surface. This is called total internal reflection. **Total internal reflection** happens when light traveling from one medium to another is completely reflected at the boundary between the two materials. For total internal reflection to happen, light must travel slower in the first material than in the second material. It also must hit the boundary at an angle greater than the critical angle.

Air

Water

Normal   Normal

Total internal reflection

Critical angle

Applying Math

3. **Use Geometry** An acute angle is an angle less than 90°. An obtuse angle is an angle greater than 90° and less than 180°. Is the critical angle an acute angle or an obtuse angle? Explain.

_____

_____

_____

Picture This

4. **Conclude** Use a protractor to measure the angle labeled "Total internal reflection." Is this an acute angle or an obtuse angle? Explain.

_____

_____

_____

5. Apply What allows light to reflect off the surface of an optical fiber without leaking out?

_____

_____

_____

Picture This

6. Identify Highlight the boundary where total internal reflection takes place.

7. Determine What are sounds converted into before they can be sent through optical fibers?

_____

_____

How does light travel through an optical fiber?

Total internal reflection allows light to be carried through optical fibers. Look at the figure below. Light enters one end of the fiber. It is reflected continuously until it comes out the other end. Almost no light is lost or absorbed. An optical fiber is almost like a water pipe, only it carries light instead of water.

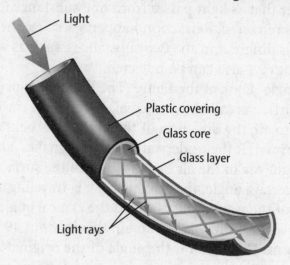

How are optical fibers used?

Optical fibers are most often used to communicate information. They are used as telephone lines, television cable lines, and computer cables. Signals can't leak from optical fibers, so they don't interfere with other nearby fibers. This allows the messages to be transmitted very clearly.

To send telephone conversations through optical fibers, sounds are changed to digital signals. Those signals are pulses of light from a light-emitting diode or a laser. The light pulses are then sent through the optical fiber. The light pulses are changed back into sounds at a receiver. ✔

Some systems have more than one laser. Each laser uses a different wavelength. Signals from each laser can be sent through the same fiber without interfering with one another.

Optical Scanners

You may have wondered how the machine at a supermarket checkout counter can read the prices of the items you purchase. The machine is an optical scanner. It reads the intensity, or brightness, of reflected light. Then it converts this information to a digital signal. The scanner shines light onto a barcode that is printed on an item. The reflection is then read by the scanner and turned into a signal. The signal goes to a computer, which finds the price of the item.

After You Read

Mini Glossary

holography: a method that produces a hologram, which is a complete, three-dimensional photographic image of an object

linearly polarized light: light in which the waves vibrate in only one direction

optical scanner: reads reflected light intensity and converts it to digital signals

total internal reflection: when light traveling from one medium to another is completely reflected at the boundary between the two materials

1. Review the terms and their definitions in the Mini Glossary. What is polarized light?

   _____

   _____

2. On the lines below, list three things that must happen for total internal reflection to take place.

   Total Internal Reflection

   1. Light traveling from one medium to another is completely reflected at the _____. between two materials.

   2. Light must travel _____ in the first material than in the second material.

   3. Light must hit the boundary at an angle _____ than the _____ angle.

3. Study Coach Think about what you have learned. How did making flash cards help you learn about how light is used?

   _____

   _____

End of Section

section ❶ Mirrors

What You'll Learn

- how three kinds of mirrors work
- the difference between real and virtual images
- uses of plane, concave, and convex mirrors

◀ **Mark the Text**

Locate Information Many headings in this section are questions. Underline the answers to these questions.

FOLDABLES

A **Find Main Ideas** Make a Foldable like the one shown below to write down main ideas about light and vision.

How I use light to see...

● Before You Read

Turn off the lights and try to read this page. Turn the lights back on and try again. On the lines below, explain why you think it is easier to read a printed page with the lights on.

_____

_____

_____

● Read to Learn

How do you use light to see?

Light comes from sources like lightbulbs, candles, and the Sun. When light reflects, or bounces, off of an object and travels to your eye, you see the object. For example, light from the lightbulbs in your classroom reflects off this book. The light travels from the book to your eyes. As a result, you see the words on the page. The more light there is to reflect off of objects, the easier it is to see them. When there is no light to reflect off objects, you cannot see anything. This is why it is hard to read a book in the dark.

What is a light ray?

First, think about light as waves. A light source like the Sun or a lightbulb sends out waves of light in all directions. Now, think about light as being many straight lines of light coming out from a source. Each line is called a light ray. Light rays refract and reflect just like light waves. The figure shows how a candle puts out rays of light in all directions. You can change the direction of light rays by reflecting them off a shiny object, like a mirror.

Light rays

Seeing Reflections with Plane Mirrors

Imagine a lake with trees growing along the shore. If you look at the surface of the water, you will see an image of the trees. An image is the picture you see when light reflects off of a smooth surface. You might see an image in a store window or a shiny car hood.

A very common place to see an image is in a flat, smooth mirror called a **plane mirror**. This is the kind of mirror you see in most bathrooms and dressing rooms. ☑

How does a plane mirror work?

Suppose you stand in front of a mirror and turn on a light as shown below. The light bulb puts out rays of light. Some of the light rays hit you. Then the rays reflect off of you in all directions. Some of these rays hit the mirror. The mirror reflects these rays in all directions. Some of the reflected light rays hit your eyes. You see your image in the mirror. If there are no light rays to reflect, there is no image for you to see.

Light rays

Mirror

Image

What does an image in a plane mirror look like?

Your image in a plane mirror appears right side up. Your head is at the top of the image and your feet are at the bottom. But the image that faces you in the mirror is reversed, or opposite. Your left side appears on the right side of the image. Your right side appears on the left side of the image. Suppose you stand 1 m in front of a plane mirror. Your image will look like it is standing 1 m behind the mirror. To you, your image appears to be 2 m away.

Copyright © Glencoe/McGraw-Hill, a division of The McGraw-Hill Companies, Inc.

1. **Define** What is a plane mirror?

_____

_____

Picture This

2. **Infer** What will happen to the image in a plane mirror if the lights are turned out? Why?

_____

_____

_____

3. Apply Look at the figure. Which action would make the image look like it is farther behind the mirror?

   a. Turn out the lights.

   b. Turn on more lights.

   c. Move closer to the mirror.

   d. Move away from the mirror.

What is a virtual image?

The figure below shows why an image in a plane mirror looks like it is behind the mirror.

Suppose you see your reflection in a plane mirror. The arrows show light rays from you hitting a mirror. The arrows show these light rays reflecting off of the mirror. When these reflected rays hit your eyes, you see your image in the mirror.

What Your Brain Sees Your brain interprets light rays as if they travel in a straight line. It does not understand that the light rays changed direction when they reflected off of the mirror. Imagine extending the reflected light rays back behind the mirror. The dashed lines in the figure show that the rays would meet at one point. Your brain thinks that the light rays are coming from this point. That is why the image of the person looks like it is behind the mirror.

However, no light really is coming from behind the mirror. The image you see is called a virtual image. A **virtual image** is an image that is not real, even though it looks real. No light rays actually meet to create the image. The virtual image always appears to be as far behind the mirror as the object is in front of it. ☑

Concave Mirrors

A plane mirror is flat. It produces a right side up, virtual image. However, not all mirrors are flat. Some mirrors have curved surfaces. Mirrors with curved surfaces form different kinds of images. One kind of mirror with a curved surface is a concave mirror.

✔ **Reading Check**

4. Describe What is a virtual image?

_____

_____

_____

What is a concave mirror?

A __concave mirror__ has a surface that curves in. The edges of a concave mirror are closer to you than the center of the mirror. Many shaving mirrors and makeup mirrors are concave. Have you ever looked at your image inside the bowl of a shiny spoon? The inside of the bowl of the spoon is a concave mirror, too.

How does a concave mirror work?

Every concave mirror has an optical axis. The __optical axis__ is an imaginary line perpendicular to the surface of the mirror. It also passes through the center of the mirror. The figure below shows a side view of a concave mirror and the location of its optical axis.

Parallel Light Rays Some light rays travel parallel to the optical axis on their way to a concave mirror. These rays reflect off the mirror. As the figure below shows, the reflected rays all cross each other at the same point. This point is called the __focal point__. The distance from the center of the mirror to the focal point is called the __focal length__. ☑

Intersecting Light Rays As shown below, some light rays travel through the focal point on their way to a concave mirror. The mirror reflects these rays parallel to the optical axis.

Picture This

5. **Draw** Connect the top and bottom of the concave mirror with a line. Is this line parallel or perpendicular to the optical axis?

_____

☑ Reading Check

6. **Identify** What is the focal point of a concave mirror?

_____

_____

_____

Picture This

7. **Identify** In the last two figures, use a highlighter to trace the light rays and the paths of their reflection.

What does an image in a concave mirror look like?

The image of an object in a concave mirror does not always look the same. It depends on how close the object is to the mirror.

Suppose you put a candle less than one focal length from the mirror. Look at the first figure. Like an image in a plane mirror, the image of the candle in the concave mirror is a virtual image. The image is right side up. But, the image is larger than the candle. Have you ever seen a concave shaving mirror or makeup mirror? When you stand less than one focal length away, the image of your face looks larger.

What are real images?

What happens when the distance from the candle to the mirror is between one and two focal lengths? This time the mirror will produce a real image. A **real image** is formed when rays of light converge to form the image. Rays of light really pass through the location of the image. The image is larger than the candle. And, the image is upside down.

How do mirrors decrease size?

If the distance between the candle and the mirror is more than two focal lengths, you get another kind of image. Look at the figure below. The mirror produces a real image. The image is also upside down. However, the image is smaller than the candle.

Picture This

8. Infer What will happen to the image if the candle is moved closer to the mirror?

_____

_____

Picture This

9. Highlight the part of the figure that shows why the mirror does not produce an image.

How are light beams created?

What happens if you put an object right at the focal point? Look at the next figure. All the rays of light start at the focal point. The mirror reflects them parallel to the optical axis. The rays never cross, even if you extend them backward. So, the mirror does not produce an image. You see all those parallel light rays as a bright beam of light. This is how flashlights, spotlights, lighthouses, and headlights create light beams.

Focal point

Mirror

Optical axis

Picture This

10. **Determine** where 2 focal lengths are and place a mark at that point on the optical axis.

Convex Mirrors

Concave mirrors are only one kind of curved mirror. Another kind of mirror with a curved surface is a convex mirror. Convex mirrors have some special uses in everyday life.

What is a convex mirror?

Have you seen a side-view mirror on a car that says "Objects in mirror are closer than they appear?" A **convex mirror** has a surface that curves out, like the back of a spoon. The center of a convex mirror is closer to you than the edges of the mirror. The security mirrors in banks and stores are convex mirrors. Some rear-view mirrors and side-view mirrors in cars are convex, too. ☑

✔ Reading Check

11. **Define** What is a convex mirror?

_____

_____

_____

12. Infer Why might a store-owner use convex mirrors, instead of plane or concave mirrors, to watch the store?

_____

_____

_____

Picture This

13. Compare and Contrast How are the images in a plane mirror and a convex mirror alike? How are they different?

_____

_____

_____

_____

_____

How does a convex mirror work?

When a convex mirror reflects light rays, it spreads the light rays apart from each other. Because of this, a convex mirror can show you the image of a large area. This is called a wide field of view. For example, the wide field of view of the convex mirrors in cars lets drivers see more of the traffic around them.

What does an image in a convex mirror look like?

So why do side-view mirrors say that the objects "are closer than they appear?" The light rays that reflect off of a convex mirror do not cross each other. Like a plane mirror, a convex mirror produces a virtual image. The image of an object in a convex mirror is right side up. But, the image is smaller than the object. If an image appears smaller, then we think it is farther away.

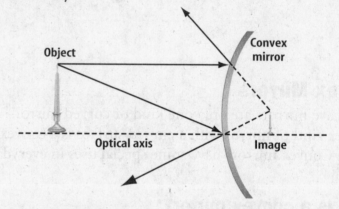

The table below summarizes the images formed by plane, concave, and convex mirrors.

	Images Formed by Mirrors			
Mirror Shape	**Position of Object**	**Virtual/Real**	**Image Created Upright/Upside Down**	**Size**
Plane		virtual	upright	same as object
Concave	Object more than two focal lengths from mirror	real	upside down	smaller than object
	Object between one and two focal lengths	real	upside down	larger than object
	Object at focal point	none	none	none
	Object within focal length	virtual	upright	larger than object
Convex		virtual	upright	smaller than object

● After You Read

Mini Glossary

concave mirror: a mirror with a curved surface so that the edges are closer to you than the center of the mirror is

convex mirror: a mirror with a curved surface so that the center is closer to you than the edges of the mirror are

focal length: the distance from the center of a concave mirror to the mirror's focal point

focal point: if light rays parallel to the optical axis hit a concave mirror, the reflected rays cross at this point

optical axis: an imaginary line perpendicular to the center of a concave mirror

plane mirror: a flat, smooth mirror

real image: an image formed when rays of light converge to form the image

virtual image: an image that looks real, but no light rays actually pass through it

1. Review the terms and their definitions in the Mini Glossary. Write a sentence that uses the terms *concave mirror*, *optical axis*, and *focal point*.

_____

_____

2. Complete the chart below to organize the information from this section.

Kind of Mirror	Position of Object	Is image right side up or upside down?	Is image larger, smaller or the same size as the object?
Plane			
Concave	object more than two focal lengths from mirror		
Concave	object between one and two focal lengths		
Concave	object at focal point		
Concave	object within one focal length		
Convex			

3. **Mark the Text** How did underlining the answers to questions asked in this section's headings help you learn about mirrors and images?

_____

_____

End of Section

Mirrors and Lenses

chapter 13

section ❷ Lenses

What You'll Learn

- what convex and concave lenses look like
- how convex and concave lenses form images
- how lenses can help people see better

 Study Coach

Make a Chart Make a two-column chart. As you read, list the main ideas in the left column. In the right column, list the details that support each main idea.

Picture This

1. **Predict** A light ray travels along the optical axis to a convex lens. Draw a line on the figure to show what will happen to the ray as it passes through the lens.

● Before You Read

Tools used to look at objects contain at least one lens. On the lines below, make a list of tools that have lenses in them.

_____

_____

● Read to Learn

What is a lens?

What do your eyes have in common with cameras, eyeglasses, and microscopes? They all contain at least one lens. A lens is made of transparent material. Remember, transparent means that almost all the light rays that hit the material pass through it. A lens has at least one curved surface. The curved surface makes light rays refract, or bend, as they pass through the lens.

Like mirrors, lenses produce images. What an image looks like depends on the shape of the lens. A lens can be convex or concave.

Convex Lenses

A <u>convex lens</u> is thicker in the middle than at the edges. The optical axis of a convex lens is an imaginary line perpendicular to the thickest point on the lens. The figure shows a convex lens.

Convex lens

Optical axis

How does a convex lens work?

Some light rays travel parallel to the optical axis on their way to a convex lens. The lens refracts these light rays in a special way. The refracted rays all cross each other at the same point, called the focal point. The focal length of the lens is the distance from the center of the lens to the focal point.

(side margin, rotated) Copyright © Glencoe/McGraw-Hill, a division of The McGraw-Hill Companies, Inc.

The focal length of a convex lens depends on the shape of the lens. Look at the figures below. If the curved sides of the lens are very rounded, the light rays bend sharply. The focal point is close to the lens. So, the focal length is short. If the sides of the lens are flatter, the light rays bend less. The focal point is farther from the lens. So, the focal length is longer.

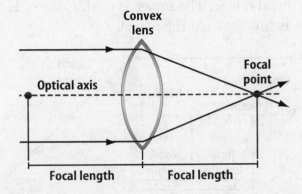

What does an image from a convex lens look like?

The image of an object from a convex lens does not always look the same. It depends on how close the object is to the lens.

Suppose you put an object less than one focal length from a convex lens. The lens will produce a virtual image of the object. The image is right side up. But, the image is larger than the object, but magnified. Have you ever used a magnifying glass? This tool is a convex lens.

Copyright © Glencoe/McGraw-Hill, a division of The McGraw-Hill Companies, Inc.

B **Organize** Take notes about lenses in a 2-tab Foldable such as the one shown below.

| Concave and Convex Lenses | Lenses and Eyesight |

Picture This

2. **Compare** Extend the top horizontal line in each figure. Compare the angles formed by these lines and the bent rays. Which angle is greater, the one in the figure of more curved lens or the one on the figure with the flatter lens?

_____

_____

Picture This

3. **Apply** Rob has a convex lens with a focal length of 5 in. He wants to produce a larger, right-side-up image of a pencil. Which figure shows where Rob should put the pencil? Circle your answer.

a. inside one focal length

b. between one and two focal lengths

c. outside two focal lengths

4. Describe Suppose you need to create an upside down image of an object. You have a convex lens. Where should you put the object?

_____

_____

What would happen if you put the object between one and two focal lengths from the lens? This time the lens will produce a real image. The image is larger than the object. And, the image is upside down.

If the distance between the object and the lens is more than two focal lengths, you get another kind of image. The lens produces a real image. The image is upside down. However, the image is smaller than the object.

Concave Lenses

A <u>concave lens</u> is thinner in the middle and thicker at the edges, as shown below. The optical axis of a concave lens is an imaginary line perpendicular to the thinnest point on the lens.

When a concave lens refracts light rays, it bends the rays outward, away from the optical axis. The rays spread out and never meet at a focal point. Concave lenses are used in some eyeglasses and some telescopes. Often they are used in combination with other lenses.

A concave lens creates a virtual image. The image is right side up. But, the image is smaller than the object.

Think it Over

5. Classify Is a concave lens most like a concave mirror or a convex mirror? Explain your choice.

_____

_____

Lenses and Eyesight

What determines how well you can see the words on this page? If you do not need eyeglasses, it is because the parts of your eyes work together to let you see clearly.

How do your eyes work?

Light enters your eye through the cornea. The **cornea** (KOR nee uh) is a transparent covering over your eyeball. The cornea bends light rays to bring them together. Then the light goes through an opening in your eye called the pupil. Behind the pupil is a convex lens. The lens also brings the light rays together. Then the rays form an image on your retina. The **retina** is the inner lining of your eye. It has cells that can change the image into electrical signals. The optic nerve sends the signals to your brain. ☑

✔ **Reading Check**

6. **Explain** On what part of the eye does the image form?

_____

_____

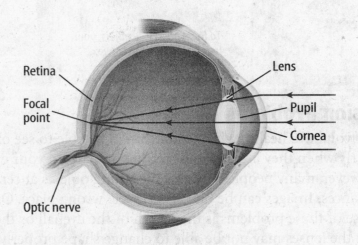

Retina

Focal point

Optic nerve

Lens

Pupil

Cornea

How do you clearly see both near and far objects?

You can see the watch on your wrist clearly. Then you can look at a clock across the room and see it clearly, too. How is this possible? For you to see an object clearly, its image must form on your retina. If the image forms in front of the retina or behind the retina, the object will look blurry.

Remember that the location of an image from a convex lens depends on the focal length and the location of the object. For an image to be formed on the retina, the focal length of the lens must change as the object's distance changes.

Picture This

7. **Find** Highlight the two parts of the eye that bring light rays together. What are these two parts?

_____

_____

How does a lens change shape?

Fortunately, the lens is made of material that can change shape. Muscles are attached to the lens of an eye. They can pull the lens into rounder or flatter shapes. This changes the lens' focal length.

If the object is far away, the muscles make the lens flatter. This increases the focal length to keep the image on the retina. If the object moves closer to you, muscles make the lens rounder. This shortens the focal length to keep the image on the retina.

When an object is far away, your lens is less convex.

To focus on a close object, your lens becomes more convex.

Vision Problems

If you have healthy vision, you should be able to see objects clearly when they are 25 cm or farther away from your eyes. However, many people have trouble seeing objects at certain distances. Images can be blurry or in the wrong place. One cause of these problems is the length of the eyeball or the fact that the lenses may not be able to change shape properly. Diseases of the retina can cause vision problems. But, eyeglasses, contact lenses, and surgery often can correct vision problems and help people see more clearly.

What does it mean to be nearsighted?

If you have nearsighted friends, you know that they can see clearly only when objects are nearby. Their eyeballs may be too long which means their lenses may not be able to flatten enough. This means that they cannot make the focal length of their lenses long enough to form an image on the retina. Instead, the image forms in front of the retina. Eyeglasses with concave lenses can help increase the focal length. Nearsighted people see objects that are near better than objects that are far away.

FOLDABLES™

C **Compare** Make three note cards out of quarter sheets of paper to take notes and compare different vision problems.

What does it mean to be farsighted?

Some people can see objects clearly that are far away. But, nearby objects look blurry to them. These people are farsighted. Their eyeballs may be too short which means their lenses may not be able to curve enough. This means that they cannot make the focal length of their lenses short enough to form an image on the retina. To correct this problem, they may wear glasses with convex lenses. As you can see in the figure, the convex lenses help shorten the focal length. Farsighted people see objects that are far away better than objects that are near.

Farsighted eye

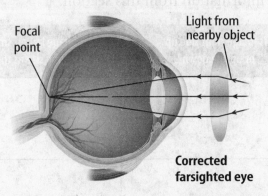

Corrected farsighted eye

Farsightedness is common as people get older. The lenses in the eyes of older people often become less flexible. The muscles cannot make the lens curve enough to form an image on the retina. ☑

What is astigmatism?

Another vision problem, called astigmatism (uh STIHG muh tih zum), occurs when the surface of the cornea is curved unevenly. When people have astigmatism, their corneas are more oval than round in shape. Astigmatism causes blurry vision at all distances. Eyeglasses with unevenly curved lenses can help people with this condition. The lenses cancel out the effects of the uneven cornea.

Think it Over

9. **Compare and Contrast** What are the differences between farsightedness and nearsightedness?

_____

_____

_____

Picture This

10. **Measure** Use a metric ruler to measure the distance from the outer edge of the cornea to the focal point in each figure. By how much did this length change?

_____

Reading Check

11. **List** What are the causes of farsightedness?

_____

_____

_____

After You Read

Mini Glossary

concave lens: a lens that is thinner in the middle than at the edges

convex lens: a lens that is thicker in the middle than at the edges

cornea: the transparent covering over the eyeball

retina: the inner lining of the eye

1. Review the terms and their definitions in the Mini Glossary. In the first box, sketch a convex lens. In the second box, sketch a concave lens. Add light rays to show how the lenses affect them.

2. Complete the chart below to organize the information from this section.

Kind of Lens	Position of Object	Is image right side up or upside down?	Is image larger, smaller, or the same size as the object?
Convex			
Convex			
Convex			
Concave			

End of Section

chapter 13 Mirrors and Lenses

section ❸ Optical Instruments

● Before You Read

Astronomers study stars that are too far away to be seen by using just their eyes. Biologists study cells that are too small to be seen by using just their eyes. On the lines below, describe how you think these scientists look at these objects.

_____

_____

_____

● Read to Learn

Telescopes

Have you ever had a hard time seeing an object that is far away? When you see an object, only some of the light reflected from its surface enters your eye. The figure shows that as an object gets farther away, less light from the object enters your eye. So, the farther away an object gets, the dimmer and less detailed it looks.

A telescope uses a lens or a combination of lenses and mirrors to gather light from distant objects. A telescope can gather more light than your eye can. This is because the lens or mirror in a telescope is much larger than your eye. In fact, the largest telescopes can gather more than a million times more light than the human eye. So, when you look at a distant galaxy through a telescope, more light enters your eye. The galaxy appears brighter and more detailed.

Copyright © Glencoe/McGraw-Hill, a division of The McGraw-Hill Companies, Inc.

What You'll Learn

- about refracting and reflecting telescopes
- why a telescope in space would be useful
- how a microscope uses lenses to magnify small objects
- how a camera creates an image

Study Coach

Summarize Read the section. Close your book and make a list of the section's main ideas. Look at the section again to check your list. Then, write a summary of the section's main ideas.

FOLDABLES

D Compare and Contrast
Make a 3-tab Foldable to organize and compare information about optical instruments.

Telescopes

Microscopes

Cameras

1. **Describe** What type of
image do you see when
using both types of
telescopes shown on
this page?

_____

_____

How does a refracting telescope work?

One common type of telescope is the refracting telescope. A simple <u>refracting telescope</u> uses two convex lenses to gather light and form an image of a distant object. Light from a distant object, like a planet, passes through the first lens. This is the objective lens. Since the light rays are from very far away, they enter the lens nearly parallel to the optical axis. This first convex lens refracts these rays to form a real image. The second convex lens is the eyepiece. When you look through the eyepiece lens, you see a larger, upside down, virtual image of the real image.

There are disadvantages to refracting telescopes. To see distant objects like planets, the objective lens must be as large as possible. The larger the lens is, the heavier it is. The weight of a large lens can make it sag in the middle. This distorts the image. Also, heavy glass lenses are expensive and difficult to make.

How does a reflecting telescope work?

Most large telescopes today are reflecting telescopes. A <u>reflecting telescope</u> uses a concave mirror, a plane mirror, and a convex lens to collect light and form an image of a distant object. Light from a distant object enters one end of the telescope. The light hits a concave mirror at the other end. The light reflects off of this mirror. The rays start to come together. But, before the rays reach the focal point, they hit a plane mirror. The plane mirror reflects the light toward the telescope's eyepiece. A real image of the object forms. Just as in a refracting telescope, a convex lens in the eyepiece then magnifies this image. ✓

✔ **Reading Check**

2. **Explain** To where does the
plane mirror reflect the light
in a reflecting telescope?

_____

_____

Space Telescopes

Earth is surrounded by a layer of atmosphere that refracts light from space. To solve this problem, telescopes have been put into space where there is no atmosphere. The Hubble Space Telescope was the first to be put in space in 1990. It has a 2.4 m mirror which produces clear and detailed images. In 2014 the James Webb telescope will be transported to space with a 6.5 m mirror. Other space telescopes use other wavelengths of energy to produce images. The Chandra X ray Telescope captures X rays and the Spitzer Space Telescope captures infrared waves. ☑

Microscopes

A telescope works well for looking at distant objects. But, if you need to look at tiny objects, like the cells in a butterfly wing, you will need a microscope. A simple **microscope** uses two convex lenses with short focal lengths to magnify objects. These lenses are the objective lens and the eyepiece lens.

The figure shows a simple microscope. You set the object you want to look at on a transparent slide under the objective lens. The microscope shines light from under the object. The light passes by or through the object. Then it travels through the objective lens. The objective lens is a convex lens. It forms a larger, real image of the object because the distance from the object to the lens is between one and two focal lengths. The eyepiece lens is also a convex lens. The eyepiece lens creates a larger, virtual image of the real image. This final image can be hundreds of times larger than the original object.

Eyepiece lens

Magnified real image

Objective lens

Object

Light source

Mirror

Digital Cameras

How does a camera make an image of a person or a scene? A camera works by gathering and bending light with a lens. Then the lens forms an image on a sensor. The sensor responds to light and digitally records the image.

3. **Explain** Why is the *Hubble Space Telescope* able to produce clearer images than telescopes on Earth?

_____

_____

_____

Picture This

4. **Predict** How does light reach the objective lens in this microscope?

_____

_____

_____

5. Circle the part of the figure that has the same function as the retina in an eye.

Sensor

Image

Shutter

Object

Lens

6. Describe How does a camera put an image on a sensor?

_____

_____

_____

When you take a picture with a camera, a shutter opens and closes. The shutter stays open long enough to let light enter the camera. The light reflecting off the person enters the camera through an opening called the aperture. The light passes through the camera lens. The lens forms the image on the sensor. The image is real, upside down, and smaller than the actual person. The size of the image depends on the focal length of the lens and how close the lens is to the sensor. ✔

Wide-Angle Lenses

Suppose you and a friend stand next to each other. You each take a photograph of the same object. If your cameras have different lenses, your pictures might look different. For example, some lenses have short focal lengths. These lenses produce a smaller image of the object but have a wide field of view. These lenses are called wide-angle lenses. A wide-angle lens must be close to the sensor in the camera to form a clear image.

Telephoto Lens

Telephoto lenses have longer focal lengths. They produce larger images. But, they have a narrower field of view than wide-angle lenses do. A camera with a telephoto lens is easy to recognize. The telephoto lens usually sticks out from the front of the camera. This increases the distance between the lens and the sensor in the camera.

● After You Read

Mini Glossary

microscope: an instrument that uses combinations of lenses with fairly short focal lengths to magnify extremely small objects

reflecting telescope: a telescope that uses a concave mirror, a plane mirror, and a convex lens to collect light and form an image of a distant object

refracting telescope: a telescope that uses two convex lenses to gather light and form an image of a distant object

1. Review the terms and their definitions in the Mini Glossary. Choose one term. In the space below, draw a picture to represent the term. Label each part of the drawing.

2. Place a check mark in each box to show what is found in each item.

	Plane Mirror	Convex Lens	Concave Mirror
Refracting Telescope			
Reflecting Telescope			
Camera			

3. **Study Coach** Think about what you have learned. How did summarizing your reading help you understand optical instruments?

_____

_____

End of Section

Solids, Liquids, and Gases

section ❶ Matter and Thermal Energy

What You'll Learn

- the kinetic theory of matter
- how particles move in the four states of matter
- how particles behave at the melting and boiling points

Mark the Text

Identify Concepts Highlight each heading in this section that asks a question. Then use a different color to highlight the answers to the questions.

FOLDABLES

A Build Vocabulary
Make the following Foldable to help you learn the vocabulary terms in this section.

Kinetic Theory
Melting Point
Heat of Fusion
Heat of Vaporization
Boiling Point
Diffusion
Plasma
Thermal Expansion

● Before You Read

Everywhere you go, you are surrounded by the three states of matter—solids, liquids, and gases. Look around you and find a substance in each of the three states of matter. List these substances on the lines below.

_____

_____

● Read to Learn

States of Matter

An everyday activity such as drinking tea may include solids, liquids, and gases. Think about heating water on the stove for a cup of hot tea. The boiling water is in the liquid state. The water vapor moving into the air is in the gaseous state. You might drop an ice cube into the tea to cool it. Ice is in the solid state. How are these states alike and different?

What is the kinetic theory of gases?

The **kinetic theory** is an explanation of how particles in gases behave. To explain the behavior of particles, you have to make some basic assumptions. An assumption is an idea that most people accept as true. The assumptions of the kinetic theory are:

1. All matter is composed of small particles—atoms, molecules, and ions.

2. These particles are in constant motion. The motion has no pattern.

3. These particles are colliding or crashing into each other and into the walls of their container.

4. Particles lose some energy when they bump into other particles. The amount of energy the particles lose is very small and can be ignored.

Copyright © Glencoe/McGraw-Hill, a division of The McGraw-Hill Companies, Inc.

A model that you picture in your mind can help you understand the kinetic theory and how particles in matter move. Think of each particle as a tiny ball. These balls constantly are bouncing and bumping into one another.

What is thermal energy?

One assumption of the kinetic theory is that particles are in constant motion. Think about an ice cube. Does an ice cube seem to move? How can a frozen, solid ice cube have motion? Remember that the kinetic theory is describing particles. Atoms in solids are held in place by the attraction between the particles. This attraction gives solids a definite shape and volume. However, the thermal energy in the particles causes them to vibrate, even when the object stays in one place.

Thermal energy is the total energy of the particles in a material. Thermal energy includes kinetic energy and potential energy. Even though the ice cube does not appear to be moving, its particles are moving because of thermal energy. When the temperature of a substance is lowered, its particles will have less thermal energy and will vibrate more slowly. ☑

What is average kinetic energy?

The word *temperature* is used to explain how much thermal energy something has. Temperature is also the average kinetic energy of particles in a substance. On average, molecules of frozen water at 0°C will move slower than molecules at 100°C. Therefore, water molecules at 0°C have lower average thermal and kinetic energy than the molecules at 100°C. Molecules have kinetic energy at all temperatures, even at absolute zero. Absolute zero is about –273°C. Scientists theorize that at absolute zero, particles move so slowly that no more thermal energy can be removed from a substance.

How are particles in a solid arranged?

The figure shows how the particles of a solid are packed together. The particles vibrate constantly. The particles of most materials always will be in the same arrangement in the solid. The type of ordered arrangement formed by a solid is important. It gives the solid its chemical and physical properties.

Solid

Reading Check

1. **Determine** What two kinds of energy are part of thermal energy?

_____

_____

Picture This

2. **Use a Model** Describe materials you might use to make a model of the particles in a solid.

_____

_____

How do solids become liquids? Gases?

Think about the ice cube you dropped into the hot tea. The moving particles in the tea bump into the vibrating particles in the ice. These collisions transfer energy from the tea to the ice cube. The particles on the ice cube surface vibrate faster and transfer energy to particles inside the ice cube. When the particles have enough energy to overcome the attractive forces, they slip out of their ordered arrangement and the ice melts. The temperature at which a solid melts is the **melting point**. The thermal energy required to change a substance from a solid to a liquid is the **heat of fusion**.

Under certain conditions, a solid can go straight to a gas without becoming a liquid first. The process is known as **sublimation**. Dry ice (frozen carbon dioxide) is a common substance that undergoes sublimation. ☑

Why do liquids flow?

Particles in a liquid have more kinetic energy than particles in a solid. This extra kinetic energy lets particles overcome some of their attraction to other particles. The particles can slide past each other. This is why liquids flow and take the shape of their container. But the particles in a liquid have not overcome all of the attractive forces among them. The particles still cling together. Because of the attractive forces, liquids have a definite volume. A certain amount of liquid will always take up the same amount of space regardless of the shape of the container.

What is a gas state?

Gas particles have enough kinetic energy to overcome the attractions among them. Gases do not have a fixed volume or shape. They can spread far apart or move close together to fill a container. How does a liquid become a gas? The particles in a liquid are moving constantly. Some particles are moving faster than other particles. They have more kinetic energy. Particles that move fast enough can escape the attractive forces of other particles. They enter the gas state. This process is called vaporization. Vaporization can occur in two ways—evaporation and boiling.

How do liquids evaporate?

Vaporization occurs when liquids evaporate. Evaporation is vaporization that occurs at the surface of a liquid. It can occur at temperatures below the liquid's boiling point. To evaporate, particles must have enough kinetic energy to escape the attractive forces of the liquid. The particles must be at the surface of the liquid. They also must be traveling away from the liquid.

✔ **Reading Check**

3. **Explain** What is the temperature at which a solid begins to turn into a liquid called?

💡 **Think it Over**

4. **Infer** Which particles have the least kinetic energy? Circle the correct answer.

   a. solid particles

   b. liquid particles

   c. gas particles

How does boiling vaporize liquids?

Another way that a liquid vaporizes is by boiling. Boiling occurs at a specific temperature. The temperature depends on the pressure on the surface of the liquid. Air exerts pressure on the surface of a liquid. This external pressure keeps particles from escaping from the liquid. The **boiling point** of a liquid is the temperature at which the pressure of the vapor in the liquid is equal to the external pressure on the surface of the liquid. Particles need energy to overcome the force of pressure. **Heat of vaporization** is the amount of thermal energy needed for the liquid at its boiling point to become a gas. ☑

Why do gases fill their containers?

Gas particles move so quickly and are so far apart that they overcome the attractive forces among them. Therefore, gases do not have a definite shape or a definite volume. The movement of particles and the collisions among them cause gases to diffuse. Diffusion is the spreading of particles throughout a given volume until they are evenly distributed. Diffusion occurs in solids, liquids, and gases, but it happens most quickly in gases. Imagine that you spray air freshener in one corner of a room. It is not long before you smell the scent all over the room. Gases will fill the container that they are in even if the container is rather large. The particles continue to move and collide in a random motion within their container.

What is the heating curve of a liquid?

The heating curve graph shows water being heated from –20°C to 100°C. It shows the temperature change as heat is added. In areas **a** and **c** the graph slopes upward because the water's kinetic energy is increasing. In areas **b** and **d** the graph is a horizontal line. The temperature of the water does not change.

State Changes of Water

5. Apply What do particles need in order to overcome the force of pressure and become a gas?

_____

💡 **Think it Over**

6. Sequence List the following substances in the order in which their particles diffuse, slowest to fastest: liquid water, water vapor, ice.

a. _____

b. _____

c. _____

Applying Math

7. Interpret a Graph At what temperature is the water turning into gas?

_____

Melting Point At 0°C, ice is melting. All of the energy put into the ice at this temperature is used to overcome the attractive forces among the particles in the solid. The temperature stays the same during melting. After the attractive forces are overcome, the particles move more freely. Their temperature increases.

Boiling Point At 100°C, water is boiling or vaporizing. The temperature stays the same again. All of the energy that is put into the water goes to overcoming the remaining attractive forces among the water particles. When all the attractive forces in the water have been overcome, the energy returns to increasing the temperature of the particles.

What is the plasma state?

Solids, liquids, and gases are three familiar states of matter. However, scientists estimate that much of the matter in the universe is plasma. **Plasma** is matter made up of positively and negatively charged particles. The overall charge of plasma is neutral. Plasma is neutral because it contains equal numbers of positive and negative particles.

The faster the particles move, the greater the force when they collide. The forces produced from high-energy collisions are so large that electrons from the atoms are ripped off. This state of matter is called plasma. Plasma is found in stars, lightning bolts, neon and fluorescent tubes, and auroras.

Thermal Expansion

The kinetic theory explains other characteristics of matter. Recall that particles move faster and separate as temperature rises. The separation of the particles in a substance increases the size of the substance. **Thermal expansion** is an increase in the size of a substance when the temperature is increased.

The kinetic theory also explains contraction in objects. When the temperature of a substance is lowered, its particles slow down. The particles move closer together, and the substance shrinks. This is known as contraction. Expansion and contraction occur in most solids, liquids, and gases.

How does expansion occur in solids?

Have you ever noticed the lines in a concrete sidewalk? A gap often is left between the sections of concrete; these are called expansion joints. When concrete absorbs heat, its particles move faster and get farther apart. The concrete expands. When the concrete cools, it contracts. Expansion joints prevent concrete cracks when temperatures change.

Think it Over

8. **Explain** How do expansion joints in a sidewalk keep the concrete from cracking?

_____

_____

_____

_____

How does expansion occur in liquids?

An example of expansion in liquids occurs in a thermometer. One kind of thermometer, as shown below, is a narrow glass tube with a liquid in it. The addition of energy makes the particles of the liquid in the thermometer move faster. The particles in the liquid move farther apart. Even if the liquid expands only a little, it will show a large change on the thermometer's temperature scale.

Picture This

9. **Infer** What will happen to the liquid in the thermometer as the temperature cools?

_____

_____

How does expansion occur in gases?

A hot-air balloon is an example of thermal expansion in gases. Hot-air balloons are able to rise because of the thermal expansion of air. When the air in the balloon is heated, the distance between the particles in the air increases. This makes the balloon expand. As the balloon expands, the number of particles per cubic centimeter inside the balloon decreases. The density of the hot air inside the balloon decreases also. The density of the air in the hot-air balloon is lower than the density of the cooler air outside the balloon. This makes the balloon rise. ☑

✔ **Reading Check**

10. **Infer** What happens to the density of the air inside a hot-air balloon as the balloon expands?

_____

Why does water behave in a different way?

Why does ice float in water? Most substances contract as the temperature is lowered because the particles move closer together. However, water molecules are unusual. Water expands when it freezes. Water has some areas that have a highly positive charge and other areas that have a highly negative charge. The diagram shows these regions. The charged regions affect the behavior of water.

Partial negative charge

Partial positive charge

Picture This

11. **Identify** Circle the atom(s) that contribute(s) to the positive charged end of the molecule.

12. Infer What will happen if you completely fill an ice tray with water?

_____

_____

_____

Unlike charges attract each other. As the temperature of water drops, the particles move closer together. The unlike charges attract each other. They line up so that only positive and negative areas are near each other. Because the water molecules line up according to charge, there are empty spaces in the structure. These empty spaces are larger in ice than in liquid water. So water expands when it goes from a liquid to a solid state. Solid ice is less dense than liquid water.

Solid or a Liquid?

Some other substances also have unusual behavior when they change states. Two kinds of materials that react unusually are amorphous solids and liquid crystals. It can be difficult to tell if these materials are in a solid state or a liquid state.

What are amorphous solids?

Ice melts at 0°C. Not all materials have a definite temperature when they change from solid to liquid. Some solids just get softer and slowly turn to liquid over a range of temperatures. These solids do not have the ordered structure of crystals. They are called amorphous (uh MOR fus) solids. Amorphous comes from the Greek word for "without form."

You are familiar with two amorphous solids—glass and plastics. The particles in amorphous solids are like long chains. They can get jumbled and twisted instead of having an ordered structure. As a result, amorphous solids have different properties than crystalline solids. ☑

 Reading Check

13. Describe What do the particles in amorphous solids resemble?

_____

What are liquid crystals?

Liquid crystals are groups of materials that do not change states in the usual way. When a substance changes from a solid to a liquid, it usually loses its ordered structure. Liquid crystals start to flow in the melting phase, like most liquids. But liquid crystals keep their ordered structure. Liquid crystals respond to temperature changes and electric fields. Scientists use these properties to make liquid crystal displays. LCDs are used in watches and calculators.

● After You Read

Mini Glossary

boiling point: the temperature at which the pressure of the vapor in the liquid is equal to the external pressure acting on the surface of the liquid

heat of fusion: the amount of energy needed to change a substance from a solid to a liquid at its melting point

heat of vaporization: the amount of energy needed for the liquid at its boiling point to become a gas

kinetic theory: an explanation of how particles in matter behave

melting point: the temperature at which a solid begins to turn into a liquid

plasma: matter consisting of positively and negatively charged particles

sublimation: the change from a solid to a gas without the liquid state

thermal expansion: an increase in the size of a substance when the temperature is increased

1. Read the vocabulary terms and their definitions in the Mini Glossary. How are the heat of fusion and the heat of vaporization similar? How are they different?

_____

_____

_____

_____

2. Complete the graphic organizer about the arrangement and behavior of particles in solids, liquids, and gases.

Solid	Liquid	Gas
Arrangement: Closely packed together	**Arrangement:**	**Arrangement:** Overcome attractions
Behavior:	**Behavior:** Slip out of arrangement and flow	**Behavior:**

End of Section

Solids, Liquids, and Gases

section ❷ Properties of Fluids

What You'll Learn
- Archimedes' principle
- Pascal's principle
- Bernoulli's principle

● Before You Read

Suppose that you have a piece of aluminum foil and have rolled it into a ball. You take another piece of foil the same size and make it into a flat, open box. Then you put both of these objects in a tub of water. What do you think would happen?

_____

_____

Mark the Text

Identify Key People
Underline the name of each person discussed in this section. Say the name out loud. Then highlight the sentences that show how that person helped explain the properties of fluids.

FOLDABLES

C **Describe** Make the following Foldable from a sheet of notebook paper to describe the three principles of fluids.

Archimedes' Principle

Pascal's Principle

Bernoulli's Principle

● Read to Learn

How do ships float?

Some ships are so huge that they are like floating cities. Even though ships are heavy, they are able to float. Ships float because the force pushing up on the ship is greater than the force of the ship pushing down. This force is called buoyancy. **Buoyancy** (BOY un see) is a fluid's ability to exert an upward force on an object in it. The fluid can be a liquid or a gas. If the buoyant force is equal to the object's weight, the object floats. If the buoyant force is less than the object's weight, the object sinks.

What is Archimedes' principle?

Almost 2,300 years ago, a Greek mathematician named Archimedes made a discovery about buoyancy. Archimedes found that the buoyant force on an object is equal to the weight of the fluid displaced by the object. _Displaced_ means "pushed away". For example, think about placing a block of wood in water. The block displaces water as it begins to sink. But the block will sink only until the displaced water's weight equals the block's weight. When this happens, the block floats.

Do equal-sized objects float?

Would a steel block the same size as the wood block float in water? They both displace the same volume and weight of water. So, the buoyant force on the blocks is the same. But the steel block sinks. What is different?

What does density have to do with buoyancy?

The volume of the two blocks and the volume of the water displaced is equal. But each volume has a different mass. Recall that density is mass per unit volume. If the three equal volumes have different masses, they must have different densities. The density of the steel block is greater than the density of water. The density of the wood block is less than the density of water. An object will float if its density is less than the density of the fluid it is placed in.

What would happen if you formed the steel block into the shape of a boat filled with air? Now the same mass takes up a larger volume. The density of the steel boat together with the air inside it is less than the density of water. The boat will float.

Weight

Wood

Buoyant force

Weight

Steel

Buoyant force

Pascal's Principle

When you are under water, you can feel the pressure of the water all around you. **Pressure** is force exerted per unit area or $P = \frac{F}{A}$. Did you know that Earth's atmosphere is a fluid? Earth's atmosphere exerts pressure all around you. Because you are used to the pressure of air, you do not feel the pressure.

Blaise Pascal was a French scientist who lived from 1623 to 1662. Pascal discovered a useful property of fluids. According to Pascal's principle, pressure applied to a fluid is transmitted or sent throughout the fluid. For example, when you squeeze a toothpaste tube, pressure is transmitted through the fluid toothpaste all around the tube. ☑

Picture This

1. **Compare** the lengths of the two "weight" arrows. Then compare the lengths of the two "buoyant force" arrows.

_____

_____

_____

_____

✔ **Reading Check**

2. **Explain** According to Pascal's principle, what is transmitted throughout fluid?

_____

_____

Picture This

3. **Determine** What does each letter in the figure stand for?

$P \rightarrow$ _____

$F \rightarrow$ _____

$A \rightarrow$ _____

How is Pascal's principle used?

Hydraulic machines move heavy loads using Pascal's principle. The figure shows how a hydraulic lift works. A pipe filled with fluid connects a small cylinder and a large cylinder. Pressure is applied to the small cylinder. The pressure is transferred through the fluid to the large cylinder. Pressure stays the same throughout the fluid. The greater surface area in the large cylinder provides more force. With a hydraulic machine, you can use your weight to lift something much heavier than you are. In the figure, P stands for pressure, F for force, and A for surface area.

Bernoulli's Principle

Daniel Bernoulli was a Swiss scientist who lived from 1700 to 1782. Bernoulli studied the properties of moving fluids such as water and air. He published his discovery in 1738. According to Bernoulli's principle, as the velocity of a fluid increases, the pressure exerted by the fluid decreases.

You can test this principle. Hold a sheet of paper near your mouth as shown. Blow across the top of the paper. The paper will rise. Why? The velocity of the air you blew over the top of the paper was greater than the velocity of the air below the paper. As a result, the paper moved upward as air pressure above decreased. Engineers use this principle in designing aircraft wings and piping systems.

Picture This

4. **Identify** Highlight the area that has the greatest air pressure.

_____

Fluid Flow

Another property of fluid is its tendency to flow. The resistance to flow by a fluid is called **viscosity**. Viscosity measures how much force is needed for one layer of fluid to flow over another layer. Some fluids tend to flow more easily than others. Water has a low viscosity because it flows easily. Syrup has a high viscosity because it flows slowly.

● After You Read

Mini Glossary

buoyancy: the ability of a fluid to exert an upward force on an object that is placed in the fluid

pressure: force exerted per unit area

viscosity: the resistance by a fluid to flow

1. Read the terms and definitions in the Mini Glossary. Circle the term that is explained by Archimedes' principle. On the lines below, write a sentence using the term.

_____

_____

2. Match the concepts in Column 1 to the definitions and explanations in Column 2.

Column 1

_____ 1. Viscosity

_____ 2. Bernoulli's principle

_____ 3. Pressure

_____ 4. Archimedes' principle

_____ 5. Buoyancy

_____ 6. Pascal's principle

Column 2

a. As the velocity of a fluid increases, the pressure exerted by the fluid decreases

b. The ability of a fluid to exert an upward force on an object

c. Pressure applied to a fluid is transmitted throughout the fluid

d. The resistance by a fluid to flow

e. Force exerted per unit area

f. The buoyant force on an object is equal to the weight of the fluid displaced by the object

3. ◄ Mark the Text You underlined the name of each person discussed in this section. You also highlighted how those people helped you understand the properties of fluids. How did this strategy help you learn the material in this section?

_____

_____

End of Section

Solids, Liquids, and Gases

section ❸ Behavior of Gases

What You'll Learn

- how a gas exerts pressure on its container
- how changing pressure, temperature, or volume affect a gas

Study Coach

Writing for Understanding As you read this section, write down each question heading. After each heading, write the answer to the question.

FOLDABLES

ⓓ **Understanding Cause and Effect** Make the following Foldable to help you understand the cause and effect relationship of gases.

Gas: Change in Pressure

Gas: Change in Temperature

Gas: Change in Volume

● Before You Read

The air that you breathe is a gas. List other gases that you have heard about in a science class.

_____

_____

_____

● Read to Learn

Pressure

You learned from the kinetic theory that gas particles are moving constantly and bumping into anything in their path. The collisions of these particles in the air result in pressure. Pressure is the amount of force exerted per unit of area. Pressure can be expressed by the formula $P = F/A$.

Often, gases are placed in containers. A balloon and a bicycle tire are gas containers. The balloon and the tire stay inflated because of collisions the air particles have with the walls of the container. The collisions of the particles cause a group of forces to push the walls of the container outward in every direction. If more air is pumped into the balloon, the number of air particles inside the balloon increases. This causes more collisions with the walls of the container and the balloon expands even more. Bicycle tires cannot expand as much as balloons. When more air is pumped into a bicycle tire, the tire's pressure increases.

How is pressure measured?

Pressure is measured in a unit called pascal, the SI unit of pressure. The abbreviation for pascal is Pa. Because pressure is the amount of force divided by area, one pascal of pressure is one Newton per square meter, or $1 N/m^2$. Because this is a very small pressure unit, most pressures are given in kilopascals, or 1,000 pascals. The abbreviation for kilopascals is kPa.

At sea level, atmospheric pressure is 101.3 kPa. This means that at Earth's surface at sea level, the atmosphere exerts a force of about 101,300 N on every square meter. This is about the weight of a large truck. The air in the atmosphere gets thinner as distance from sea level increases. This means that the air pressure decreases as you travel higher above sea level.

Boyle's Law

You have learned how gas creates pressure in a container. Suppose the container gets smaller. What happens to the gas pressure? The pressure of a gas depends on how often its particles hit the walls of the container. If you squeeze gas into a smaller space, its particles will strike the walls of their container more often. This increases the pressure. The opposite is true, too. If you give the gas particles more space, they will hit the walls less often. Gas pressure is reduced. ☑

Robert Boyle was a British scientist who lived from 1627 to 1691. Boyle explained the property of gases described above. According to **Boyle's law**, if you decrease the volume of a container of gas and keep the temperature the same, the pressure of the gas will increase. If you increase the volume of the container and keep the temperature the same, the pressure of the gas will decrease.

Pressure Outside an Object Affects Volume

Boyle's law explains the behavior of weather balloons. Weather balloons are large balloons that carry instruments to high altitudes. The instruments detect weather information and send the data back to Earth. The balloons are inflated near Earth's surface with a low-density gas. As a balloon rises, the atmospheric pressure outside the balloon decreases. This allows the balloon to slowly get bigger. It can reach a volume of 30 to 200 times its original size. Eventually, the balloon expands so much it breaks. Boyle's law states that as pressure outside an object is decreased, the volume increases. The weather balloon demonstrates this law. The graph on the next page shows that the opposite is also true. As the pressure outside an object is increased, the volume will decrease.

FOLDABLES

E Organize Make the following Foldable to help you organize the important information about each law.

Boyle's Law

Pressure-Temperature Relationship

Charles's Law

☑ **Reading Check**

1. **Apply** If you increase the size of a container, will the pressure of the gas inside it increase or decrease?

_____

2. Interpret a Graph What is the pressure of the gas at 300 L?

Volume v. Pressure for a Fixed Amount of Gas at Constant Temperature

3. Solve an Equation Use the equation $P_iV_i = P_fV_f$. Assume that $P_i = 101$ kPa and $V_i = 10.0$ L. If $P_f = 43.0$ kPa, what is V_f? Show your work.

Using Boyle's law

Using Boyle's law, you will find that the pressure inside an object multiplied by the volume of the object is always equal to the same number, or a constant. This is true only if the temperature stays the same. As the pressure and volume change, the constant number remains the same. You can use the equations $P_iV_i = $ constant $= P_fV_f$ to express this mathematically. In this equation, P_i and V_i stand for the initial pressure and volume. P_f and V_f stand for the final pressure and final volume. The equation shows us that the product of the initial pressure and volume is equal to the product of the final pressure and volume. If you know three of these values, you can find the unknown fourth value.

The Pressure-Temperature Relationship

Have you ever seen the words "keep away from heat" on a spray can? If you heat a gas inside a closed container, the particles of gas will strike the walls of the container more often. Because the container is rigid, its volume cannot increase. Instead, its pressure increases. The pressure can become greater than the container can hold. If this happens, the container will explode. If the volume stays the same, an increase in temperature results in an increase in pressure. The opposite is also true. That is, a decrease in temperature results in a decrease in pressure, if the volume stays the same. Suppose the same spray can is placed in a very low temperature. The gas particles will slow down and decrease in pressure. The pressure decrease can buckle or crumple the container.

Charles's Law

Have you ever seen a hot-air balloon being inflated? You know that gases expand when they are heated. Because particles in the hot air are farther apart than particles in the cool air, the hot air is less dense than the cool air. This difference in density allows the hot-air balloon to rise. Jacques Charles was a French scientist who studied gases. He lived from 1746 to 1823.

According to Charles's law, the volume of a gas increases with increasing temperature. This only happens if pressure does not change. The graph shows that the volume of a gas increases as temperature increases, but only at a constant pressure. As with Boyle's law, the reverse is true, also. The volume of a gas decreases as temperature decreases at a constant pressure.

Temperature v. Volume for a Fixed Amount of Gas at Constant Pressure

Applying Math

4. **Interpret a Graph** About how much did the volume of Gas B increase when the temperature rose from 100°C to 200°C?

Charles's law can be explained using the kinetic theory of matter. As a gas is heated, its particles move faster and faster and its temperature increases. Because the gas particles move faster, they begin to strike the walls of their container more often and with more force. In the hot-air balloon, the walls have room to expand. Instead of increased pressure, the volume of the gas increases. ☑

How is Charles's law used?

In the same way that there is a formula for the relationship of pressure and volume, there is a formula for the relationship of temperature to volume. The temperature is given in kelvins. The formula for that relationship is $V_i/T_i = V_f/T_f$. In this formula, V_i and T_i stand for initial volume and temperature. V_f and T_f stand for final volume and temperature. The pressure must be kept constant when you use Charles's law.

✔ **Reading Check**

5. **Infer** When the temperature of the air in a hot-air balloon increases, what happens to the volume of the gas?

◉ After You Read

Mini Glossary

pascal: the SI unit of pressure

1. Review the term *pascal* and its definition in the Mini Glossary. Use the term in a sentence that shows that you understand its definition.

   _____

   _____

   _____

   _____

2. Complete the graphic organizer to compare Boyle's law and Charles's law.

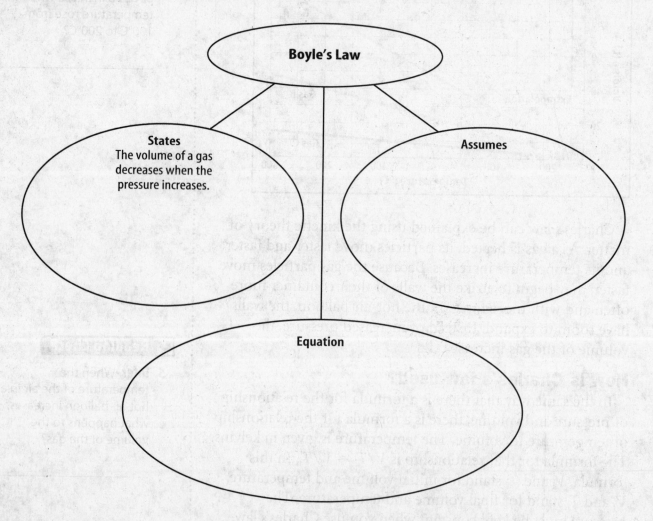

Boyle's Law

States
The volume of a gas decreases when the pressure increases.

Assumes

Equation

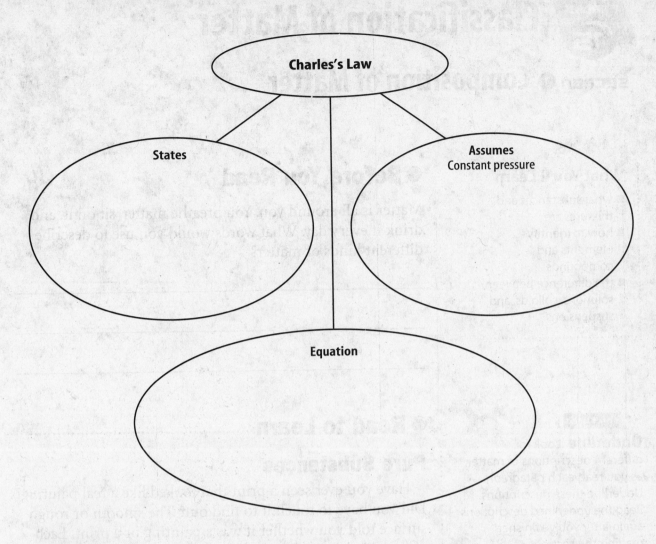

Charles's Law

States

Assumes
Constant pressure

Equation

3. **Study Coach** As you read this section, you wrote down the question headings and their answers. How could you use this to study for a quiz on this section?

_____

_____

_____

End of
Section

section ❶ Composition of Matter

What You'll Learn

- what substances and mixtures are
- how to identify elements and compounds
- the difference between solutions, colloids, and suspensions

● Before You Read

Matter is all around you. You breathe matter, sit on it, and drink it every day. What words would you use to describe different kinds of matter?

_____

_____

_____

Mark the Text

Underline Look for different descriptions of matter as you read each paragraph. Underline these descriptions. Read the underlined descriptions again after you've finished reading the section.

● Read to Learn

Pure Substances

Have you ever seen a print that looked like a real painting? Did you have to touch it to find out? The smooth or rough surface told you whether it was a painting or a print. Each material has its own properties. The properties of materials can be used to classify them into categories.

Each material is made of a pure substance or of a mix of substances. A **substance** is a type of matter that is always made of the same material or materials. A substance can be either an element or a compound. Some substances you might recognize are helium, aluminum, water, and salt.

What are elements?

All substances are made of atoms. A substance is an **element** if all the atoms in the substance are the same. The graphite in your pencil is an element. The copper coating on most pennies is an element, too. In graphite, all atoms are carbon atoms. In copper, all atoms are copper atoms. The metal under the copper coating of a penny is another element, zinc. There are about 90 elements found on Earth. More than 20 other elements have been made in laboratories. Most of the 20 human-made elements are unstable. They exist for only a short time in laboratories. You may recognize the elements that are shown in the figure on the next page. ☑

✔ Reading Check

1. **Explain** Why is graphite considered an element?

_____

_____

_____

Atoms in Elements

Copper

Nitrogen

Picture This

2. Determine What are the elements in each object shown in the figure?

_____

_____

_____

What are compounds?

A **compound** is a pure substance in which two or more elements are combined in a fixed proportion. A common compound is water. Water is made up of the elements hydrogen and oxygen. Each individual particle of water contains two hydrogen atoms and one oxygen atom. A molecule is the smallest particle of a compound. Chalk is another compound. It contains calcium, carbon, and oxygen. Each particle of chalk contains one calcium atom, one carbon atom, and three oxygen atoms.

Do compounds look like their elements? Can you imagine putting a silvery metal and a greenish-yellow poisonous gas on your food? Table salt is a compound made from elements that fits this description. Another name for table salt is sodium chloride. This common compound is made up of sodium, a silvery metal, and chlorine, a greenish-yellow poisonous gas. Many compounds look different from the elements in them.

What are molecules?

A particle that consists of two or more atoms bonded together is a molecule. As an example, oxygen in the air is a molecule made of two atoms of oxygen. A molecule is the basic unit of a molecular compound. Molecules are all around you. The simple sugars that you eat and the proteins in your body are molecules. The wool and cotton fibers in the clothes you wear all are molecules. They are formed from atoms that are bonded together.

Think it Over

3. Infer Could a molecule contain only one atom? Explain your answer.

_____

_____

_____

Mixtures

Is pizza one of your favorite foods? Do you like soft drinks? If so, you like two foods that are mixtures. A mixture is a material made up of two or more substances that can be separated physically. There are many different kinds of mixtures.

4. Explain How can you tell that a pizza is a heterogeneous mixture?

_____

_____

_____

FOLDABLES

A Classify Make the following Foldable to help you classify heterogeneous and homogeneous mixtures, colloids, and suspensions.

✔ **Reading Check**

5. Explain Why is soda in an unopened bottle a homogeneous mixture?

_____

_____

_____

What are heterogeneous mixtures?

Mixtures are different from compounds in several ways. One difference is that mixtures do not always contain the same proportions of the substances that make them up. For example, a pizza chef might put more cheese on one pizza than on another. In a pizza, you can see the different ingredients. A <u>heterogeneous</u> (he tuh ruh JEE nee us) <u>mixture</u> is one in which different materials can be identified easily. Granite, concrete, and pizza are some heterogeneous mixtures.

Recognizing Heterogeneous Mixtures You may be wearing a heterogeneous mixture. Some fabrics are labeled as permanent-press. These fabrics resist wrinkles. Permanent-press fabric contains fibers of two materials. The materials are polyester and cotton. The amounts of polyester and cotton can change from one piece of fabric to another. Look at the labels on some of your clothes. Do they contain different amounts of polyester and cotton?

You probably cannot tell that permanent-press fabric is a heterogeneous mixture by looking at it. It looks like it is made up of only one material. However, you might be able to see the mixture with a microscope. Under a microscope, the polyester fibers probably would look different from the cotton fibers.

Many substances around you are heterogeneous mixtures. Some have materials that are easy to see, such as those in pizza. Others have materials that are not easy to tell apart, such as the fibers in permanent-press fabrics. In fact, some parts of heterogeneous mixtures can be mixtures themselves. The cheese in pizza is a mixture, but you cannot see the materials. Cheese contains many compounds, such as milk, proteins, butterfat, and, sometimes, food coloring.

What are homogeneous mixtures?

Soft drinks are mixtures. They contain water, sugar, flavorings, coloring, and carbon-dioxide gas.

Soft drinks in sealed bottles are homogeneous mixtures. A <u>homogeneous</u> (hoh moh JEE nee us) <u>mixture</u> contains two or more substances blended evenly throughout. You cannot see the different substances in a homogeneous mixture. ✔

When a soft drink is poured into a glass, the carbon-dioxide gas forms bubbles. You then can see that the gas is separate from the other ingredients. When this happens, the soda becomes a heterogeneous mixture. However, after the carbon dioxide has bubbled out, the drink is flat. It is a homogeneous mixture of water, sugar, coloring, and flavoring.

What is a solution?

Vinegar is another homogeneous mixture. It looks clear, but it contains particles of acetic acid mixed with water. Homogenous mixtures, such as bottled soft drinks and vinegar, are called solutions. A **solution** is a homogeneous mixture of particles so small that they cannot be seen with a microscope and will never settle to the bottom of their container. Solutions stay evenly mixed. The diagram below shows the difference between substances and mixtures.

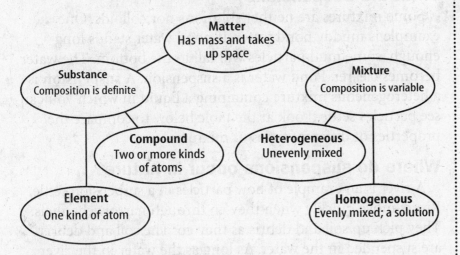

Matter
Has mass and takes up space

Substance
Composition is definite

Mixture
Composition is variable

Compound
Two or more kinds of atoms

Heterogeneous
Unevenly mixed

Element
One kind of atom

Homogeneous
Evenly mixed; a solution

Picture This

6. **Identify** In the diagram, highlight the words that explain the difference between a substance and a mixture.

What are colloids?

Milk is an example of a mixture called a colloid. A **colloid** (KAH loyd) is a type of mixture with particles that are larger than those in solutions but not heavy enough to settle to the bottom of their container. Remember how a pizza is still a pizza even if the proportion of its ingredients are changed? Milk is similar to a pizza in that way. Milk contains water, fat, and proteins, but like any mixture, these substances can be in different proportions. What makes milk a colloid is that these ingredients form large particles, but they are not heavy enough to settle.

Paint is an example of a liquid colloid. Gases and solids can also be colloids. For example, fog and smoke are colloids. Fog is made up of liquid water particles suspended in air. Smoke contains solids suspended in air.

Do colloids and solutions look the same?

One way to tell the difference between a colloid and a solution is by how each looks. Fog looks white because its particles are large enough to scatter light. Sometimes it is not easy to tell that a liquid is a colloid. For example, some shampoos and gelatins are colloids called gels that look almost clear.

Think it Over

7. **Give an Example** Name a colloid that you can see through.

_____

_____

How do you identify colloids?

You can tell if a liquid is a colloid by shining a beam of light through it. You cannot see a light beam as it passes through a solution. But you easily can see a light beam in a colloid because its large particles scatter light. Small particles in solutions do not scatter light. Have you ever noticed how, at night, the fog scatters the light from a car's headlights? The **Tyndall effect** is the scattering of light by particles in a colloid. ☑

What are suspensions?

Some mixtures are neither solutions nor colloids. One example is muddy pond water. If pond water stands long enough, some mud particles will fall to the bottom. The water becomes clearer. Pond water is a suspension. A **suspension** is a heterogeneous mixture containing a liquid in which you can see particles settle. Look at the table below to compare the properties of different types of mixtures.

Where do suspensions occur in nature?

A river is an example of how particles in a suspension settle. Rivers move quickly when they go through narrow channels. They pick up soil and debris as they go. The soil and debris are suspended in the water. As long as the water in the river moves fast enough, the suspended soil does not settle. When the river slows, the particles fall out of the suspension and settle on the bottom of the river. This also happens when a river flows into a large body of water, such as an ocean. After many years, a delta forms made up of mud and debris.

Comparing Solutions, Colloids, and Suspensions

Description	Solutions	Colloids	Suspensions
Settle upon standing?	no	no	yes
Separate using filter paper?	no	no	yes
Particle size	0.1–1 nm	1–100 nm	>100 nm
Scatter light?	no	yes	yes

✔ **Reading Check**

8. Define What is the scattering of light by particles in a colloid called?

_____

_____

_____

Applying Math

9. Comparison Which type of mixture has the largest particles? Explain how you know.

_____

_____

_____

_____

● After You Read

Mini Glossary

colloid: a type of mixture with particles that are larger than those in solutions, but not heavy enough to settle to the bottom of their container

compound: a substance with two or more elements that are combined in a fixed proportion

element: a substance in which all the atoms are the same

heterogeneous mixture: a mixture in which different materials can be identified easily

homogeneous mixture: a mixture that contains two or more substances blended evenly throughout

solution: a homogeneous mixture of particles so small that they cannot be seen with a microscope and that will never settle to the bottom of their container

substance: a type of matter that is always made of the same material or materials

suspension: a heterogeneous mixture containing a liquid in which you can see particles settle

Tyndall effect: the scattering of light by particles in a colloid

1. Review the terms and their definitions in the Mini Glossary. The oxygen that you breathe is made up of tiny particles that are actually two atoms of oxygen bonded together. Would you say that oxygen is an element or a compound? Explain.

_____

_____

2. Fill in the blanks with an example of each type of matter.

End of Section

Classification of Matter

section ❷ Properties of Matter

What You'll Learn

- to identify substances using physical properties
- differences between physical and chemical changes
- how to identify chemical changes
- the law of conservation of mass

● Before You Read

When you see someone, how do you identify that person as a friend or a stranger? How do you identify a friend on the phone? What are some things about people that help you recognize them? On the lines below, list some things you use to identify people.

_____

_____

Study Coach

Make Flash Cards For each heading in this section, make a flash card. The flash card should contain the main point of the paragraphs below the heading. When you finish the section, review the flash cards.

● Read to Learn

Physical Properties

You can stretch a rubber band, but you can't stretch a piece of string very much. You can bend a piece of wire, but you can't bend a matchstick easily. The rubber band and the wire change shape, but the substances that they are made of do not change.

The ability to stretch or bend is a physical property. A **physical property** is a feature or characteristic that describes an object or substance. Some examples of physical properties are color, shape, size, density, melting point, and boiling point.

How do physical properties describe appearance?

How would you describe a tennis ball? You could describe some of its physical properties, such as shape and color. You could say it is a solid, not a liquid or a gas. For example, you might describe a tennis ball as a brightly colored, hollow sphere. You could also measure some physical properties of the ball. You could measure its diameter with a tape measure. You could measure its mass with a balance. You could measure how high it will bounce.

To describe a soft drink in a cup, you could start by saying it is a brown liquid. You could measure the volume and the temperature of the soft drink. Each of these characteristics is a physical property of that soft drink.

FOLDABLES

B Compare and Contrast Make the following Foldables to help you understand how physical and chemical properties are different, and how physical and chemical changes are different.

How do physical properties describe behavior?

Some physical properties describe the behavior of a material or substance. You recall that a magnet attracts objects that contain iron, such as a safety pin. Attraction to a magnet is a physical property of iron. Every substance has physical properties that make it useful for certain tasks.

Some metals, such as copper, are useful because they bend easily and can be pulled into wires. Other metals, such as gold, are useful because they easily can be pounded into sheets as thin as 0.1 micrometers. This property of gold makes it useful for decorating picture frames and other things. Flattened gold is called gold leaf.

Think again about the soft drink. If you knock over the cup, the drink will spill. If you knock over a jar of honey, however, it does not flow as quickly. The ability to flow is a physical property of liquids. ☑

How are physical properties used to separate materials?

You can use the physical properties of size and hardness to separate some substances. Removing the seeds from a watermelon is easy. The seeds are small and hard, and the flesh of the large watermelon is soft and juicy.

Using Magnetism The dish in the figure below contains sand mixed with iron filings. You probably would not be able to sift out the iron filings. They are about the same size as the sand particles. However, if you pass a magnet through the mixture, it will attract only the iron filings. A magnet does not attract sand. This is an example of using the physical property of magnetism to separate substances in a mixture.

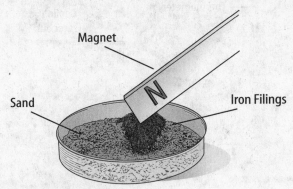

Magnet

Sand

Iron Filings

Physical Changes

When you break a stick of chewing gum, you change its size and shape. You do not change the identity of the materials that make up the gum. A physical change does not change identity.

☑ **Reading Check**

1. **Explain** Why does honey flow more slowly than water?

_____

_____

Picture This

2. **Observe** Why would it be difficult to sift the iron filings from the sand?

_____

_____

Why doesn't the identity change?

When a substance freezes, boils, evaporates, or condenses, it undergoes a physical change. A **physical change** is a change in size, shape, or state of matter. Heat might be added or removed during a physical change. Changes in energy do not change the identity of the substance being heated or cooled. All substances have distinct properties that are constant, or never change. The properties of density, specific heat, boiling point, and melting point are constant for substances. These properties can be used to identify unknown substances in a mixture. ☑

Iron is a substance that changes states when it absorbs or releases energy. At high temperatures, iron melts. However, iron has the same physical properties that identify it as iron, whether it is in the liquid or solid state.

What is distillation?

Distillation is the process of separating substances in a mixture by evaporating a liquid and condensing its vapor. A laboratory distillation process is shown below.

To distill a liquid, it is heated until it vaporizes. Then, the vapors are cooled until they condense. All solid material is left behind. Distillation is used to make drinking water out of salt water.

Liquids with different boiling points can be separated by distilling. The mixture is heated slowly until it begins to boil. Vapors of the liquid with the lowest boiling point form first. They are condensed and collected. As temperature increases, the second liquid boils. Its vapors are condensed and collected. Distillation is used often in industry. Natural oils such as mint are distilled.

Picture This

4. **Observe** Where would you expect to find the solid material left behind in the distillation process?

_____

_____

Chemical Properties and Changes

A <u>chemical property</u> is a characteristic of a substance that indicates whether it can undergo a change that results in a new substance. A warning on a can of liquid paint thinner or lighter fluid states that the liquid is flammable (FLA muh buhl). If a substance is flammable, it can burn. When a substance burns, it produces new substances during a chemical change. Therefore, whether or not a substance is flammable is a chemical property. Knowing which substances are flammable helps you to use them safely.

Another chemical property is how compounds react to light. Some medicines are kept in dark bottles because the compounds will change chemically in the light.

Detecting Chemical Change

Your senses tell you when a chemical change has happened. Leave a pot of chili cooking on the stove too long and you will smell it burning. The smell tells you a new substance formed.

How does the identity of a substance change?

You smell a rotten egg or see rust on a bike. These are signs that chemical changes have taken place. A <u>chemical change</u> is a change of one substance to another. When you drop antacid tablets in a glass of water, they foam up. After a lightning strike the air smells different. These are signs that new substances have been produced. Heat, light, and sound are often signs of a rapid energy release and of a chemical change.

The only proof that a chemical change has taken place is that a new substance is formed. For example, when hydrogen and oxygen combine in a rocket engine, their chemical changes produce heat, light, and sound. But there are no such clues when iron combines with oxygen to form rust. Rust forms slowly. The only clue that iron has changed to a new substance is the presence of rust. Burning and rusting are chemical changes because new substances are formed. ☑

How can a chemical change be used to separate substances?

You can separate substances using a chemical change. One example is cleaning silver. Silver chemically reacts with sulfur compounds in the air to form silver sulfide, or tarnish. A different chemical reaction changes the tarnish back to silver using warm water, baking soda, and aluminum foil.

Think it Over

5. Infer How can knowing a chemical property such as flammability help you to use a product safely?

_____

_____

_____

Reading Check

6. Explain What is the only proof that a chemical change has taken place?

_____

_____

Weathering—Chemical or Physical Change?

The forces of nature continuously shape Earth's surface. Rocks split, rivers carve deep canyons, sand dunes shift, and interesting formations develop in caves. These changes are known as weathering. Weathering changes are both physical and chemical changes.

What is physical weathering?

As a stream cuts through rock to form a canyon, small particles of rock are carried downstream. The large rocks and the particles of rock have the same properties. Their properties are not changed, so this weathering is physical.

What is chemical weathering?

Limestone is made up mostly of a chemical called calcium carbonate. Calcium carbonate does not dissolve easily in water. But if the water is even slightly acidic, calcium carbonate reacts. A new substance, calcium hydrogen carbonate, is formed. This substance dissolves in water. This change in limestone is a chemical change. The calcium carbonate changes to calcium hydrogen carbonate in the chemical reaction. Rainwater can dissolve limestone because of this reaction. This chemical change leads to weathering. Chemical changes like this one create caves and the rock formations that are found in them.

The Conservation of Mass

Wood is combustible, which means it can burn. As you have learned, this is a chemical property of wood. Think about a log burning in a fireplace. After you burn a piece of wood, there is nothing left but a small pile of ashes. During the fire, the wood gives off heat, light, and smoke. These changes in the wood are all signs of a chemical reaction.

Where did all the matter in the log go as it burned? At first, you might think that matter was lost as the log burned, since the pile of ashes is so small. The ashes have a smaller mass than the wood you started with. But imagine that you could collect all the smoke and gases that escaped from the log while it burned. If you added this all up, you would find that no mass was actually lost.

Mass is not lost during burning. In the same way, mass is not gained or lost in any chemical change. In other words, matter is not created or destroyed in a chemical change. The **law of conservation of mass** states that the mass of all the substances before a chemical change equals the mass of substances after a chemical change. ✔

Copyright © Glencoe/McGraw-Hill, a division of The McGraw-Hill Companies, Inc.

Think it Over

7. **Apply** How are shifting sand dunes an example of physical weathering?

_____

_____

_____

Reading Check

8. **Explain** what the law of conservation of mass means.

_____

_____

● After You Read

Mini Glossary

chemical change: change of one substance to another

chemical property: characteristic of a substance that indicates whether it can undergo a certain chemical change

distillation: the process of separating substances in a mixture by evaporating a liquid and condensing its vapor

law of conservation of mass: the mass of all substances that are present before a chemical change equals the mass of all the substances that remain after the change

physical change: a change in size, shape, or state of matter

physical property: a feature or characteristic that describes an object or substance

1. Review the terms and their definitions in the Mini Glossary. What is the main difference between a physical change and a chemical change?

_____

_____

2. Complete the table below by giving an example of the property or change.

Physical property	Example:
Chemical property	Example:
Physical change	Example:
Chemical change	Example:
Separation using physical change	Example:
Separation using chemical change	Example:

3. ◀ Study Coach Imagine explaining physical and chemical changes to a group of elementary school students. Describe some items around your house to use as examples of physical and chemical changes.

_____

_____

End of Section

Properties of Atoms and the Periodic Table

section ❶ Structure of the Atom

What You'll Learn

- the names and symbols of common elements
- what subatomic particles and quarks are
- how to describe the atom
- how electrons are arranged in an atom

Study Coach

Create a Quiz As you read this section, write a quiz question for each paragraph. After you finish reading the section, answer your quiz questions.

FOLDABLES

Ⓐ Organize Information
Make the following Foldable to help organize information about scientific shorthand, atomic components, and quarks.

Scientific Shorthand	Atomic Components	Quarks

● Before You Read

You use symbols to make it easier to write certain things, such as $25.08 instead of twenty-five dollars and eight cents. On the following lines, write some symbols you may use to make writing easier.

_____

_____

● Read to Learn

Scientific Shorthand

Do you have a nickname? Do you use abbreviations for long words or the names of states? Scientists also do this. In fact, scientists have developed their own shorthand, a way to shorten long, complicated names.

C, Al, Ne, and Ag are all chemical symbols for different elements. A chemical symbol is shorthand for the name of an element. Chemical symbols make writing names of elements easier. Chemical symbols are either one capital letter or a capital letter plus one or two small letters. The table shows the chemical symbols for some elements. For some elements, the symbol is the first letter of the element's name. For example, C is for carbon. For other elements, the symbol is the first letter plus another letter from its name. For example, Ca is for calcium. Some symbols come from the Latin names of elements. *Argentum* is Latin for "silver." Silver's symbol is Ag.

Symbols of Some Elements

Element	Symbol	Element	Symbol
Aluminum	Al	Iron	Fe
Calcium	Ca	Mercury	Hg
Carbon	C	Nitrogen	N
Chlorine	Cl	Oxygen	O
Gold	Au	Potassium	K
Hydrogen	H	Sodium	Na

How have elements been named?

Elements have been named in many different ways. Elements have been named to honor scientists, for places, or for the elements' properties. Other elements have been named using rules made by an international committee. No matter what the origin of the name, scientists worldwide use the same system of element names and chemical symbols. People everywhere know that H means hydrogen, O means oxygen, and H_2O means dihydrogen monoxide, or water. ☑

Atomic Components

An element is matter that is made up of one type of atom. An **atom** is the smallest piece of matter that still has the properties of the element. For example, the element silver is made up of only silver atoms. The element hydrogen is made up of only hydrogen atoms.

The figure below shows the structure of the atom. Atoms are made up of protons, neutrons, and electrons. **Protons** are particles with an electrical charge of $1+$. **Neutrons** are particles with no electrical charge. **Electrons** are particles with an electrical charge of $1-$. The **nucleus** is the small, positively charged center of the atom. It is made up of protons and neutrons. The nucleus is surrounded by a cloud containing electrons. The number of protons in an atom determines which element it is. For example, all atoms with 47 protons are silver atoms. All atoms with 1 proton are hydrogen atoms.

☑ Reading Check

1. **Identify** What is the symbol for dihydrogen monoxide?

_____

Picture This

2. **Label** Write a plus sign on each proton shown in the nucleus.

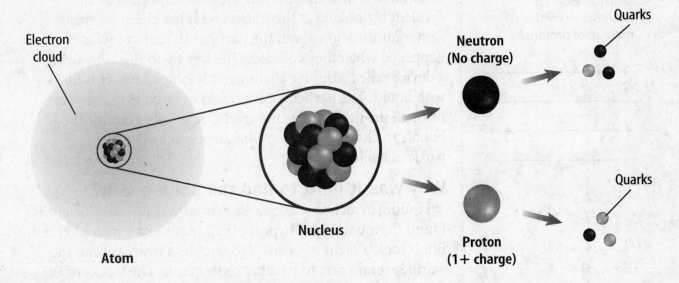

Electron cloud

Atom

Nucleus

Neutron (No charge)

Quarks

Proton (1+ charge)

Quarks

Quarks—Even Smaller Particles

Are protons, neutrons, and electrons the smallest particles that exist? Scientists hypothesize that electrons are not made up of smaller particles. If this is true, electrons are one of the most basic types of particles. But protons and neutrons are made up of smaller particles called **quarks**. So far, scientists have discovered six different quarks. Scientists theorize that protons are made up of three quarks. The quarks in a proton are held together with a force called the strong nuclear force. Neutrons are made up of another arrangement of three quarks. Scientists are still studying protons and neutrons to better understand them. ☑

How do scientists find quarks?

To study quarks, scientists accelerate, or speed up, charged particles until they are moving extremely fast. Then they force the particles to collide with—or smash into—protons. The collision causes the protons to break apart. The Fermi National Accelerator Laboratory in Illinois has a machine that can accelerate particles fast enough to smash protons. This machine, called a Tevatron, is in a circular tunnel. The tunnel is 6.4 km in circumference. Scientists use electric and magnetic fields in the Tevatron to accelerate and smash particles.

How do scientists study quarks?

Scientists use different kinds of devices to detect the new particles that are made when particles are smashed together. Just as traffic investigators can tell what happened at an accident by looking at tire tracks and other clues, scientists gather information about the particles to find out what happened when they collided. One way to do this is by using a device called a bubble chamber. A bubble chamber is filled with liquid. The particles leave tracks of bubbles as they pass through the liquid—like tire tracks. Scientists examine the bubble tracks to determine what happened when the particles collided.

Why was it hard to find the sixth quark?

Finding evidence that quarks existed was not easy. Scientists found five quarks and hypothesized that a sixth quark existed. But it took a team of nearly 450 scientists from around the world several years to find the sixth quark. The tracks of the sixth quark were hard to detect. They were hard to detect because there was evidence of the sixth quark in only about one billionth of a percent of proton collisions. The sixth quark is called the *top* quark.

☑ **Reading Check**

3. **Compare** Which is smaller, a proton or a quark?

_____

_____

💡 **Think it Over**

4. **Describe** How do scientists study the makeup of protons?

_____

_____

_____

_____

Models—Tools for Scientists

Scientists use models to represent things that are difficult to visualize—or picture in your mind. Scaled-down models let you visualize something that is too large to see. Models of buildings, the solar system, and airplanes are scaled-down models. Scaled-up models are used to represent things that are too small to see. Scientists have developed scaled-up models to help them study the atom. To give you an idea of how small the atom is, it would take about 50,000 atoms stacked on top of each other to equal the thickness of a sheet of aluminum foil.

For a model of the atom to be useful, it must accurately represent everything we know about matter and how the atom behaves. As they learn more about atoms, scientists must change their models to include the new information.

How has the atomic model changed?

People have not always known that matter is made up of atoms. Around 400 B.C., a Greek philosopher named Democritus came up with the idea that atoms make up all substances. Another famous Greek philosopher, Aristotle, did not agree with Democritus' theory. Aristotle believed that each kind of matter was uniform, or the same all the way through, and not made of smaller particles. Aristotle's incorrect theory was accepted for about two thousand years. But in the 1800s, an English scientist named John Dalton was able to prove that atoms existed.

Dalton Model Dalton's model of the atom was a solid sphere, as shown in the first figure on the right. Dalton's model helped scientists explain why chemical reactions occur. Scientists then could use chemical symbols and equations to describe these reactions.

Dalton's Model

Thomson Model In 1904, English physicist Joseph John Thomson decided from his experiments that atoms contained small, negatively charged particles. He thought these "electrons" were spread out evenly throughout a positively charged sphere. His model, shown in the second figure, looks like a ball of chocolate chip cookie dough.

Ball of positive charge

Negatively charged electron

Thomson's Model

Copyright © Glencoe/McGraw-Hill, a division of The McGraw-Hill Companies, Inc.

Rutherford Model In 1911, another British physicist, Ernest Rutherford, thought that almost all the mass of an atom and all its positive charge were concentrated in the nucleus of an atom. He also thought the nucleus of an atom was surrounded by electrons, as shown in the first figure.

Bohr Model In 1913, Danish physicist Neils Bohr hypothesized that electrons travel in fixed orbits around the nucleus of the atom, as the second figure shows. One of Bohr's students, James Chadwick, found that the nucleus contained positive protons and neutral neutrons. ☑

Positively charged nucleus

"Empty space" containing electrons

Rutherford's Model

Nucleus of protons and neutrons

Bohr's Model

✔ **Reading Check**

7. **Explain** What did Chadwick discover about the nucleus of an atom?

_____

_____

_____

What is the electron cloud model?

The model of the atom has changed over time. By 1926, scientists had developed the electron cloud model of the atom in use today. An **electron cloud** is the area around the nucleus of an atom where its electrons are most likely found. The electron cloud is 100,000 times larger than the diameter of the nucleus. However, each electron in the cloud is much smaller than a single proton.

Scientists do not really know where in the electron cloud the electrons might be. Electrons move so fast and have such a small mass that it is impossible to describe exactly where they might be. It is best to describe their location as somewhere in the cloud. Think of the spokes on a spinning bicycle wheel. The spokes are moving so quickly that you can't tell exactly where any one spoke is. All you see is a blur. The spokes lie somewhere in the blur. An electron cloud is similar. It is a blur containing all of the electrons somewhere within it. The figure illustrates what an electron cloud might look like.

Picture This

8. **Compare and Contrast** How are Dalton's model (see figure on previous page) and electron cloud models of the atom similar? How are they different?

Similar:

_____

_____

Different:

_____

_____

_____

Electron cloud

Nucleus

Electron Cloud Model

● After You Read

Mini Glossary

atom: the smallest piece of matter that still has the properties of the element

electron: particle with an electrical charge of 1−

electron cloud: the area around the nucleus of an atom where its electrons are most likely found

neutron: particle with no charge

nucleus: the small, positively charged center of the atom

proton: particle with an electrical charge of 1+

quark: smaller particle that makes up protons and neutrons

1. Review the terms and their definitions in the Mini Glossary. Write a sentence describing what parts make up an atom.

   _____

   _____

2. Below is a model of an atom. Label and describe each part of the atom. If any particles are made up of even smaller particles, list these also.

3. **Study Coach** As you read this section, you created a quiz question for each paragraph. Did answering these quiz questions after you read the section help you learn the material? Why or why not?

   _____

   _____

End of Section

Properties of Atoms and the Periodic Table

section ❷ Masses of Atoms

What You'll Learn

- the difference between the atomic mass and the mass number of an atom
- how to identify components of isotopes
- how to interpret the average atomic mass of an element

Mark the Text

Highlight As you read the text under each heading, highlight the main ideas. After you finish reading the section, review the highlighted main ideas to help you learn the important topics of the section.

Applying Math

1. **Comparing Decimals** Which has a larger mass, a proton or a neutron?

● Before You Read

Which metric unit do you use to measure the amount of gas that a car's gas tank holds? Which metric unit would you use to measure the distance to the next town? Explain why you would use these units and not smaller units.

● Read to Learn

Atomic Mass

Neutrons and protons are much more massive than electrons. Since the nucleus contains the neutrons and protons, it contains most of the mass of an atom. The mass of a proton is about the same as the mass of a neutron—about 1.67×10^{-24} g, as shown in the table. The mass of a proton or a neutron is about 1,800 times greater than the mass of an electron. The mass of an electron is so small that it is not even considered when finding the mass of an atom.

Subatomic Particle Masses	
Particle	**Mass (g)**
Proton	1.6726×10^{-24}
Neutron	1.6749×10^{-24}
Electron	9.1093×10^{-28}

What is the atomic mass unit?

What unit would you use to estimate the height of your school building? Kilometers would be difficult to use. You probably would use a more appropriate unit, such as meters. Just as the kilometer is not the right unit for measuring the height of a building, scientists found that the gram was not the right unit for measuring the mass of an atom.

A useful unit gives numbers that are easy to work with. The unit used for measuring atomic particles is called the atomic mass unit (amu). The mass of a proton or neutron is almost equal to 1 amu. This is not a coincidence. The amu was defined that way. The amu is one-twelfth the mass of a carbon atom. A carbon atom contains six protons and six neutrons, or twelve particles. Since most of the mass of an atom is in the nucleus, each proton and neutron has a mass nearly equal to 1 amu.

How do protons identify elements?

Remember that atoms of different elements have different numbers of protons. In fact, the number of protons tells you what type of atom you have and vice versa. For example, every carbon atom has six protons. Also, every atom with six protons is carbon.

The **atomic number** of an element is the number of protons in an atom of the element. Since carbon has six protons, the atomic number of carbon is six. If you are given any one of the following for an element—its name, number of protons, or atomic number—you can find the other two.

What is the mass number?

The **mass number** of an atom is the sum of the number of protons and the number of neutrons in the nucleus of an atom. The table below shows this.

FOLDABLES™

B Find the Main Idea

Make the following Foldable to help take notes on the main ideas from this section.

Atomic Number
Mass Number
Isotope
Average Atomic Mass

Picture This

2. **Complete** the table by finding the mass numbers for oxygen, sodium, and copper.

Mass Numbers of Some Atoms						
Element	Symbol	Atomic Number	Protons	Neutrons	Mass Number	Average Atomic Mass*
Boron	B	5	5	6	11	10.81 amu
Carbon	C	6	6	6	12	12.01 amu
Oxygen	O	8	8	8		16.00 amu
Sodium	Na	11	11	12		22.99 amu
Copper	Cu	29	29	34		63.55 amu

*The atomic mass units are rounded to two decimal places.

How is the number of neutrons found?

If you know the mass number and atomic number of an atom, you can find the number of neutrons it contains.

number of neutrons = mass number − atomic number

Atoms of the same element with different numbers of neutrons can have different properties. For example, carbon with a mass number of 12 is called carbon-12. Carbon-14, with a mass number of 14 is radioactive. Carbon-12 is not radioactive.

Isotopes

Not all atoms of an element have the same number of neutrons. Atoms of the same element that have different numbers of neutrons are called <u>isotopes</u>. For example, boron atoms can have mass numbers of 10 or 11. To find the number of neutrons in an isotope, you can use the formula above. Look at the table on the previous page. Notice that boron has an atomic number of five. That means it has five protons. Substitute these numbers into the formula to get $11 - 5 = 6$ and $10 - 5 = 5$. So, boron isotopes have either five or six neutrons.

How can isotopes be used?

Atoms can be used to find the age of bones and rocks that are millions of years old. Radioactive isotopes release nuclear particles and energy as they decay into another element. The time it takes for half of the radioactive isotopes in a piece of rock or bone to change into another element is called its half life. Scientists use half lives of radioactive isotopes to measure time.

The table below lists the half-lives of some radioactive elements. It also lists the elements that the radioactive elements decay into. For example, it would take 5,715 years for half of the carbon-14 atoms in a rock to change into atoms of nitrogen-14. After another 5,715 years, half of the remaining carbon-14 atoms will change, and so on. These radioactive "clocks" can be used to measure different periods of time.

Half-Lives of Radioactive Isotope

Radioactive Element	Changes to This Element	Half-Life
uranium-238	lead-206	4,460 million years
potassium-40	argon-40, calcium-40	1,260 million years
rubidium-87	strontium-87	48,800 million years
carbon-14	nitrogen-14	5,715 years

Applying Math

3. **Apply** The element uranium has a mass number of 238, and an atomic number of 92. How many neutrons does an atom of uranium have?

_____

Think it Over

4. **Determine** What is the same in two isotopes of an element? What is different?

_____

_____

_____

Picture This

5. **Apply** How many years would it take half of the atoms in uranium-238 to change into lead-206?

_____

How do you identify isotopes?

The figure shows models of the two isotopes of boron. Because the numbers of neutrons in the isotopes is different, their mass numbers are different. To identify an isotope, use the name of the element followed by the mass number of the element. For example, the isotopes of boron are boron-10 and boron-11, because boron isotopes have mass numbers of either 10 or 11.

Boron-10
5 Electrons
5 Protons
5 Neutrons
Nucleus

Boron-11
5 Electrons
5 Protons
6 Neutrons
Nucleus

Most elements have more than one isotope. Because of this, each element has an average atomic mass. The **average atomic mass** of an element is the weighted-average mass of the mixture of its isotopes. For example, four out of five atoms of boron are boron-11. That means one out of five atoms is boron-10. To find the average atomic mass of boron, solve the following equation:

$$\frac{4}{5}(11 \text{ amu}) + \frac{1}{5}(10 \text{ amu}) = 10.8 \text{ amu}$$

The average atomic mass of boron is 10.8 amu. You round the average atomic mass to the nearest whole number to find the most abundant isotope of an atom. For example, the average atomic mass of boron, 10.8, rounds to 11. So, the most abundant isotope of boron is boron-11.

Picture This

6. Draw and Label Carbon-12 is an isotope with 6 protons and 6 neutrons. Draw a model of carbon-12. Label the protons and neutrons.

Applying Math

7. Apply The element magnesium has an average atomic mass of 24.305. What is the most abundant isotope of magnesium?

_____

● After You Read
Mini Glossary

atomic number: a number equal to the number of protons in an atom

average atomic mass: the weighted-average mass of the mixture of isotopes of an element

isotopes: atoms of the same element that have different numbers of neutrons

mass number: the sum of the number of protons and the number of neutrons in an atom

1. Review the terms and their definitions in the Mini Glossary. How can you find the number of protons and neutrons in an atom of an element from the atomic number and the mass number?

   _____

   _____

   _____

2. Complete the Venn diagram by writing the given phrases in the correct area.

   • determines which isotope

   • equals the atomic number

   • equals the mass number

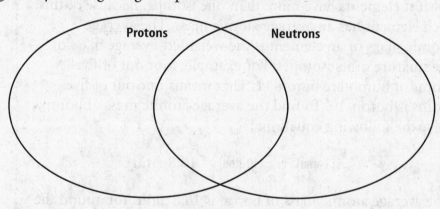
Protons   Neutrons

3. ◄ **Mark the Text** Tell how you could use a set of red and blue marbles to teach a friend about the atomic number and mass number of an element.

   _____

   _____

   _____

   _____

End of Section

section ❸ The Periodic Table

● Before You Read

Many parts of our lives are affected by repeated patterns. For example, a calendar shows the patterns of weeks. Name some repeated patterns that you see happening all the time. How could you keep track of a pattern?

_____

_____

What You'll Learn

- the composition of the periodic table
- how to get information from the periodic table
- what metal, nonmetal, and metalloid mean

● Read to Learn

Organizing the Elements

When you look at the Moon, does it always appear the same? Each month, the Moon grows larger until it is full, then grows smaller until it seems to disappear. This type of change is called periodic. The word *periodic* means "repeated in a pattern." The days of the week are periodic because they repeat every seven days. Think of the calendar as a periodic table of days and months.

Who was Dmitri Mendeleev?

In the late 1800s, a Russian chemist named Dmitri Mendeleev wanted to find a way to organize the elements. He organized the elements known at the time into a table. He placed the elements in the table in order of increasing atomic mass.

Mendeleev discovered a pattern in his table. The properties of some lighter elements seemed to repeat in heavier elements. Because this pattern repeated, the pattern was considered to be periodic. Today, this arrangement is called a periodic table of elements. In the **periodic table**, the elements are arranged by increasing atomic number and by changes in physical and chemical properties. ☑

Mark the Text

Identify the Main Point Look for the main point of the paragraph or paragraphs under each heading in this section. When you have found the main point, write it down on a piece of paper. After you read the section, look over the main points again to help you learn the content of the section.

✔ Reading Check

1. **Determine** Who was the first person to organize the elements into a periodic table?

_____

_____

How did Mendeleev's table predict properties?

Mendeleev left blank spaces in his table so that he could line up the elements. He looked at the elements surrounding the blank spaces. He predicted the properties and atomic masses of unknown elements to fit in the blank spaces.

The table shows Mendeleev's predictions for the element germanium. He called the element ekasilicon. His predictions proved to be accurate. Scientists eventually found all of the elements that were missing from Mendeleev's periodic table. The properties of these "missing" elements turned out to be extremely close to what Mendeleev predicted.

Applying Math

2. **Use Decimals** Find the difference between the predicted density of germanium and its actual density.

_____

_____

Mendeleev's Predictions	
Predicted Properties of Ekasilicon (Es)	**Actual Properties of Germanium (Ge)**
Existence Predicted—1871	*Actual Discovery—1886*
Atomic mass = 72	Atomic mass = 72.61
High melting point	Melting point = 938°C
Density = 5.5 g/cm^3	Density = 5.323 g/cm^3
Dark gray metal	Gray metal
Density of EsO$_2$ = 4.7 g/cm^3	Density of GeO$_2$ = 4.23 g/cm^3

How has the periodic table been improved?

Mendeleev's periodic table was very good for its time. However, scientists eventually found some problems with it. The elements on Mendeleev's table increased in atomic mass from left to right. Look at the modern periodic table at the back of this book. You will find examples, such as cobalt and nickel, that decrease in mass from left to right. However, notice that the atomic number always increases from left to right.

In 1913, the arrangement of the periodic table was changed. Instead of being arranged by increasing atomic mass, it was arranged by increasing atomic number. This change was due to the work of an English scientist named Henry G. J. Moseley. The new arrangement seemed to correct some of the problems of the old table. The current periodic table uses Moseley's arrangement and is shown in the back of this book. ✓

3. **Observe** How did Moseley arrange the periodic table?

_____

_____

_____

The Atom and the Periodic Table

Objects are often sorted or grouped according to the properties they have in common. Elements on the periodic table are grouped according to their chemical properties. The vertical columns on the periodic table are called **groups**. The groups are numbered 1 through 18. Sometimes they are called families. Elements in each group have similar properties. For example, in Group 11, copper, silver, and gold have similar properties. Each is a shiny metal. Each is a good conductor of electricity and heat. Why are the elements in a group similar? Look at the structure of the atom to answer this question.

What is the structure of the electron cloud?

Where are the electrons located in an atom? How many are there? In a neutral atom, the number of electrons is equal to the number of protons. Carbon has an atomic number of six, which means it has six protons and six electrons. These electrons are located in the electron cloud that surrounds the nucleus.

Scientists have found that electrons in the electron cloud have different amounts of energy. Look at the figure. It shows a model of the energy differences that scientists use. Electrons fill the energy levels from the inner levels to the outer levels. The inner levels are closer to the nucleus and the outer levels are farther from the nucleus. Inner levels have less energy than the outer levels. Imagine that the nucleus is like a floor. Each energy level is a step up a flight of stairs above the floor. Each stair step represents an increase in energy. The figure shows the maximum number of electrons that will fit in each energy level of an atom. Not all atoms will have all levels filled. This depends on the number of electrons in an atom of that element.

Copyright © Glencoe/McGraw-Hill, a division of The McGraw-Hill Companies, Inc.

4. **Draw Conclusions** Neon is a gas. Do you think neon is an element in Group 11? Explain.

_____

_____

_____

Picture This

5. **Apply** The element magnesium has 12 electrons. In how many energy levels are the electrons of magnesium?

_____

Energy Levels

Step 4 = energy level 4 | 32 electrons

Step 3 = energy level 3 | 18 electrons

Step 2 = energy level 2 | 8 electrons

Step 1 = energy level 1 | 2 electrons

Floor (nucleus)

Energy

How are electrons arranged in energy levels?

Elements that are in the same group have the same number of electrons in their outer energy level. The number of electrons in the outer energy level determines the chemical properties of the element. It is important to understand the link between the location on the periodic table, chemical properties, and the structure of the atom.

These energy levels are named using numbers one to seven. Electrons fill the energy levels starting with the inner level. For example, the element sulfur has 14 electrons. Two electrons will be in energy level 1 and eight electrons will be in energy level 2. The rest of the electrons will be in energy level 3.

Look again at the diagram on the previous page. Notice that energy levels 3 and 4 have increasingly large numbers of electrons. However, a stable outer energy level has eight electrons. How is this possible? In elements that have three or more energy levels, more electrons can be added to inner energy levels as long as the outer level contains eight electrons.

How are rows on the periodic table arranged?

Remember that the atomic number found on the periodic table is equal to the number of electrons in an atom. Look at the partial periodic table below.

Top Row The top row has hydrogen with one electron and helium with two electrons. Both of these electrons are in energy level 1. Energy level 1 is the outermost level in these elements. So, hydrogen has one outer electron and helium has two. Recall from the figure on the previous page that energy level 1 can hold only two electrons. Therefore, helium has a full outer energy level.

Picture This

6. Identify What is hydrogen's outermost energy level?

_____

Hydrogen 1 **H**					Helium 2 **He**
Lithium 3 **Li**	Beryllium 4 **Be**				
Sodium 11 **Na**	Magnesium 12 **Mg**				

Boron 5 **B**	Carbon 6 **C**	Nitrogen 7 **N**	Oxygen 8 **O**	Fluorine 9 **F**	Neon 10 **Ne**
Aluminum 13 **Al**	Silicon 14 **Si**	Phosphorus 15 **P**	Sulfur 16 **S**	Chlorine 17 **Cl**	Argon 18 **Ar**

Second Row The second row of the periodic table begins with lithium. Lithium has three electrons, two in energy level 1, and one in energy level 2. Next is beryllium with two outer electrons and boron with three outer electrons. The pattern continues until you reach neon. Neon has eight outer electrons. Look at the figure on the previous page. Energy level 2 can hold eight electrons. So, neon has a full outer energy level. Notice how a row in the table ends when an outer energy level is filled. The third row of elements, electrons begin filling energy level 3. The row ends with argon, which has a stable outer energy level.

What are electron dot diagrams?

Elements in the same group have the same number of electrons in their outer energy level. Outer electrons are used to determine the chemical properties of an element.

American chemist G. N. Lewis invented the electron dot diagram to show the outer electrons of an element. An **electron dot diagram** is the symbol of an element with dots representing the number of electrons in the outer energy level. The diagram for the elements sodium (Na) and chlorine (Cl) is shown below. These diagrams show how electrons in the outer energy level bond when elements combine to form compounds.

How are elements in the same group similar?

Elements in Group 17 are called halogens. They all have electron dot diagrams similar to chlorine, shown below. You can see that chlorine has seven electrons in its outer energy level. So do the other halogens. Since all elements in a group have the same number of electrons in their outer levels, those elements undergo chemical reactions in similar ways. ☑

How do halogens form compounds?

All halogens can form compounds with elements in Group 1. Group 1 elements, like sodium, all have one electron in their outer energy level. The figure shows an example of a compound formed by a reaction between sodium and chlorine. Sodium combines with chlorine to give each element a complete outer energy level. The result is the compound sodium chloride (NaCl), ordinary table salt.

Sodium	Chlorine	Sodium Chloride
$\text{Na}\cdot$	$\cdot\ddot{\ddot{\text{Cl}}}:$	$[\text{Na}]^{+}\,[:\ddot{\ddot{\text{Cl}}}:]^{-}$

Not all elements form compounds with other elements. Group 18 elements have completely filled outer energy levels. This makes group 18 elements unreactive.

Copyright © Glencoe/McGraw-Hill, a division of The McGraw-Hill Companies, Inc.

7. **Use Numbers** How many electrons would fluorine need to gain in order to have a stable outer energy level?

_____

_____

☑ Reading Check

8. **Apply** What do all of the elements in a group have in common?

_____

_____

_____

Picture This

9. **Draw Conclusions** What seems to happen to the one electron in the outer level of sodium when it combines with chlorine to form sodium chloride?

_____

_____

_____

C Compare and Contrast
Divide your paper into four
quarters to compare and
contrast metals, nonmetals,
and metalloids.

Metal:

Nonmetal:

Metalloid:

Think it Over

10. Draw Conclusions Silicon
conducts electricity under
some conditions but not
under other conditions. In
which region would you
place silicon?

_____

_____

Picture This

11. Apply On which side of
the periodic table would
you look for an element
that definitely will not
conduct electricity?

_____

_____

Regions on the Periodic Table

The periodic table has several regions with specific names.
The horizontal rows of elements are called **periods**. Recall
that the elements increase by one proton and one electron as
you go from left to right across a period. Also, each period
represents a higher electron energy level.

All of the elements in the white squares in the diagram are
metals. Iron, zinc, and copper are some examples of metals.
Most metals are solids at room temperature. They usually are
shiny. They can be drawn into wires and pounded into sheets.
They are good conductors of heat and electricity.

The elements on the right side of the diagram in dark gray
are classified as nonmetals. Oxygen, bromine, and carbon are
nonmetals. Most nonmetals are gases. The elements in this
region that are solids are brittle. Nonmetals are also poor
conductors of heat and electricity at room temperature. The
elements in light gray are metalloids, or semimetals. They
have some properties of metals and some properties of
nonmetals. Boron and silicon are examples of metalloids.

Regions on the Periodic Table

Elements in the Universe

Scientists have found the same elements throughout the
universe. Many scientists hypothesize that hydrogen and
helium are the building blocks of other elements. Atoms join
within stars to form elements with atomic numbers greater
than those of hydrogen and helium. Exploding stars, called
supernovas, spread their mixture of elements throughout the
universe. Scientists have made new elements in laboratories.
These elements may have life spans less than a second.

● After You Read

Mini Glossary

electron dot diagram: the symbol of an element with dots representing the number of electrons in the outer energy level

group: a vertical column of elements on the periodic table

period: a horizontal row of elements on the periodic table

periodic table: an arrangement of the elements by increasing atomic number and by changes in physical and chemical properties

1. Review the terms and their definitions in the Mini Glossary. Write a sentence using one of the terms that shows that you understand the term.

   _____

   _____

2. Below is a blank periodic table of elements. On this table, label the different sections as *metals, metalloids, nonmetals, period,* or *group.*

3. Mark the Text Explain what is important about the rows and columns on the periodic table.

   _____

   _____

   _____

End of Section

Elements and Their Properties

section ❶ Metals

What You'll Learn

- the properties of metals
- how to identify alkali metals and alkaline earth metals
- the difference among transition elements

 Mark the Text

Underline As you read each paragraph, underline the main idea. Also underline other important information, such as definitions.

● Before You Read

Think about what you are wearing. On the lines below, list all of the metal objects that are on your body.

_____

_____

● Read to Learn

Properties of Metals

People have used metals for thousands of years. The first metal used was gold, about 6,000 years ago. People started using copper and silver a few thousand years later. Then came tin and iron. Aluminum wasn't refined—separated from other material—until the 1800s.

On the periodic table, metals are elements found to the left of the stair-step line. Metals usually have many common properties. <u>**Metals**</u> are good conductors of heat and electricity, and all but mercury are solid at room temperature. Metals reflect light. This property is called luster. Metals are <u>**malleable**</u> (MAL ee uh bul), which means they can be hammered or rolled into sheets. Metals are also ductile. <u>**Ductile**</u> (DUK tul) means they can be stretched into wires. The properties of metals make them good materials for making many objects, ranging from eyeglass frames to computers to building materials.

How do metals form ionic bonds?

When a metal atom gives one or more electrons in its outer energy level to a nonmetal atom in a chemical reaction, the atoms bond. This electron loss and gain results in both atoms becoming ions. The ions form ionic bonds, such as shown in the figure. Both atoms become more chemically stable because each ion has a full outer energy level.

Picture This

1. **Circle** the atom that gave away one or more electrons.

Metal   Nonmetal

Ag⁺   I⁻

What is metallic bonding?

The atoms in metals have a special property called metallic bonding. In **metallic bonding**, a cloud of electrons surrounds positively charged metallic ions. Metallic bonding happens because electrons in the outer energy level of a metal atom are not held tightly to the nucleus. This property lets the electrons move freely among many positively charged metal ions. The electrons form a cloud around the metal ions. ☑

Metallic bonding explains many of the properties of metals. For example, metals can be hammered, rolled, and stretched without breaking because of metallic bonding. The ions are in layers that slide past each other without losing their attraction to the electron cloud. Metallic bonding also makes metals good conductors of electricity. Electrons in an electric current flow easily through the electron cloud.

Where are metals on the periodic table?

How many elements are classified as metals? On the periodic table, all the elements in Groups 1 through 12, except hydrogen, are metals. Also, the elements under the stair-step line in Groups 13 through 15 are metals.

The Alkali Metals

The elements in Group 1 on the periodic table shown in the figure are called alkali (AL kuh li) metals. Like other metals, Group 1 metals are shiny, malleable, and ductile. They are also good conductors of electricity. However, alkali metals are softer than most other metals. They also are the most reactive of all the metals. They react rapidly with oxygen and water. Because they are so reactive, alkali metals are stored in substances that are not reactive, such as oil.

Alkali in Compounds Alkali metals are always found in chemical compounds in nature. Each alkali metal atom has one electron in its outer energy level. Alkali metals give up the electron and form ionic bonds with other elements. Ionic compounds with alkali metals include sodium chloride, NaC1, and potassium bromide, KBr.

☑ **Reading Check**

2. **Infer** Why does metallic bonding happen?

_____

_____

_____

Picture This

3. **Determine** Circle the element in Group 1 that is not an alkali metal. What is the name of that element?

_____

A Organize Information
Make the following Foldable to help you organize information about the differences among the alkali metals, alkaline earth metals, and transition elements.

Alkali metals and their compounds have many uses. All living things need potassium and sodium compounds to stay healthy. Doctors use lithium compounds to regulate bipolar disorder. Cesium and francium are radioactive elements. A **radioactive element** is an element in which the nucleus of the atom breaks down and gives off particles and energy.

The Alkaline Earth Metals

The alkaline earth metals are the elements in Group 2 on the periodic table shown in the figure. Alkaline earth metals have properties similar to most other metals. Like alkali metals, they form ionic bonds easily with other elements and are not found as free elements in nature. Each alkaline earth metal atom has two electrons in its outer energy level.

What are some uses of alkaline earth metals?

Some alkaline earth metals give fireworks their colors. Strontium compounds create bright red flashes and magnesium produces brilliant white sparks. Magnesium is very light and strong, so it is used in cars, airplanes, spacecraft, and even baseball bats. The magnesium compound chlorophyll makes it possible for plants to make food. Calcium carbonate is a common calcium compound found in chalk and limestone.

Are alkaline earth metals used in your body?

Do you take a vitamin with calcium? Calcium is important to your health. Calcium phosphate in your bones helps make them strong. Patients who need X rays of their digestive systems swallow a barium compound. It absorbs X-ray radiation and gives doctors better pictures of the digestive system. Radioactive radium is used in treating cancer.

Transition Elements

Transition elements are the elements in Groups 3 through 12 on the periodic table. They are called transition elements because they are in the middle, or transition, between Groups 1 and 2 and Groups 13 through 18. Titanium bike frames and tungsten lightbulb filaments are some of the things that are made from transition elements. ☑

Picture This

4. **Interpret a Table** How many elements are classified as alkaline earth metals?

_____

_____

✔ Reading Check

5. **Identify** Where are the transition elements located on the periodic table?

_____

_____

_____

Transition elements are often found in nature in their pure, or elemental, form. The elements we think of as being typical metals are transition elements. Iron, copper, and gold are some examples. Transition elements often form colored compounds. Cadmium yellow and cobalt blue paints are made from compounds of transition elements.

What are the uses of iron, cobalt, and nickel?

Iron, cobalt, and nickel are the first elements in Groups 8, 9, and 10 on the periodic table. They are sometimes called the iron triad. These elements are used in the process to create steel and other metal mixtures. Iron is the main ingredient in steel

and is the most often used metal. Iron is the second most common metallic element found in Earth's crust, after aluminum. Some steels contain nickel and cobalt. Nickel is added to some metals to give them strength or a shiny protective coating.

How are copper, silver, and gold used?

The three elements in Group 11 on the periodic table are copper, silver, and gold. These elements are stable, malleable, and found as free elements in nature. They are known as the coinage metals because they were once used to make coins. Gold and silver are too expensive to be used in most coins today, but copper is still used. Copper also is used for electrical wiring. It is inexpensive and conducts electricity very well. Silver compounds are used to make photographic film. Silver and gold have an attractive color, do not corrode or wear away, and are rare, so they often are used to make jewelry.

What are uses for zinc, cadmium, and mercury?

Zinc, cadmium, and mercury are the elements in Group 12 on the periodic table. Zinc and cadmium are often used to coat other metals. Cadmium also is used in rechargeable batteries. Since mercury is a liquid at room tempera-

ture, it is used in thermometers, thermostats, switches, and batteries. Mercury is poisonous and can accumulate in the body.

Picture This

6. Explain Use the periodic table to explain why iron, cobalt, and nickel are not in the same group of elements.

_____

_____

_____

Think it Over

7. Apply Pennies used to be called "coppers." How do you suppose they got this nickname?

_____

_____

_____

_____

Picture This

8. Infer On the periodic table to the left, highlight the group that contains the coinage metals. What are the atomic numbers of copper, silver, and gold?

_____

Lanthanide series   Actinide series

Copyright © Glencoe/McGraw-Hill, a division of The McGraw-Hill Companies, Inc.

Picture This

9. Locate On the periodic table, highlight the lanthanides in one color. Highlight the actinides in a second color. Make the key match your colors.

✔ Reading Check

10. Determine How are metals obtained from ores?

_____

_____

_____

_____

The Inner Transition Metals

Two rows of elements seem to be apart from the rest on the periodic table shown on the figure. These are the inner transition elements, listed below the main periodic table to save room.

Lanthanides The first row of inner transition elements are called lanthanides (LAN thuh nidez) because they follow lanthanum on the table. Europium, gadolinium, and terbium compounds are used to produce the colors on a TV screen.

Actinides In the second row are the actinides (AK tuh nidez) because they follow actinium. All actinides are radioactive and unstable. Thorium and uranium are the only actinides found in Earth's crust in usable quantities. Thorium is used to make high-quality glass for camera lenses. Uranium is best known for its use in nuclear reactors and weapons.

Metals in the Crust

Earth's crust contains a few pure metals, such as gold and copper. The rest of the metals are combined with other elements in ores. Most ores are made up of a metal, or mineral, mixed with clay and rock. Ores are mined from Earth's crust. Then the mineral is separated from the clay and rock. The mineral is then changed, often with heat, to another physical form. The metal is refined into a pure form. This process can be expensive. Sometimes it costs more to separate a metal from the waste rock than the metal is worth. When this is true, the mineral is no longer called an ore. ✔

● After You Read

Mini Glossary

ductile: the ability to be stretched into a wire

malleable: the ability to be hammered or rolled into a sheet

metal: an element that is a good conductor of heat and electricity and is usually a solid at room temperature

metallic bonding: occurs because a cloud of electrons moves freely among positively charged metallic ions

radioactive element: an element in which the nucleus of the atom breaks down and gives off particles and energy

transition element: one of the elements in Groups 3 through 12 on the periodic table

1. Review the terms and their definitions in the Mini Glossary. Write a sentence that shows your understanding of one of the properties of metals.

_____

_____

2. In the graphic organizer below, list characteristics that are common to all metals. Then list characteristics of the three types of metals. Give examples of each type of metal.

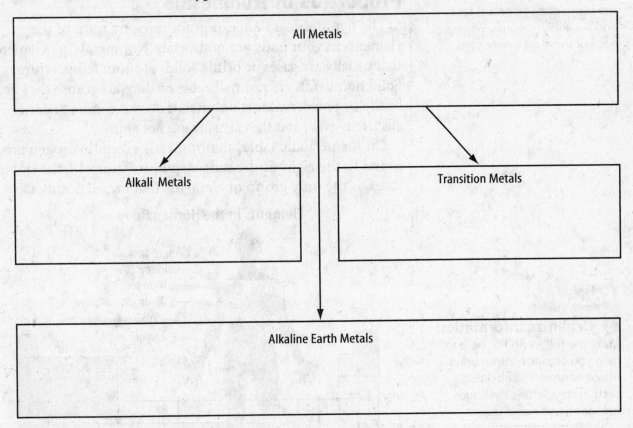

All Metals

Alkali Metals

Transition Metals

Alkaline Earth Metals

End of Section

Elements and Their Properties

chapter 17

section ❷ Nonmetals

What You'll Learn

- recognize hydrogen as a nonmetal
- the similarities and differences between the halogens
- the properties and uses of noble gases

Study Coach

Make Flash Cards As you read each paragraph, identify the main idea, important facts, and vocabulary words. Write each of these on a note card.

FOLDABLES™

B Organize Information Make the following Foldable to help you organize information about nonmetals. Label the sections properties, hydrogen, halogens, and noble gases.

● Before You Read

Think of an object that is metal. Now think of an object that is not metal. What are some differences between these objects? How do they look different? How do they feel different?

_____

_____

● Read to Learn

Properties of Nonmetals

The figure shows that, except for calcium, most of the elements in your body are nonmetals. **Nonmetals** are elements that usually are gases or brittle solids at room temperature. Solid nonmetals are not malleable or ductile because they are brittle or powdery. Most nonmetals do not conduct heat or electricity well, and they usually are not shiny.

On the periodic table, all nonmetals except hydrogen are found to the right of the stair-step line. Group 18, the noble gases, is the only group of elements that are all nonmetals.

Elements in the Human Body

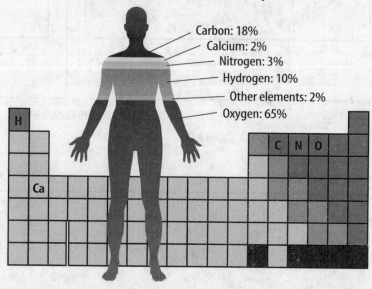

Carbon: 18%
Calcium: 2%
Nitrogen: 3%
Hydrogen: 10%
Other elements: 2%
Oxygen: 65%

How do nonmetals bond?

Most nonmetals can form ionic and covalent compounds. When nonmetals gain electrons from metal atoms, the nonmetals become the negative ions in ionic compounds. One example is sodium chloride (NaCl), or table salt, combined from the metal sodium and the nonmetal chlorine.

Many covalent compounds are made up of nonmetals. When nonmetals bond with other nonmetals, the atoms usually share electrons to form covalent compounds. Ammonia, NH_3, is a covalent compound. Ammonia produces a strong odor that you may have noticed in some household cleaning products.

Hydrogen

About 90 percent of all the atoms in the universe are hydrogen atoms. Most hydrogen on Earth is found in the compound water. As an element, hydrogen is a gas that is made up of diatomic molecules. A **diatomic molecule** consists of two atoms of the same element in a covalent bond. ☑

Hydrogen is very reactive. A hydrogen atom has a single electron. It shares this electron when it combines with other non-metals, such as oxygen to form water, H_2O. Hydrogen also can gain an electron when it combines with alkali and alkaline earth metals. These compounds are called hydrides. An example is sodium hydride, NaH.

The Halogens

The Group 17 elements are called halogens. They are very reactive and their compounds have many uses. Halogen atoms have seven electrons in their outer energy level. Only one electron is needed to fill the outer level. If a halogen gains an electron from a metal, an ionic compound, called a salt, is formed. An example is NaCl. Halogens also can share electrons to form covalent compounds that can be identified by their colors. Chlorine is greenish yellow, bromine is reddish orange, and iodine is violet, or purple.

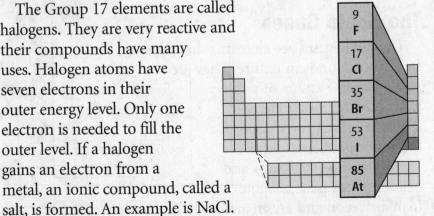

Copyright © Glencoe/McGraw-Hill, a division of The McGraw-Hill Companies, Inc.

☑ **Reading Check**

1. **Identify** About what percent of the atoms in the universe are hydrogen?

_____

_____

Picture This

2. **Explain** In which group is hydrogen located on the periodic table?

_____

_____

💡 Think it Over

3. **Conclude** In a salt, what charge does the halogen element have when it gains an electron?

_____

_____

What are some uses of halogens?

Fluorine The halogen fluorine is the most chemically active of all the nonmetals. Hydrofluoric acid, a mixture of hydrogen fluoride and water, is used to etch glass and frost the inner surface of lightbulbs. It is also used when making semiconductors. Fluorine compounds are added to toothpaste and drinking water to prevent tooth decay.

Chlorine The most common halogen is chlorine. It is obtained from seawater. Chlorine compounds are added to swimming pools and drinking water to disinfect the water. Many bleaches also contain chlorine compounds. Chlorine bleaches are used to whiten clothing, flour, and paper.

Bromine The only nonmetal compound that is liquid at room temperature is bromine. Bromine also is produced from compounds found in seawater. Bromine compounds are used as dyes and in cosmetics.

Iodine The halogen iodine is a shiny solid at room temperature. When heated, iodine changes to a vapor without first becoming a liquid. The process of a solid changing directly to a vapor without first forming a liquid is called sublimation. ✔

Iodine is an important part of your diet. Without it, your thyroid gland cannot produce the hormone thyroxin. If you do not get enough iodine in your diet, your thyroid gland will enlarge due to a condition called goiter. To prevent goiter, potassium iodide is often added to table salt.

Astatine The last element in Group 17 is astatine. It is rare and radioactive. Astatine has properties similar to those of the other halogens. Because it is so rare, it has no known uses.

The Noble Gases

The noble gases are elements that do not form compounds in nature. They are stable because atoms of noble gases have full outer energy levels. Because they are stable, noble gases are useful. Helium is used in blimps and balloons because it is lighter than air. Neon and argon are used in "neon" lights. Argon and krypton are used in lightbulbs and in some lasers.

✔ **Reading Check**

4. Define What is sublimation?

_____

_____

_____

Picture This

5. Infer What can you tell about the noble gases from their position on the periodic table?

_____

_____

2 He
10 Ne
18 Ar
36 Kr
54 Xe
86 Rn

● After You Read

Mini Glossary

diatomic molecule: a molecule consisting of two atoms of the same element in a covalent bond

nonmetals: elements that usually are gases or brittle solids at room temperature

1. Review the terms and their definitions in the Mini Glossary. In the space below, draw a diagram that will help you remember the definition of a diatomic molecule.

2. Complete the graphic organizer below to organize properties of nonmetals.

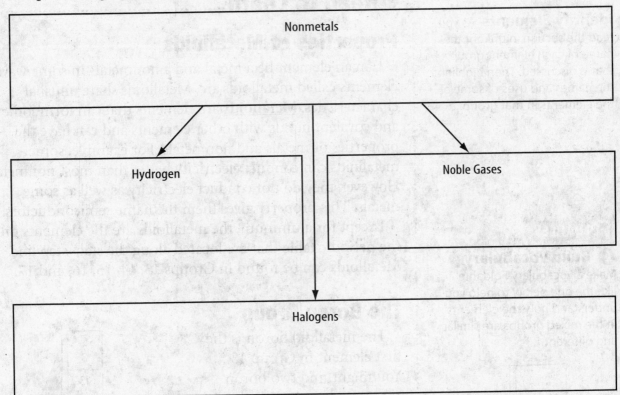

Nonmetals

Hydrogen

Noble Gases

Halogens

End of Section

Elements and Their Properties

section ❸ Mixed Groups

What You'll Learn

- the differences between metals, nonmetals, and metalloids
- what allotropes are
- the different crystal structures of carbon
- why synthetic elements are important

Mark the Text

Identify Elements As you read the section, highlight the name of each element group that is discussed. Then highlight the names and uses of some of the elements in that group.

FOLDABLES

C Build Vocabulary
Make a vocabulary Foldable like the one below to help you understand how the elements in the mixed groups are similar and different.

Metalloids

Boron Group

Carbon Group

Nitrogen Group

Oxygen Group

Synthetic Elements

● Before You Read

Have you ever seen a mixed-breed dog? What characteristics of different breeds did the dog have?

_____

_____

_____

● Read to Learn

Properties of Metalloids

Can an element be a metal and a nonmetal? In some ways, elements called metalloids are. Metalloids share unusual characteristics. **Metalloids** are elements that can form ionic and covalent bonds with other elements and can have the properties of metals and nonmetals. For example, some metalloids can conduct electricity better than most nonmetals. However, they do not conduct electricity as well as some metals. This property gives them the name semiconductors.

Except for aluminum, the metalloids are the elements on the periodic table that are located along the stair-step line. Metalloids can be found in Groups 13, 14, 15, 16, and 17.

The Boron Group

The metalloid boron is the first element in Group 13. You might find two boron compounds in your house. The first is borax. Borax is added to laundry detergents to soften water. The other is boric acid, an antiseptic. Compounds called boranes are used for jet and rocket fuel.

Aluminum is a metal in Group 13. It is the most common metal in Earth's crust. You have seen aluminum in soft drink cans, foil wrap, and cooking pans. Aluminum is strong and light and is used in making airplanes.

The Carbon Group

Atoms of the elements in Group 14 have four electrons in their outer energy levels. Other than this, the elements in the carbon group are very different. Carbon is a nonmetal, silicon and germanium are metalloids, and tin and lead are metals.

Picture This

1. **Identify** Circle the metalloids in the figure.

Carbon Carbon occurs as an element in coal. Compounds of carbon make up oil and natural gas. Carbon can combine with oxygen to produce carbon dioxide, CO_2. Plants use CO_2 and sunlight to make food. All organic compounds contain carbon, but not all carbon compounds are organic.

Silicon and Germanium The metalloid silicon is the second most common element in Earth's crust. Most silicon is found in sand (SiO_2). It also is found in almost all rocks and soil. <u>Allotropes</u> are different forms of the same element that have different molecular structures. Molecular structure is how the atoms in a molecule are arranged.

Silicon has two allotropes. One allotrope of silicon is a hard, gray substance. The other is a brown powder. Because it is a semiconductor, silicon is used in many electronic devices, such as transistors and computer chips. A <u>semiconductor</u> is an element that conducts electric current under certain conditions. Germanium is the other metalloid in the carbon group. It is used with silicon to make semiconductors. ☑

Tin and Lead Tin is used to coat other metals to prevent corrosion. It also is mixed with other metals to make bronze and pewter. Tin cans are made of steel coated with tin. Lead is a soft metal that was once used to make paints. Lead is no longer used in paint because it is toxic.

What are the allotropes of carbon?

What do a diamond ring and a pencil have in common? The diamond in a ring and the graphite in a pencil are carbon. There are three known allotropes of carbon. They are diamond, graphite, and buckminsterfullerene (BUK mihn stur ful ur een).

☑ **Reading Check**

2. **Explain** What are semiconductors?

_____

_____

Copyright © Glencoe/McGraw-Hill, a division of The McGraw-Hill Companies, Inc.

3. **Apply** Why do you think some saw blades are coated in diamond powder?

_____

_____

Diamonds Diamonds are clear and extremely hard. In fact, they are the hardest things in the world. In a diamond, each carbon atom is bonded to four other carbon atoms. The bonded carbon atoms form a geometric shape called a tetrahedron. Many of these tetrahedrons join together to form a giant molecule. In this molecule, the atoms are held tightly together in a strong crystalline structure. This strucure is why diamonds are so hard.

Graphite Graphite is a black powder that is made of hexagonal layers of carbon atoms. In the hexagons, each carbon atom is bonded to three other carbon atoms. The fourth electron in each atom is bonded weakly to the layer above or below it. These weak bonds let the layers of carbon atoms slide easily past each other. This makes graphite very slippery.

Buckminsterfullerene Buckminsterfullerene was discovered in the 1980s. It forms molecules that are shaped like soccer balls. It is named after R. Buckminster Fuller, who designed structures with similar shapes. Scientists have used buckministerfullerene to make tiny tubes called nanotubes. They are one-billionth of a meter in diameter. You can stack tens of thousands of nanotubes to get the thickness of one sheet of paper. Nanotubes might be used some day to make smaller, faster computers.

Applying Math

4. **Use Percentages** Assume you inhale in about 6 L of air per minute. How much nitrogen do you breathe in one minute?

_____

The Nitrogen Group

Group 15 is known as the nitrogen family. Atoms of each element in the group have five electrons in their outer energy levels. Group 15 elements usually share their electrons in covalent bonds with other elements.

About 80 percent of the air you breathe is nitrogen. In the air, nitrogen exists as diatomic molecules, N_2. All organisms need nitrogen compounds to live. Even though you breathe it, your body cannot use nitrogen in its diatomic form. It must be in the form of a nitrogen compound. Nitrogen often is used to make compounds called nitrates and ammonia, NH_4. Nitrates are compounds that contain the nitrate ion, NO_3. Nitrates and ammonia are important ingredients of fertilizers.

How are elements of the nitrogen group used?

Phosphorus is a nonmetal that has three allotropes. Its compounds are used to make many things, including fertilizers, water softeners, match heads, and fine china. Antimony is a metalloid, and bismuth is a metal. These elements are added to metals to lower their melting points. Automatic fire sprinkler heads sometimes contain bismuth. The metal melts from the heat of a fire and turns on the sprinkler.

The Oxygen Group

Oxygen Group 16 on the periodic table is the oxygen group. Oxygen, a nonmetal, makes up about 21 percent of air. Oxygen exists in the air as a diatomic molecule, O_2. During electrical storms, some oxygen molecules, O_2, change into ozone, O_3. Nearly all living things on Earth need O_2 to live. Living things also depend on a layer of O_3 in the atmosphere, called the ozone layer. The ozone layer protects living things from the Sun's radiation.

Sulfur The second element in the oxygen group is sulfur. It is a nonmetal and has several allotropes. Sulfur combines with metals to form compounds called sulfides. Some sulfides are colorful and are used as pigments in paints.

Other Elements Selenium, a nonmetal, and tellurium and polonium, metalloids, are the other elements in Group 16. You need a tiny amount of selenium in your body, so it is included in multivitamins. But too much selenium is toxic. Selenium also is used in photocopiers.

Synthetic Elements

Scientists have created elements that usually do not exist on Earth. These are called synthetic elements. To create them, scientists smash existing elements with particles from a heavy ion accelerator. Except for some isotopes of natural elements, all synthetic elements have more than 92 protons. Many synthetic elements disintegrate, or fall apart, soon after they are made. It may seem strange to make elements that fall apart, or disintegrate, but scientists are learning about atoms this way. ☑

Picture This

5. Use a Diagram What is the atomic number of oxygen?

_____

☑ Reading Check

6. Apply What happens to many of the synthetic elements after they are made?

_____

How are synthetic elements used?

Scientists smash protons into uranium to make neptunium, element 93. In about two days, half of neptunium atoms disintegrate. When neptunium atoms disintegrate, they form plutonium, element 94. Plutonium has been produced in nuclear reactors and is used in bombs. Plutonium also can be changed to americium, element 95. You probably have some americium in your home. There is a small amount of the element in smoke detectors. An electric plate in smoke detectors attracts some of the charged americium particles and makes an electric current. A lot of smoke will break the current and set off an alarm.

What are transuranium elements?

Transuranium elements have more than 92 protons, the atomic number of uranium. The transuranium elements are located toward the bottom of the periodic table. Some are in the actinide series. Others are on the bottom row of the main periodic table. All of the transuranium elements are synthetic. They also are unstable and many disintegrate quickly.

Why make elements?

For centuries, scientists discovered new natural elements. Now, all new elements are created in laboratories. When these atoms disintegrate, they are said to be radioactive. Radioactive elements can be useful. For example, technetium has medical uses.

If most transuranium elements break down quickly, why make them? Scientists make new elements to study the forces that hold the nucleus together. In the 1960s, scientists theorized that stable synthetic elements exist. Maybe scientists will one day find a way to make transuranium elements that do not break down. Perhaps a transuranium element will be found that has everyday uses.

Picture This

7. Infer The atomic number of einsteinium is 99. Is einsteinium a transuranium element? Why or why not?

_____

_____

Think it Over

8. Draw Conclusions Why do you think scientists have not discovered new natural elements in many years?

_____

_____

● After You Read

Mini Glossary

allotropes: different forms of the same element that have different molecular structures

metalloids: elements that can form ionic and covalent bonds with other elements and can have the properties of metals and nonmetals

semiconductor: an element that can conduct electricity under certain conditions

transuranium element: an element with more than 92 protons

1. Review the terms and their definitions in the Mini Glossary. Write a sentence that tells what semiconductors are used for.

   _____

   _____

2. Use the graphic organizer below to list the properties of metalloids. Then list examples of some important metalloids from each group.

Metalloids

Boron Group

Oxygen Group

Carbon Group

Nitrogen Group

3. **Mark the Text** As you read the section, you highlighted the element groups that were discussed and the names and uses of some of the elements in each group. Was this a good strategy for learning the information? Why or why not?

   _____

   _____

   _____

End of Section

Chemical Bonds

section ❶ Stability in Bonding

What You'll Learn

- about elements in a compound
- chemical formulas
- how electric forces help form compounds
- why a chemical bond forms

Study Coach

Make Flash Cards As you read, make a flash card of each element mentioned in the text. Include the name of the element, its chemical symbol, and its electron dot diagram.

FOLDABLES

Ⓐ **Find Main Ideas** Make a Foldable like the one below. As you read this section, write down the main ideas about combined elements, formulas, and atomic stability.

Combined Elements

Formulas

Atomic Stability

● Before You Read

Have you ever seen an object with rust on it? What was the object made of? Why do you think rust formed on it?

_____

_____

_____

● Read to Learn

Combined Elements

Have you ever noticed that the Statue of Liberty is green? Was it painted green? Is it made of green stone? Actually, the Statue of Liberty is made of the metal copper. Copper is an element. Pennies are also made of copper. If copper usually has a shiny, metallic color, then why is the Statue of Liberty green?

Do compounds look like their elements?

Some matter around you is in the form of elements. For example, oxygen, copper, and sulfur are all elements. All three exist by themselves. They can also combine with other elements in a chemical reaction to form compounds.

When copper, sulfur, and oxygen combine in a chemical reaction, they produce the green coating like on the Statue of Liberty. This compound is copper sulfate, which is a green solid. It is nothing like the elements that combined to make it. Copper is a shiny, copper-colored solid. Sulfur is a yellow solid. Oxygen is a colorless, odorless gas. When elements combine, they form compounds with their own properties.

Do compounds have new properties?

Compounds may have different properties from the elements that form them. Sodium chloride, or table salt, is a compound made from the elements sodium and chlorine. Sodium is a shiny, soft, silvery metal that reacts violently with water. Chlorine is a greenish-yellow, poisonous gas. These two elements combine to form the salt that people use in their food.

Formulas

The formula for sodium chloride is NaCl. Na is the chemical symbol for the element sodium. Cl is the chemical symbol for the element chlorine. Written together, they make up the formula for sodium chloride. A **chemical formula** tells what elements are in a compound and how many atoms of each element are in one unit of the compound. ☑

What is the formula for water?

Let's look at the formula for a compound you use every day. H_2O is the chemical formula for water. H is the symbol for the element hydrogen. O is the symbol for the element oxygen. The number 2 in the formula is called a subscript. Subscript means "written below." A subscript written after a symbol tells how many atoms of that element are in one unit of the compound. In H_2O, the 2 tells you there are two atoms of hydrogen in one unit of water. If there is no subscript after a symbol in a formula, there is only one atom of that element in the compound. So, in one unit of H_2O there are two hydrogen atoms and one oxygen atom.

The table shows some familiar compounds and their formulas. Notice that each kind of atom can be combined with other atoms in many different kinds of compounds. For instance, oxygen is a part of the compounds of sand, milk of magnesia, cane sugar, and vinegar.

Some Familiar Compounds		
Familiar Name	**Chemical Name**	**Formula**
Sand	Silicon dioxide	SiO_2
Milk of magnesia	Magnesium hydroxide	$Mg(OH)_2$
Cane sugar	Sucrose	$C_{12}H_{22}O_{11}$
Vinegar	Acetic acid	CH_3COOH

Atomic Stability

Recall that protons have a positive charge and electrons have a negative charge. These opposite electric forces attract each other. They are the forces that hold atoms and molecules together. The kinds of electric forces that hold atoms together also bring atoms together to form compounds.

✔ **Reading Check**

1. **Explain** What does a chemical formula tell you?

_____

_____

_____

Picture This

2. **Describe** The formula for cane sugar, or sucrose, is $C_{12}H_{22}O_{11}$. Describe what one unit of cane sugar is made of.

_____

_____

_____

Copyright © Glencoe/McGraw-Hill, a division of The McGraw-Hill Companies, Inc.

Picture This

3. Interpret Scientific Illustrations How many electrons are in the outer energy level of an atom of carbon?

Why do atoms form compounds?

Look at the periodic table on the inside back cover of this book. It lists all the known elements. Most of these elements can combine with others to form compounds that are more stable. Notice the six elements in Group 18. These elements are gases called the noble gases. Atoms of the noble gases are very stable. They are different from the other elements because they almost never combine to form compounds. Compounds that are formed with a noble gas are less stable than the original noble gas atom.

Why are the noble gases stable?

A helpful way to picture the stability of the noble gases is to look at the electron dot diagrams of those elements. An electron dot diagram shows the symbol of the element. It also shows the electrons in the outer energy level of an atom. The number of electrons in the outer energy level of an atom determines if that atom will combine to form a compound.

How do you know how many dots to put in an electron dot diagram? For elements in Groups 1 and 2 and 13 through 18, you can use the periodic table. The above shows two rows of the periodic table. Look at the outer energy level of the elements in Group 1. All elements in Group 1 have one outer electron. The elements in Group 2 have two outer electrons. Group 13 elements have three outer electrons, Group 14 elements have four, and so on. The noble gases in Group 18 have eight outer electrons.

What makes an atom stable?

An atom is chemically stable when its outermost energy level has all the electrons it can hold. If an atom is chemically stable, it does not easily form compounds with other atoms.

Stable Noble Gases The figure to the right shows electron dot diagrams of the noble gases. The first electron dot diagram is for helium. Notice that there are two electrons around the symbol for helium, He. Hydrogen and helium need only two electrons in their outer level to be stable. The outer energy levels of all the other elements are stable when they contain eight electrons.

Now look at the electron dot diagram for neon, Ne. There are eight electrons in neon's outer energy level. Neon is stable as an atom. Neon does not become more stable if it forms a compound. The noble gases are stable because they already have the maximum number of electrons in their outer energy levels. ☑

Picture This

4. Describe How many electrons are in the outer energy level of xenon?

_____

_____

What elements have incomplete outer energy levels?

Hydrogen and helium are the only elements in the first row, or period, of the periodic table. Both hydrogen and helium need only two outer electrons to be stable. Helium has two electrons in its outer energy level and is stable. But hydrogen has only one electron. Its outer level is not full. Therefore, hydrogen is more stable when it is part of a compound.

Look at the periodic table again. You can see that none of the elements in Group 13 through Group 17 have full outer energy levels. These elements are more stable when they form compounds.

✔ **Reading Check**

5. Infer What makes noble gases stable?

_____

_____

How do atoms become more stable?

As you have learned, hydrogen is an element that does not have a full outer energy level. How does hydrogen, or any other element become stable? Atoms that do not have a stable outer energy level can do one of three things to complete their outer level. They can gain electrons, lose electrons, or share electrons to make a full outer level. Atoms combine with other atoms that also do not have complete outer levels. This way, each atom becomes more stable.

The figure on the next page shows electron dot diagrams for sodium and chlorine. Sodium has one outer electron and chlorine has seven. When these two elements combine, sodium gives its outer electron to chlorine. Chlorine now has eight electrons in its outer level. This is a full, stable energy level. But what about sodium?

Picture This

6. Use a Model Draw an electron dot diagram of sodium showing its new outer energy level after it has given its outer electron to chlorine.

Sodium Combines with Chlorine to form Sodium Chloride

Na· → ·C̈l:

Sodium   Chlorine

How can sodium become more stable?

Sodium gave up the only electron in its outer level when it combined with chlorine to form sodium chloride. If sodium loses the electron in the outer energy level, what is left? The next energy level of the sodium atom has eight electrons. When the one outer electron is removed, the next level becomes the new outer level. Sodium now has a complete, stable outer energy level. Sodium and chlorine exchanged an electron and are now both stable in the compound they formed.

What is another way atoms can become stable?

Remember that the formula for the compound water is H_2O. Hydrogen atoms need one electron in their outer level to be stable. Oxygen atoms need two electrons to be stable. In this case, neither atom can give up an electron. Instead, they share electrons. Each hydrogen atom shares one electron from an oxygen atom to complete its outer level. In turn, the oxygen atom shares each of the two hydrogen atoms' electrons to complete its outer level. The compound water is more stable than a hydrogen atom or an oxygen atom.

Atoms gain and lose electrons to become more stable. Recall that electrons have a negative charge and protons have a positive charge. Positive charges attract negative charges like the opposite ends of a magnet. The force that holds atoms together is called a **chemical bond**. ☑

✔ Reading Check

7. Explain What happens to an atom's charges when it gains or loses an electron?

_____

● After You Read

Mini Glossary

chemical bond: the force that holds atoms together in a compound

chemical formula: a formula that tells what elements and how many atoms of each element are in a compound

1. Review the terms and their definitions in the Mini Glossary. Find one chemical formula you learned about in this section and explain the parts of the formula.

   _____

   _____

   _____

2. Complete the concept web to organize the information from this section.

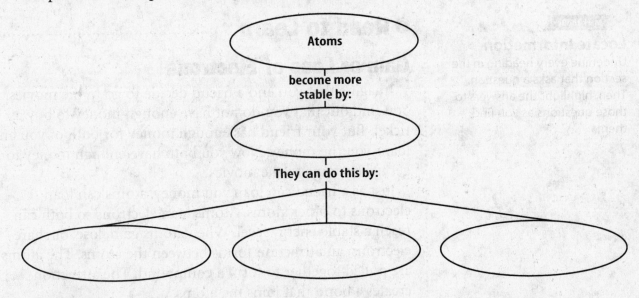

3. Study Coach As you read this section, you made flash cards for the elements mentioned in the text. Do you think that making flash cards would be a good way to learn the symbols of all the elements in the periodic table? Why or why not?

   _____

   _____

   _____

End of Section

section ❷ Types of Bonds

What You'll Learn

- what ionic bonds and covalent bonds are
- what particles are produced by ionic and covalent bonding
- about polar and nonpolar covalent bonds

▶ Mark the Text

Locate Information
Underline every heading in the section that asks a question. Then, highlight the answers to those questions as you find them.

☑ **Reading Check**

1. **Explain** What is a compound?

_____

_____

● Before You Read

Some atoms share electrons and become more stable. Describe a situation in which people share something and everyone benefits.

_____

_____

● Read to Learn

Gain or Loss of Electrons

Imagine that you and a friend decide to go to the movies. You find out that you do not have enough money to buy a ticket. But your friend has enough money for both of you and loans you the money. Now you both have enough money to get tickets to go to the movies.

Just as a friend can loan you money, atoms can loan electrons to other atoms. Atoms loan electrons so both can reach a stable energy level. When atoms gain, lose, or share electrons, an attraction forms between the atoms. The atoms are pulled together to form a compound. The attraction creates a bond that joins the atoms.

A compound is completely different from the atoms that make it up. A compound has different physical and chemical properties from the atoms that are in it. A compound is two or more atoms bonded together. ☑

What are some common compounds?

Ions form when atoms gain or lose electrons and become changed particles with more or less electrons than protons. Many common compounds are formed by the loss or gain of just one electron. These compounds are made from a Group 1 element, such as sodium, and a Group 17 element, such as chlorine. Examples are sodium chloride, also known as table salt, and potassium iodide, an ingredient in iodized salt.

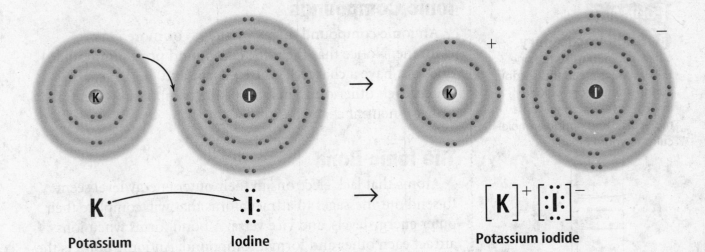

Potassium   **Iodine**   **Potassium iodide**

How is potassium iodide formed?

What happens when potassium and iodine bond to form potassium iodide? The figure above shows the electron dot diagrams and electron distribution for this process.

An atom of potassium has one electron in its outer level. This is not a stable outer energy level. When a potassium atom combines with an iodine atom, the potassium atom loses the one electron in its outer level. The third energy level now becomes a complete outer level. The potassium atom now has one less electron than it has protons. The positive and negative charges are no longer equal. Recall than an ion is a charged particle that is formed when an atom gains or loses electrons. The potassium atom is now a positive ion. It has more protons with positive charges than electrons with negative charges.

The potassium ion has a 1+ charge. The symbol for a positive potassium ion is K^+. The plus sign shows its positive charge. In the symbol K^+, the plus sign is a superscript. *Superscript* means "written above."

How does the iodine atom change?

The iodine atom also changes when it combines to form potassium iodide. An iodine atom has seven electrons in its outer energy level. A stable outer energy level has eight electrons. When the iodine atom reacts with the potassium atom, the iodine atom gains one electron from potassium. There are now eight electrons in iodine's outer energy level.

The iodine atom now has a stable outer energy level, but it has one more electron than it has protons. It has more negative particles than positive particles. The iodine atom has become a negative ion with a charge of 1−. Now it is called an iodide ion and its symbol is written as I^-.

Picture This

2. Describe How many electrons are there in potassium's outer energy level *after* the bond has formed?

_____

Picture This

3. Think Critically Explain why magnesium chloride cannot be made from one atom of magnesium and one atom of chlorine.

_____

_____

_____

Ionic Compounds

An ionic compound forms when two or more ions combine. Notice that the ionic compound potassium iodide does not have a charge. The 1+ charge and the 1− charge cancel each other out. This means the compound is neutral or has a neutral charge.

The Ionic Bond

Atoms that lack electrons in their outer energy level seem to send out messages to attract atoms that will complete their outer energy levels, and vice versa. A bond forms when ions attract each other and form a compound. An **ionic bond** is the force of attraction between a positive ion and a negative ion in an ionic compound.

In an ionic compound, one atom gives up electrons and the other atom takes them. Electrons are not shared in ionic compounds. When an atom loses electrons, one or more atoms must gain the same number of electrons that were lost. That way, the compound stays neutral. When atoms form an ionic compound, a large amount of energy is released. This occurs because of the transfer of electrons.

What if many electrons are transferred?

Potassium iodide forms an ionic bond when one electron is transferred. What happens when more than one electron is transferred? The figure below shows the formation of another ionic compound, magnesium chloride, $MgCl_2$.

When magnesium reacts with chlorine, a magnesium atom loses the two electrons in its outer energy level. The atom becomes a positively charged magnesium ion. The symbol for this ion is Mg^{2+} because it has lost two electrons and now has a charge of 2+.

Two chlorine atoms each take one of the electrons and complete their outer levels. Each chlorine atom becomes an ion with a 1− charge. Cl^- is the symbol for the chloride ion.

In this case, the magnesium atom has two electrons to give. However, a chlorine atom can only use one electron. So, it takes two chlorine atoms to receive the two electrons from the magnesium atom and combine to form the ionic compound magnesium chloride.

Magnesium   2 chlorine atoms   Magnesium chloride

Does an ionic compound have a charge?

The ionic compound magnesium chloride is neutral. It does not have a charge. The compound is neutral because the sum of the charges on the ions is zero. The 2+ charge on the magnesium ion is equal to two 1− charges on the chloride ions. When different atoms form an ionic compound, they transfer their electrons. The total number of electrons and protons stays the same. The compound is neutral. ☑

Metals and nonmetals usually combine by forming ionic compounds. Elements on the far left side of the periodic table tend to form ionic bonds with elements on the far right. Ionic compounds are often crystalline solids with high melting points.

Sharing Electrons

Some atoms of nonmetal elements become more stable when they share electrons. Look at the elements in Group 14 of the periodic table. These elements have four electrons in their outer levels. They would have to gain or lose four electrons to have a stable outer energy level. It takes a lot of energy for an atom to lose or gain that many electrons. Each time an ion loses an electron, the nucleus holds the remaining electrons even more tightly. Once one electron is removed, it takes more energy to remove a second electron. It takes even more energy to remove a third, and so on. For elements in Group 14, it is much easier for them to become stable by sharing electrons.

Atoms that share electrons form a covalent bond. A **covalent bond** is the force of attraction between two atoms that share electrons. A **molecule** is the neutral particle that forms when atoms share electrons.

How are single covalent bonds formed?

A single covalent bond forms when two atoms share two electrons. Usually one electron comes from each atom in the covalent bond. Look at the water molecule in the figure. There are two single covalent bonds in a water molecule. In each single bond, a hydrogen atom and an oxygen atom each give one electron, which the atoms share.

Recall that ionic bonds help atoms become more stable by filling their outer energy levels. Covalent bonds also help atoms fill their outer energy levels and become more stable.

✔ Reading Check

4. **Explain** Why is the ionic compound magnesium chloride neutral?

_____

_____

_____

Picture This

5. **Locate** Circle the electrons that could have belonged only to the hydrogen atoms before they bonded with the oxygen atom.

Picture This

6. Identify Highlight the electrons that are being shared in the diagram of the nitrogen molecule. Circle the electrons that are not being shared.

What are multiple bonds?

A covalent bond can have more than two shared electrons. Look at the dot diagram of the two nitrogen atoms. Each nitrogen atom has five electrons in its outer energy level. A nitrogen atom needs to gain three electrons to have a stable number of eight electrons in its outer energy level.

$$\cdot \ddot{N} \cdot \quad + \quad \cdot \ddot{N} \cdot \quad \rightarrow \quad : N ::: N :$$

When two nitrogen atoms combine, they share three electrons with each other. The bond between the two atoms has six electrons, or three pairs of electrons. Each electron pair is a covalent bond. Three pairs of electrons form a triple bond. By sharing the electrons, each nitrogen atom now has eight electrons in its outer energy level. The symbol for the nitrogen molecule is N_2.

Can double bonds occur?

Molecules also can have double bonds. A double bond is two pairs of electrons shared between two atoms. In the carbon dioxide molecule, CO_2, the carbon atom shares two electrons with one oxygen atom and two electrons with the other oxygen atom. In return, each oxygen atom shares two electrons with the carbon atom. In this way, all three atoms have eight electrons in their outer energy levels. This makes all three atoms stable.

When electrons are shared in covalent bonds by similar or identical atoms, such as in N_2 or O_2 molecules, the electron charge is shared equally across the bond. A **nonpolar bond** is a covalent bond in which electrons are shared equally by both atoms. Nonmetals commonly form nonpolar bonds.

Are electrons always shared equally?

Atoms in a covalent bond do not always share electrons equally. The positive charge of an atom's nucleus attracts the electrons in a bond. Some nuclei attract electrons more strongly than others. If a shared electron is closer to one nucleus in a bond, that nucleus could attract the electron more strongly.

You can see this with a magnet and a piece of metal. When the magnet is closer to the metal, it attracts the metal more strongly. Some nuclei have a greater positive charge than others. Nuclei with a greater positive charge attract electrons more strongly. In the same way, a strong magnet holds the metal more firmly than a weak magnet.

How can you picture the sharing of electrons?

Imagine a game of tug-of-war. A covalent bond is like the rope in a game of tug-of-war. The shared electrons are like the knot in the center of the rope. Look at the diagram of a molecule of HCl (hydrochloric acid) below. Each atom in the molecule of HCl is attracting the shared electrons. However, the atoms are not the same size. Then what happens?

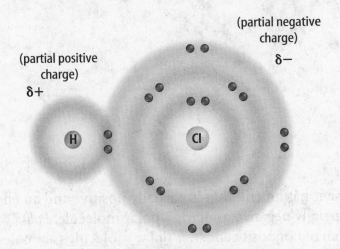

(partial positive charge)
δ+

(partial negative charge)
δ−

H

Cl

The chlorine atom pulls on the electrons more than the hydrogen atom. The chlorine atom is like the stronger team in tug-of-war. When one team is stronger, the knot in the middle of the rope moves towards the stronger team. A similar thing happens in the molecule. The shared electrons are held more closely to the atom with the stronger pull.

What is an example of unequal electron sharing?

The covalent bond in a molecule of hydrogen chloride, HCL, is an example of unequal electron sharing. When HCl mixes with water, it becomes hydrochloric acid. Hydrochloric acid is used to clean metal in factories. It is also found in your stomach where it helps digest food.

Because of its size, the chlorine atom attracts the electrons in the bond more strongly than the hydrogen atom. That means the electrons in the bond spend more time closer to the chlorine atom than to the hydrogen atom. The chlorine atom has a partial negative charge when the bonding electrons are closer to it. The hydrogen atom then has a slight positive charge when the electrons are farther away from it. The bond is said to be polar. A **polar bond** is a covalent bond in which the electrons are unequally shared, producing a molecule with a negatively-charged end and a positively-charged end. ☑

Picture This

7. **Label** Draw an arrow showing the pull of electrons toward the chlorine atom.

✔ Reading Check

8. **Apply** Why does the chlorine atom have a partial negative charge in a molecule of hydrogen chloride?

_____

_____

_____

_____

What are polar and nonpolar molecules?

The figure shows the energy levels of a water molecule. The oxygen atom forms a covalent bond with each hydrogen atom. The oxygen atom has a stronger attraction for these bonding electrons. Two electron pairs spend more time closer to the oxygen atom than to the hydrogen atom. This gives the oxygen atom a partial negative charge.

H_2O
Water

Because it has an end that is partially positive and an end that is partially negative, water is a polar molecule. *Polar* means "having opposite ends." A **polar molecule** is a molecule that has a slightly positive end and a slightly negative end, but the molecule itself is neutral.

When two atoms that are exactly alike form a covalent bond, they share electrons equally. In a **nonpolar molecule**, the electrons are shared equally in the bond. A nonpolar molecule does not have oppositely charged ends. Atoms of the same element can form nonpolar molecules.

Properties of Compounds

You have just learned about ionic bonds and covalent bonds. A covalent compound is one whose atoms are held together by covalent bonds. Sugar is a covalent compound. An ionic compound is one that is held together by ionic bonds. Table salt, NaCl, is an ionic compound.

From the way they look, you might think that sugar and salt are similar compounds. They are both little white grains, and they are both used in foods. They look like they might have the same physical and chemical properties.

Both sugar and salt are used to change the taste of foods. You add sugar to lemonade or tea to make them taste sweet. Table salt is used to enhance the taste of foods such as meat and vegetables. The reason sugar and salt have different effects on the taste of foods is because they have different physical and chemical properties. ☑

Picture This

9. **Identify** Highlight the parts of the water molecule that have a slight positive charge. Circle the part that has a slight negative charge.

✔ **Reading Check**

10. **Explain** Although sugar and salt look alike, why do they have different effects on the taste of food?

_____

_____

_____

_____

How are covalent and ionic compounds different?

Ionic and covalent compounds have different physical and chemical properties. These differences are due to differences in the attractive forces of the bonds.

What are some properties of covalent compounds?

Covalent bonds are strong, but the attraction between individual molecules of a covalent compound is weak. These weak forces between individual molecules are responsible for the properties of covalent compounds. ☑

For example, melting and boiling points of covalent compounds are low, compared to those of ionic compounds. Sugar will melt at 185° C, but table salt will melt at 801° C. Covalent compounds form soft solids. They are poor electrical and thermal conductors. They do not conduct heat and electricity well.

What are some properties of ionic compounds?

The ionic bonds between ions are relatively strong. This is why ionic compounds like table salt have high melting points. They also have relatively high boiling points. Ionic compounds are hard and brittle solids. If ionic compounds are dissolved in water, they become good conductors of electric current.

Strong attractive forces hold the ions in place in an ionic compound. Because of this strong attraction, ionic compounds are stable. When ions are drawn together, energy is released. The table below shows some of the different properties of covalent and ionic compounds.

Table 2 Comparison of Covalent and Ionic Compounds		
	Covalent Compounds	**Ionic Compounds**
Bond Type	Electron Sharing	Electron Transfer
Melting and Boiling Points	Lower	Higher
Electrical Conductivity	Poor	Good
State at Room Temperature	Solid, liquid, or gas	Solid
Forces Between Particles	Strong bonds between atoms; weak attraction between molecules	Strong attraction between positive and negative ions

✔ Reading Check

11. **Explain** Describe the strength of the forces between covalent molecules.

_____

_____

Picture This

12. **Identify** Highlight the information in the table that shows that sugar melts at a lower temperature than salt.

● After You Read

Mini Glossary

chemical bond: the force that holds atoms together in a compound

covalent bond: the force of attraction between two atoms that share electrons

ionic bond: the force of attraction between a positive ion and a negative ion in an ionic compound

molecule: the neutral particle that forms when atoms share electrons

nonpolar bond: a covalent bond where the electrons are shared equally

nonpolar molecule: a molecule in which the electrons are shared equally in a covalent bond

polar bond: a covalent bond where electrons are unequally shared producing charged molecules

polar molecule: a molecule that has a slightly positive end and a slightly negative end, but the molecule itself is neutral

1. Review the terms and definitions in the Mini Glossary. Choose two terms that are related and write a sentence that uses those two terms.

_____

_____

2. Complete the outline to help you organize what you learned about chemical bonds.

> Types of Bonds
>
>   I.   What is an ionic bond?
>
>   A. Forms between a positive and a negative ion
>
>   B. Electrons are _____
>
>   C. Usually forms between a metal and _____
>
>   D. Forms a(n) _____ compound
>
>   II. What is a covalent bond?
>
>   A. _____
>
>   B. Electrons are _____
>
>   C. _____
>
>   D. Forms a _____

3. **Mark the Text** You underlined every heading in the section that asks a question and then highlighted the answers. How did this help you learn the content of this section?

_____

End of Section

Chemical Bonds

section ❸ Writing Formulas and Naming Compounds

● Before You Read

Shakespeare asked, "What's in a name?" In this section, you are going to learn what the names of chemical compounds can tell you. On the lines below, explain what your name tells about you.

_____

_____

● Read to Learn

Binary Ionic Compounds

Alchemists were early scientists who tried to turn lead into gold. They never succeeded, but they did develop some laboratory methods and equipment that scientists still use today. The alchemists also used symbols to write formulas.

The first formulas you will learn to write are for binary ionic compounds. A **binary compound** is a compound made of two elements. An example of a binary compound is potassium iodide, a compound added to table salt.

What are oxidation numbers?

Before you can write the correct formula of a compound, you need to know which elements combine to make that compound. All elements in a certain group of the periodic table have the same number of electrons in their outer energy levels. So, elements in the same group all gain or lose the same number of electrons.

Metals always lose electrons, and nonmetals always gain electrons when they form ions. The **oxidation number** of an atom tells you how many electrons an atom gains, loses, or shares to become stable. It is the same as the charge on the ion.

The charge on the ion in an ionic compound is the same as its oxidation number. For example, a sodium ion has a charge of $1+$ and an oxidation number of $1+$. A chloride ion has a charge of $1-$ and an oxidation number of $1-$.

What You'll Learn
- how to determine oxidation numbers
- how to write formulas and names for ionic compounds
- how to write formulas and names for covalent compounds

Study Coach

Create a Quiz As you read the text under each heading, write a question that your teacher might ask on a quiz. Exchange your questions with a partner and take each other's quizzes.

❸ **Build Vocabulary**
Make two quarter-sheet Foldables as shown below. Record information about the oxidation number of the elements and rules for writing formulas as you read.

| Oxidation Number: | Ionic Compounds: |

Picture This

1. **Identify** Highlight the oxidation numbers on the periodic table.

Picture This

2. **Compare** Circle the Roman numerals and the oxidation numbers for each element in the table.

How are oxidation numbers related to the periodic table?

Look at the periodic table above. Notice the numbers above each column. These are the oxidation numbers for the elements in the column. Notice how the oxidation numbers fit with the periodic table groupings.

What are special ions?

Part of the periodic table is not included in the figure above. Some elements in this section can have more than one oxidation number. The table shows some of these elements and their oxidation numbers. Because these elements can have more than one oxidation number, you must include more information when you name them. When naming these compounds, the oxidation number is expressed in the name as a roman numeral. For example, when iron with an oxidation number of 3+ combines with oxygen, the compound they form is iron(III) oxide.

Some Special Ions	
Name	**Oxidation Number**
Copper(I)	1+
Copper(II)	2+
Iron(II)	2+
Iron(III)	3+

How many positive and negative ions must a formula have?

Remember that ionic compounds are neutral. The ions in an ionic compound have charges, but the compound itself does not. The formula for an ionic compound must have the right number of positive and negative ions to make the charges balance.

For example, sodium chloride has one sodium ion with a charge of $1+$ and one chloride ion with a charge of $1-$. The $1+$ balances the $1-$. The correct formula for sodium chloride is NaCl. The formula tells you that one positively charged ion combines with one negatively charged ion and makes a neutral compound.

What about the ionic compound calcium fluoride? A calcium ion has a charge of $2+$. A fluoride ion has a charge of $1-$. The charge $1-$ does not balance $2+$. You need to have two fluoride ions for every calcium ion to balance the charges. The formula for the neutral compound calcium fluoride, CaF_2, shows exactly that. There is one calcium ion and two fluoride ions in the compound. ☑

How do you find the correct subscripts?

Sometimes you need to use math skills to write a formula correctly. In the compound aluminum oxide, there are aluminum ions and oxygen ions. But, how many of each one? From the periodic table, you can get the information about aluminum and oxygen that you need to write the formula.

By its place on the periodic table, you can tell that aluminum has an oxidation number of $3+$. That means the aluminum ion has a charge of $3+$. Oxygen has an oxidation number of $2-$, so its ion has a charge of $2-$. You must find the least common multiple of 3 and 2 to balance the charges. The least common multiple of 3 and 2 is 6.

Multiply $3+$ by 2 to equal $6+$. You need two aluminum ions to have a charge of $6+$. Multiply $2-$ by 3 to equal $6-$. You need three oxygen ions to have a charge of $6-$. The charges balance. The correct formula for the neutral compound aluminum oxide is Al_2O_3.

How do you write the formula of an ionic compound?

You learned how to use the oxidation number to find the charge on an ion. You also learned that the charges on the ions must balance to form a neutral compound. Now you can write formulas for ionic compounds. Follow rules 1, 2, and 3 on the next page.

☑ Reading Check

3. Infer What does the formula CaF_2 tell you about the compound?

_____

_____

_____

Applying Math

4. Calculate What is the least common multiple of 5 and 2?

_____

Writing Formulas You have learned how to find oxidation numbers and their least common multiples. Now you can write formulas for ionic compounds. What is the formula for an ionic compound containing sodium and oxygen? Use these rules to figure it out:

1. Write the symbol of the element that has the positive oxidation number or charge. Sodium is a Group 1 element. It has an oxidation number of 1+.

2. Write the symbol of the element with the negative oxidation number or charge. Except for hydrogen, all nonmetals have negative oxidation numbers. Oxygen has an oxidation number of 2−.

3. The compound should be neutral. To make it neutral, the positive charges have to balance the negative charges. It takes two sodium ions to balance one oxygen ion. Thus, the formula becomes Na_2O. ☑

Now use these rules to write the formula for lithium nitride. Lithium and nitrogen are the two atoms that make this compound. Look at the periodic table in the back of this book. Lithium is in Group 1, so it forms ions with a 1+ charge. Write the symbol for lithium, Li, first.

Find the oxidation number of nitrogen. Nitrogen is in Group 15. It forms ions with a charge of 3−. You now can write LiN. Can you stop now? Look at the charges of the two ions. Do 1+ and 3− balance? No, you cannot stop yet.

Use the number of the charge of a nitrogen ion as the subscript for Li. Use the number of the charge of a lithium ion as the subscript for N. That gives Li_3N. When an element has no subscript, it means only one ion is in the compound. Do 3(1+) and 3− balance? Yes, this is the correct formula.

How do you name a binary ionic compound?

When you know the formula, you can write the name of a binary ionic compound by following these rules:

1. Write the name of the positive ion.

2. Look to see if the positive ion is listed in the Special Ions table on 404. If it is not in the table, go right to Step 3. If it is, the ion can have more than one oxidation number. To find the correct oxidation number, look at the formula of the compound. The charge of the compound is always zero. The negative ion can only have one possible charge. From the table, pick the charge of the positive ion that balances the negative charge. Write that positive ion's symbol along with the correct roman numeral in parentheses.

✔ **Reading Check**

5. Restate What will a compound be when its positive and negative charges balance?

💡 **Think it Over**

6. Apply Write the formula for calcium chloride. Use the periodic table to identify the positive and negative ions.

3. Write the root name of the negative ion. The root is the first part of the element's name. For example, the root name of chlorine is *chlor-*. The root name of oxygen is *ox-*. ☑

4. Add the ending *-ide* to the root name. For example, write *oxide*.

Do not use subscripts in the name of an ionic compound. Subscripts can be used in the formula to figure out the charge on a metal ion with more than one positive charge.

Table 4 Elements in Binary Compounds	
Element	*-ide* Name
Oxygen	Oxide
Phosphorus	Phosphide
Nitrogen	Nitride
Sulfur	Sulfide

How do you use these rules?

Use these rules to write the name of the compound CuCl. Find the name of the positive ion on the periodic table. Cu is the symbol for copper. Is copper in the Special Ions table on 404? Yes, and the copper ion can have a 1+ or a 2+ charge. To find the charge on the copper in CuCl, look at the negative ion. Cl is the symbol for chlorine. Chlorine is in Group 17 of the periodic table. That means it has an oxidation number of 1−. From the formula, you can see that there is only one chloride ion in the compound. To balance a 1− charge on the chloride ion, the copper ion must have a 1+ charge. Now you can write copper(I) as the first part of the name.

Write the root name of the negative ion. The root name for chlorine is *chlor-*. Add *-ide* to the root. That gives you chloride. The correct name of CuCl is copper(I) chloride.

Compounds with Polyatomic Ions

Not all ionic compounds are binary compounds. The formula for baking soda used in cooking is $NaHCO_3$. Baking soda is an ionic compound that is not binary. Compounds like baking soda are made of more than two elements. These compounds have polyatomic ions. A **polyatomic ion** is a charged group of atoms that are bonded together by a covalent bond. The prefix *poly-* means "many" and *polyatomic* means "many atoms." The polyatomic ion in baking soda is the bicarbonate or hydrogen carbonate ion. The symbol of this polyatomic ion is HCO_3^-.

Reading Check

7. **Determine** What is the root name of oxygen?

_____

_____

Think it Over

8. **Apply** Write the name of the compound FeI_2.

_____

_____

Picture This

9. Explain Use the table to determine what elements make up acetate.

_____

_____

Polyatomic Ions		
Charge	**Name**	**Formula**
1+	ammonium	NH_4^+
1−	acetate	$C_2H_3O_2^-$
	chlorate	ClO_3^-
	hydroxide	OH^-
	nitrate	NO_3^-
2−	carbonate	CO_3^{2-}
	sulfate	SO_4^{2-}
3−	phosphate	PO_4^{3-}

How do you name a polyatomic compound?

To write the name of a compound with a polyatomic ion, first write the name of the positive ion. If the positive ion is polyatomic, use the table above to find its name.

Next, write the name of the negative ion. Again, if it is polyatomic, look up its name in the table. What is the name of the compound K_2SO_4? K is the symbol of the positive ion. It is not polyatomic. K is the symbol for potassium. The negative ion is SO_4^{2-}. It is polyatomic. Use the table again. K_2SO_4 is potassium sulfate.

Now try naming the compound $Sr(OH)_2$. You can see that the positive ion is not polyatomic. Find the name for the symbol Sr on the periodic table. It is strontium. The negative ion is polyatomic. From the table of polyatomic ions, you can see that OH^- is named the hydroxide ion. The name of $Sr(OH)_2$ is strontium hydroxide.

How do you write the formula of a polyatomic compound?

To write the formula for a compound with a polyatomic ion, use the rules for writing the formula of a binary compound, but add one more thing. If you need to show more than one polyatomic ion, put parentheses around the formula for the ion before you write the subscript.

Write the formula for barium chlorate. First, write the symbol of the positive ion. The symbol for barium is Ba. Barium is in Group 2, so it forms a 2+ ion.

Applying Math

10. Determine How many 2+ ions does it take to balance the charge on one 4− ion?

_____

Balancing Negative Charges Now write the formula for the negative ion. The table on 408 shows you the formula for the chlorate ion is ClO_3^-. Are the charges on these two ions balanced? No, 2+ does not balance 1−. It takes two chlorate ions to balance the 2+ charge on the barium ion. The formula for barium chlorate is $Ba(ClO_3)_2$.

Compounds with Added Water

Some ionic compounds have water molecules as part of their structure. A **hydrate** is a compound that has water chemically attached to its ions and written into its formula. The word *hydrate* comes from a word that means "water."

What are common hydrates?

When a solution of cobalt chloride evaporates, pink crystals form. The crystals have six water molecules for each unit of cobalt chloride. The formula for this compound is $CoCl_2 \cdot 6H_2O$. The compound's name is cobalt chloride hexahydrate. The prefix *hexa-* means "six," so hexahydrate means "six waters."

You can remove water from these crystals by heating them. The blue compound that results from heating contains no water. It is called anhydrous, meaning "without water." When anhydrous $CoCl_2$ is exposed to water, even water in the air, it absorbs water and becomes pink again.

Plaster of paris becomes a hydrate when water is added. It becomes calcium sulfate dihydrate, which is also known as gypsum. The water added to the powder becomes part of the compound.

Formulas for Hydrates To write a formula for a hydrate, first write the formula for the compound. Then make a dot followed by the number of water molecules. The dot stands for the ratio of compound-to-water molecules. For example, the formula for calcium sulfate dihydrate is $CaSO_4 \cdot 2H_2O$, a compound of calcium sulfate that contains two water molecules.

Naming Binary Covalent Compounds

Covalent compounds form between elements that are nonmetals. Some nonmetals can form many compounds. For example, nitrogen and oxygen can form N_2O, NO, NO_2, and N_2O_5. Using the rules you learned earlier, all these compounds would be called nitrogen oxide. Now you will learn how to give each of these compounds a different name. ☑

Think it Over

11. **Analyze** How many water molecules does the hydrate $LiNO_2 \cdot H_2O$ have?

_____

_____

12. **Identify** What is the element N?

_____

What is the compound NO?

_____

How are prefixes used to name covalent compounds?

The table below lists some Greek prefixes used to name covalent compounds made with the same elements. These prefixes tell how many atoms of each element are in a compound. For example, the compound NO_2 is nitrogen dioxide. The prefix *di-* tells you that there are two oxygen atoms in the compound. N_2O is dinitrogen oxide. The compound has two nitrogen atoms.

The name of the compound N_2O_5 is dinitrogen pentoxide. This name uses two prefixes. There are two nitrogen atoms, so *dinitrogen* is used. There are also five oxygen atoms, so *pentoxide* is used.

What rules apply in naming covalent compounds?

Drop the last vowel of the prefix when the second element of the compound begins with a vowel. In pentoxide, the *a* is dropped from *penta-*.

There is a prefix to use when a compound has only one atom of an element. The prefix is *mono-*. Many times *mono-* is not used. Instead, it is understood that if no prefix is used, there is only one atom of that element in a compound. In some cases, *mono-* is used for emphasis. Carbon monoxide is one example.

Picture This

13. Use Numbers What does the prefix *tetra-* mean?

_____

Prefixes for Covalent Compounds	
Number of Atoms	**Prefix**
1	*mono-*
2	*di-*
3	*tri-*
4	*tetra-*
5	*penta-*
6	*hexa-*
7	*hepta-*
8	*octa-*

● After You Read

Mini Glossary

binary compound: a compound made of two elements

hydrate: a compound that has water chemically attached to its ions and written into its formula

oxidation number: the number that tells how many electrons an atom gains, loses, or shares to become stable

polyatomic ion: a charged group of atoms that are bonded together by a covalent bond

1. Review the terms and definitions in the Mini Glossary. Write a sentence that explains in your own words what a polyatomic ion is.

_____

_____

2. Complete the flow chart with the steps used in writing the name of a binary ionic compound.

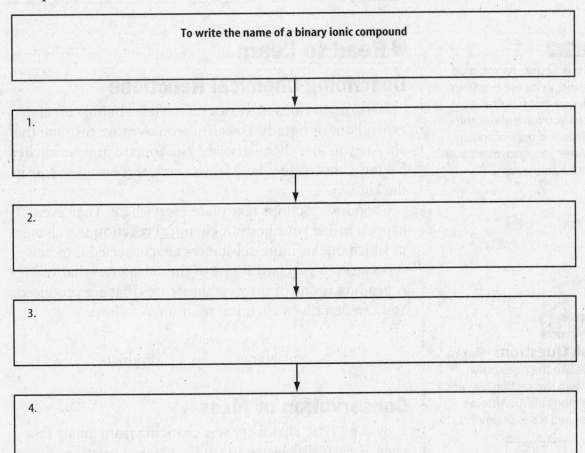

To write the name of a binary ionic compound
1.
2.
3.
4.

End of Section

chapter 19 Chemical Reactions

section ❶ Chemical Changes

What You'll Learn

- how to identify the reactants and products in a chemical reaction
- how a chemical reaction follows the law of conservation of mass
- how chemists describe chemical changes with equations

Study Coach

Create a Quiz As you read this section, write the headings that ask questions. Write questions you have about the main ideas and the vocabulary terms. After you read, make your questions into a quiz.

FOLDABLES™

A **Ask Questions** Make a Foldable like the one shown. As you read the section, answer the questions about Antoine Lavoisier and his experiments.

Who? Antoine Lavoisier
When?
What?
Where?

●Before You Read

Think about what happens when you bake a cake. On the lines below, describe how the ingredients change form throughout the process of baking a cake.

_____

_____

_____

● Read to Learn

Describing Chemical Reactions

Dark, mysterious mixtures react. Gases bubble up and expand out of liquids. Powerful aromas move through the air. Are you in a chemistry lab? No. You are in your kitchen baking a chocolate cake. Many chemical reactions occur in the kitchen.

Chemical reactions take place everywhere. They even happen inside your body. A **chemical reaction** is a change in which one or more substances are converted into new substances. A **reactant** is one of the substances that react. A **product** is one of the new substances that are produced. You can describe a chemical reaction as follows:

$$\text{reactants} \xrightarrow{\text{produce}} \text{products}$$

Conservation of Mass

By the 1770s, chemistry was changing from an art to a science. Scientists began to study chemical reactions more carefully. The French chemist Antoine Lavoisier discovered an important rule. He found that the total mass of the products of a chemical reaction always equals the mass of the reactants. This is called the conservation of mass.

Before burning

After burning

The figures above show an experiment he performed. The mass of the candle and the air in the jar (the reactants) before burning is the same as the mass of the gases and the candle (the products) after burning.

What were Lavoisier's experiments?

Lavoisier wanted to know exactly what happened when substances changed form. To answer this question, he experimented with mercury. He put solid mercury(II) oxide, a red powder, in a sealed container. He found the mass of the reactant in the container. When he heated the container, the mercury(II) oxide changed to a silvery liquid. It also gave off a gas. The silvery liquid was the metal mercury. He then found the mass of the products in the container again. It was the same as the mass before the experiment. ☑

mercury(II) oxide		oxygen	plus	mercury
10g	=	0.3g	+	9.7g

Lavoisier also figured out that the gas produced in the experiment, oxygen, was a part of air. He did this by heating mercury metal with air. He saw that a portion of the air combined with mercury to make mercury(II) oxide. He studied the effect of oxygen on living animals and humans.

Lavoisier did hundreds of experiments in his laboratory. He confirmed that in a chemical reaction, matter is not created or destroyed, but is conserved. This principle is known as the law of conservation of mass. This means that the total starting mass of the reactants of a chemical reaction always equals the total final mass of the products.

Why is Lavoisier called the father of modern chemistry?

Lavoisier's explanation of the law of conservation of mass started modern science. He also was the first to describe a chemical reaction called combustion. These discoveries are why Lavoisier is called the father of modern chemistry.

Picture This

1. **Compare** How does the height of the right side of the scale in the first figure compare to the height of the right side of the scale in the second figure?

_____

_____

☑ Reading Check

2. **Identify** What did Lavoisier find about the mass of the container with reactants and the mass of the container with products in his experiment with mercury(II) oxide?

_____

_____

_____

3. **Draw Conclusions** Imagine that chemists did not use the same rules to name compounds. How might this cause problems for a chemist who was trying to repeat an experiment done by the first chemist?

_____

_____

_____

Picture This

4. **Summarize** What does the symbol (*g*) placed next to a compound in a chemical equation mean?

_____

_____

_____

5. **Explain** What do you notice about the numbers on the left side of the arrow and the numbers on the right side of the arrow in the chemical equation?

_____

_____

Why are names important?

Scientists needed better ways to describe their ideas. Lavoisier wanted to improve the way elements and compounds were named. He knew that if all chemists used the same names for elements and compounds, they could understand one another better. In 1787, Lavoisier and several other scientists wrote the first instructions for naming compounds. Since then, the guidelines have continued to evolve. In 1919, an organization was formed to coordinate guidelines for naming compounds. It is called the International Union of Pure and Applied Chemistry (IUPAC).

Writing Equations

It is important to include all the information when you describe a chemical reaction. What were the reactants? What did you do with them? What happened when they reacted? What were the products? When you answer all these questions, the description of the reaction can be quite long.

Scientists have a shortcut for describing chemical reactions. A **chemical equation** is a way to describe a chemical reaction using chemical formulas and other symbols. Some of the symbols used in chemical equations are shown in the table.

Symbols Used in Chemical Equations			
Symbol	**Meaning**	**Symbol**	**Meaning**
→	produces or forms	(*aq*)	Aqueous; a substance is dissolved in water.
+	plus	heat →	The reactants are heated.
(*s*)	solid	light →	The reactants are exposed to light.
(*l*)	liquid	elec. →	An electric current is applied to the reactants.
(*g*)	gas		

Look at this description of a chemical reaction:

Nickel(II) chloride, dissolved in water, plus sodium hydroxide, dissolved in water, produces solid nickel(II) hydroxide plus sodium chloride, dissolved in water.

If you use a chemical equation, the same description is shorter and easier to understand as:

$$NiCl_2(aq) + 2NaOH(aq) \rightarrow Ni(OH)_2(s) + 2NaCl(aq)$$

Coefficients

Look again at the chemical equation on the previous page. What do the numbers to the left of NaOH and NaCl mean? Do you remember the law of conservation of mass? Matter is not made or lost in a chemical reaction. Atoms are rearranged, but they are never created or destroyed. The numbers in the equation are called coefficients. A **coefficient** shows the number of units of a substance taking part in a reaction. ☑

Suppose you were going to make sandwiches for a picnic. You know that each sandwich needs two slices of bread, one slice of turkey, one slice of cheese, two slices of tomato, and one leaf of lettuce. If you also know how many sandwiches you need to make, you can figure out how much bread, turkey, cheese, tomato, and lettuce you need to buy so you do not have any food left over.

Making sandwiches is like a chemical reaction. The ingredients for the sandwiches are the reactants. The finished sandwiches are the products. The number of units of bread, turkey, cheese, tomato, and lettuce are the coefficients of the reactants. The number of finished sandwiches is the coefficient of the product. However, the quantity of each ingredient is the same in the reactants and product.

How do chemists use coefficients?

When chemists know the number of units of each reactant, they are able to add the correct amounts of reactants needed for a reaction. The units or coefficients will tell how much product will form. For example, here is the chemical equation from the example on the previous page.

$$NiCl_2(aq) + 2NaOH(aq) \rightarrow Ni(OH)_2(s) + 2NaCl(aq)$$

You can see that one unit of $NiCl_2$ and two units of NaOH produce one unit of $Ni(OH)_2$ and two units of NaCl. The figure below shows you how the coefficients affect the number of molecules in the reaction.

| $NiCl_2$ | + | 2NaOH | → | $Ni(OH)_2$ | + | 2NaCl |

☑ Reading Check

6. **Explain** What does a coefficient show?

_____

_____

_____

Applying Math

7. **Apply** Suppose $NiCl_2$ reacts with NaOH. For each molecule of $NiCl_2$, how many molecules of NaOH are needed?

_____

Picture This

8. **Observe** What does the 2NaOH represent?

_____

_____

Balancing Equations

The equation below is for Lavoisier's mercury(II) oxide reaction.

$$HgO(s) \xrightarrow{heat} Hg(l) + O_2(g)$$

How many atoms of mercury (Hg) are on each side of the equation? There is one mercury (Hg) atom on the reactant side and one mercury (Hg) atom on the product side. How many atoms of oxygen (O) are on each side? Notice that there is one oxygen (O) atom on the reactant side, but the product side has two oxygen (O) atoms.

Atoms	HgO	→	Hg	+	O₂
Hg	1		1		
O	1				2

Remember that according to the law of conservation of mass, one oxygen atom cannot become two oxygen atoms. You cannot rewrite HgO as HgO_2. That would make the number of oxygen atoms balance, but HgO and HgO_2 are not the same compound. The formula in a chemical equation must accurately represent the compounds that react.

What does a balanced equation show?

A chemical equation must be balanced. Balancing only changes the way a reaction is represented. It does not change what happens in the reaction. To balance a chemical equation, you change the coefficients. A **balanced chemical equation** has the same number of atoms of each element on each side of the equation.

How do you choose coefficients?

You often can find the coefficients to balance an equation just by guessing and checking your guess. In the mercury(II) oxide equation, the number of mercury atoms is balanced. You need to balance the number of oxygen atoms. Try putting a coefficient of 2 in front of HgO on the left side of the equation. This balances the oxygen, but not the mercury.

Atoms	2HgO	→	Hg	+	O₂
Hg	2		1		
O	2				2

Think it Over

9. **Analyze Results** Why does putting a coefficient of 2 in front of HgO on the left side of the equation balance the oxygen but not the mercury?

_____

_____

_____

_____

To balance the mercury, put a 2 in front of the mercury on the right side of the equation.

Atoms	2HgO	→	2Hg	+	O₂
Hg	2		2		
O	2				2

Now the equation is balanced.

$$2HgO \rightarrow 2Hg + O_2$$

What are the steps for balancing an equation?

Magnesium burns with a very bright light. Have you ever seen a flare burning at the scene of a traffic accident? The flare probably was made of magnesium. When magnesium burns, it leaves a white powder, magnesium oxide. To write a balanced chemical equation for the burning of magnesium, follow these steps.

Step 1 Write a chemical equation using formulas and symbols. Remember that oxygen is a diatomic molecule, which means that it consists of two oxygen atoms in a covalent bond.

$$Mg(s) + O_2(g) \rightarrow MgO(s)$$

Step 2 Count the atoms in the reactants and products.

Atoms	Mg	+	O₂	→	MgO
Mg	1				1
O			2		1

The magnesium atoms are balanced, but the oxygen atoms are not. So, the equation is not balanced.

Step 3 Choose coefficients to balance the equation. Remember that you cannot change the subscripts of a formula to balance the equation. Instead, try putting the coefficient 2 in front of MgO.

$$Mg(s) + O_2(g) \rightarrow 2MgO(s)$$

Think it Over

10. **Apply** Hydrogen (H), like oxygen, is a diatomic molecule. Write the chemical formula for hydrogen.

_____

Applying Math

11. **Apply** When lithium metal is treated with water, hydrogen gas and lithium hydroxide are produced. Balance the following chemical equation that shows this reaction.

Li(s) + H₂O(g) → LiOH(aq) + H₂(g)

_____

_____

Step 4 Check the number of atoms on each side of the equation again. Now, there are two magnesium atoms on the right side of the equation and only one on the left. You need to put the coefficient 2 in front of Mg to balance the equation.

$$2Mg(s) + O_2(g) \rightarrow 2MgO(s)$$

The above chemical equation for the burning of magnesium is balanced.

Now try one on your own. Balance the equation for the following reaction:

$$Fe(s) + Cl_2(g) \rightarrow FeCl_3(s)$$

Write the correct number of atoms in the column under the each element. Then write the balanced numbers in the bottom row.

Atoms	Fe	Cl₂	→	Fe	Cl₃
Fe			→		
Cl			→		
balanced					

What are moles? To help chemists figure out how much of a substance is needed to get a certain reaction, they use a counting unit called the mole (mol). One **mole** is the amount of a substance that contains 6.022×10^{23} particles of that substance. The mass in grams of one mole of a substance is called its **molar mass**. Just as the mass of a dozen eggs is different from the mass of a dozen watermelons, different substances have different molar masses. The atomic mass of titanium (Ti), for example, is 47.87 amu, and the molar mass is 47.87 g/mol. By comparison, the atomic mass of sodium (Na) is 22.99 amu, and its molar mass is 22.99 g/mol. For a compound such as nitrogen dioxide (NO_2), the molar mass is the sum of the masses of its component atoms. The nitrogen dioxide (NO_2) molecule contains one nitrogen atom (1×14.01 amu) and two oxygen atoms (2×16.00 amu = 32.00 amu). So, NO_2 has a molar mass of 46.01 g/mol.

Given the mass of a substance, you can use the molar mass as a conversion factor to calculate the number of moles.

$$50.00 \text{ g } NO_2 \times \frac{1 \text{ mol } NO_2}{46.01 \text{ g } NO_2} = 1.087 \text{ mol } NO_2$$

Given the number of moles of a substance, you can use the molar mass as a conversion factor to calculate the mass.

$$0.2020 \text{ mol } NO_2 \times \frac{46.01 \text{ g } NO_2}{1 \text{ mol } NO_2} = 9.294 \text{ g } NO_2$$

Applying Math

12. **Calculate** What number goes in each indicated blank in the table?

Fe _____

Cl₂ _____

Cl₃ _____

● After You Read

Mini Glossary

balanced chemical equation: having the same number of atoms of each element on both sides of an equation

chemical equation: a way to describe a chemical reaction using chemical formulas and other symbols

chemical reaction: a change in which one or more substances are converted into new substances

coefficient: a number that shows how many units of a substance take part in a reaction

molar mass: the mass of one mole of substance in grams

mole: 6.022×10^{23} number of particles of a substance

products: the new substances that are produced in a chemical reaction

reactants: the substances that react in a chemical reaction

1. Review the terms and their definitions in the Mini Glossary. Write a sentence describing a chemical equation.

_____

2. Complete the concept web by writing three ways that Antoine Lavoisier helped to make chemistry into a modern science.

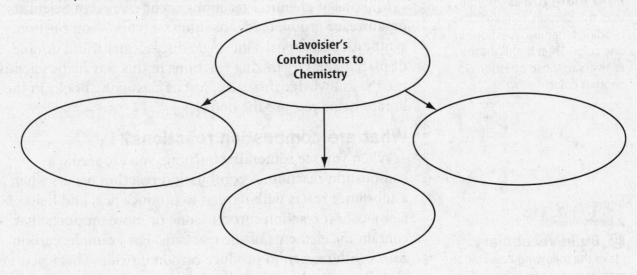

Lavoisier's Contributions to Chemistry

3. ◄ **Study Coach** You created a quiz with questions about important topics from the section. Which question was the hardest for you to answer? Why do you think this was?

_____

_____

End of Section

Chemical Reactions

section ❷ Classifying Chemical Reactions

What You'll Learn

- what the five kinds of chemical reactions are
- what oxidation and reduction are
- what a redox reaction is
- which metals replace others in compounds

FOLDABLES

ⓑ Build Vocabulary

Make the following Foldable to help you organize information about chemical reactions. Label the tabs as shown.

Combustion
Synthesis
Decomposition
Single Displacement
Double Displacement

● Before You Read

Why do you think there are different sections in your school library to organize, or classify, the books?

_____

_____

● Read to Learn

Types of Reactions

Millions of chemical reactions occur every day. Scientists organize, or group, reactions into five types—combustion, synthesis, decomposition, single displacement, and double displacement. Organizing reactions in this way helps scientists use the knowledge they gain, just as classifying books in the library helps you use the books.

What are combustion reactions?

When you see something burning, you are seeing a combustion reaction. A **combustion reaction** occurs when a substance reacts with oxygen to produce heat and light. Combustion reactions produce one or more products that contain the elements of the reactants. For example, carbon reacts with oxygen to produce carbon dioxide. This reaction describes what happens when coal burns. Many combustion reactions also fit into other types of reactions. For example, the reaction between carbon and oxygen is a combustion reaction and a synthesis reaction.

What are synthesis reactions?

In a **synthesis reaction**, two or more substances combine to form another substance. The general formula for a synthesis reaction is $A + B \rightarrow AB$. Substance A reacts with substance B to form substance AB.

An example of a synthesis reaction is hydrogen burning in oxygen to form water.

$$2H_2(g) + O_2(g) \rightarrow 2H_2O(g)$$

This reaction is used to power some rockets, including the main engines of a space shuttle. Have you ever seen a rusty car or bike? The reaction that causes rust is a synthesis reaction. When iron reacts with oxygen in the presence of water, hydrated iron(II) oxide, or rust, is formed. ☑

What are decomposition reactions?

A decomposition reaction is the reverse of a synthesis reaction. A **decomposition reaction** occurs when one substance breaks down, or decomposes, into two or more substances. The general formula for a decomposition reaction is $AB \rightarrow A + B$. Most decomposition reactions use heat, light, or electricity. For example, an electric current passed through water produces hydrogen and oxygen. The chemical equation and figure below show this.

$$2H_2O(l) \xrightarrow{\text{elec.}} 2H_2(g) + O_2(g)$$

Test Tubes

Battery

Beaker

What are single-displacement reactions?

A **single-displacement reaction** happens when one element replaces another element in a compound. The general formula for a single-displacement reaction is $A + BC \rightarrow AC + B$. Atom A displaces, or takes the place of, atom B. A new molecule, AC, forms. A single-displacement reaction occurs when a copper wire is put into a solution of silver nitrate. Copper is a more active metal than silver, so it replaces the silver. A blue copper(II) nitrate solution forms. The silver, which is not soluble, forms on the wire.

$$Cu(s) + 2AgNO_3(aq) \rightarrow Cu(NO_3)_2(aq) + 2Ag(s)$$

1. **Explain** What type of reaction is the formation of rust?

_____

Picture This

2. **Interpret Illustrations** Look at the coefficients of the products in the equation. Which test tube has hydrogen in it, the left one or the right one? How do you know this?

_____

_____

_____

What is the activity series?

You can predict which metal will replace another metal in displacement reactions. The diagram lists metals by how reactive they are. The most-active metals are at the top of the list. The least-active metals are at the bottom. A metal will replace any less-active metal. Notice that copper, silver, and gold are the least-active metals on the list. That is why these elements often occur in relatively pure deposits.

Picture This

3. Order Write these elements in order from least active to most active: lead, aluminum, copper, silver, zinc.

_____

_____

_____

_____

_____

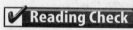

4. Identify Which type of reaction causes a precipitate to form?

_____

_____

What are double-displacement reactions?

In a **double-displacement reaction**, the positive ion of one compound replaces the positive ion of the other compound to form two new compounds. The general formula for a double-displacement reaction is AB + CD → AD + CB. A double-displacement reaction takes place if a precipitate, water, or gas forms when two ionic compounds in solution are combined. A **precipitate** is an insoluble compound (one that cannot be dissolved) that comes out of solution during a double-displacement reaction. ☑

Look at the following example of a double-replacement reaction.

$$Ba(NO_3)_2(aq) + K_2SO_4(aq) \rightarrow BaSO_4(s) + 2KNO_3(aq)$$

The reactants are barium nitrate and potassium sulfate. Both compounds are in solutions. The products are solid barium sulfate, which is the precipitate, and potassium nitrate, which is in solution.

The chemical reactions you learned about in this section are only a few examples. Thousands of reactions of each type happen all around you.

What are oxidation-reduction reactions?

In many chemical reactions, substances gain or lose electrons. Chemists use two terms to describe gaining or losing electrons. **Oxidation** is a loss of electrons during a chemical reaction. **Reduction** is a gain of electrons during a chemical reaction. Chemical reactions involving electron transfer of this sort often involve oxygen. Oxygen is very reactive. It often pulls electrons from metals. Oxidation-reduction reactions cause metals to corrode, or rust, as you can see in the figure.

During an oxidation-reduction reaction, one substance gains electrons while another loses electrons. The substance that gains electrons becomes more negative, so we say it is reduced. Another substance loses electrons and becomes more positive. We say it is oxidized. Reduction and oxidation always occur together. That is why these reactions are often called redox reactions. Redox stands for *red*uction and *ox*idation.

Copyright © Glencoe/McGraw-Hill, a division of The McGraw-Hill Companies, Inc.

Think it Over

5. **Draw Conclusions** Chlorine has seven electrons in its outer energy level. In reactions, chlorine usually gains an electron. In a redox reaction, would chlorine be oxidized or reduced? Explain.

_____

_____

_____

_____

Picture This

6. **Explain** What has happened to the electrons in the metal of the car in the figure?

_____

_____

_____

● After You Read

Mini Glossary

combustion reaction: a reaction in which a substance reacts with oxygen to produce heat and light

decomposition reaction: a reaction in which one substance breaks down, or decomposes, into two or more substances

double-displacement reaction: a reaction in which the positive ion of one compound replaces the positive ion of the other compound to form two new compounds

oxidation: a loss of electrons during a chemical reaction

precipitate: an insoluble compound that comes out of solution during a double-displacement reaction

reduction: a gain of electrons during a chemical reaction

single-displacement reaction: a reaction in which one element replaces another element in a compound

synthesis reaction: a reaction in which two or more substances combine to form another substance

1. Review the terms and their definitions in the Mini Glossary. Choose two terms that are related and write a sentence using both of them.

_____

_____

2. Write the letter of the description in Column 2 that matches the reaction in Column 1.

Column 1

_____ 1. decomposition reaction

_____ 2. single-displacement reaction

_____ 3. synthesis reaction

_____ 4. combustion reaction

_____ 5. double-displacement reaction

Column 2

a. A + B → AB

b. AB + CD → AD + CB

c. burning

d. A + BC → AC + B

e. AB → A + B

3. ◀ **Mark the Text** How did highlighting the questions and answers in different colors help you learn the material in this section?

_____

_____

End of Section

Chemical Reactions

section ❸ Chemical Reactions and Energy

● Before You Read

You have probably seen video of a building being demolished by an explosion. Describe what happened on the lines below.

_____

_____

● Read to Learn

Chemical Reactions—Energy Exchanges

When they are no longer useful, buildings are sometimes demolished with dynamite. A dynamite explosion is an example of a rapid chemical reaction.

Most chemical reactions happen more slowly than a dynamite explosion, but all chemical reactions release or absorb energy. The energy released in a chemical reaction can be in the form of heat, light, sound, or electricity. Wood burns and releases heat and light. A glow stick releases only light.

Chemical bonds are the source of this energy. Most chemical reactions break some chemical bonds in the reactants. It takes energy to break the chemical bonds. That is why many substances need heat to make them react. For products to be produced, new bonds must form. When bonds form, energy is released. The amount of energy required to break the chemical bonds in dynamite is much less than the amount of energy released when new bonds form. The result is a release of energy and sometimes a loud explosion.

More Energy Out

An **exergonic** (ek sur GAH nihk) **reaction** releases energy. In an exergonic reaction, less energy is needed to break the bonds in the reactants than is released when new bonds in the products form. Exergonic reactions give off energy, such as light or heat. An exergonic reaction produces visible light in a glow stick.

What You'll Learn

- energy change sources in chemical reactions
- the difference between exergonic and endergonic reactions
- how catalysts and inhibitors are used

Mark the Text

Underline As you read this section, underline the information you think is important. When you finish reading, look back at what you underlined.

FOLDABLES

❻ Make a Venn Diagram
Make the following Foldable to compare and contrast exergonic and endergonic reactions.

Picture This

1. Identify What type of reactions provides the energy for the items in the figures?

_____

_____

2. Determine What happens to energy in an endergonic reaction?

_____

_____

What are exothermic reactions?

In some reactions, the energy is given off as thermal energy. Have you ever used a heat pack? Heat packs release thermal energy. An **exothermic** (ek soh THUR mihk) **reaction** releases thermal energy. Burning wood and exploding dynamite are examples of exothermic reactions. Iron rusting is also exothermic. The chemical reaction that produces rust occurs so slowly that you can't detect the thermal energy.

How are exothermic reactions used?

Have you turned on a light or used a blow-dryer today? The energy you used probably came from exothermic reactions. The power plant the electricity came from probably uses fossil fuels. The carbon in the fossil fuels combines with oxygen to form carbon dioxide gas and energy. This reaction is exothermic.

Other substances in fossil fuels also react. Often, the products of these other reactions are pollutants. Sulfur in fossil fuels reacts with oxygen to form sulfur dioxide, which combines with water in the atmosphere to form acid rain.

More Energy In

An **endergonic** (en dur GAH nihk) **reaction** absorbs energy. In an endergonic reaction, it takes more energy to break the bonds in the reactants than is released when new bonds in the products form. Endergonic reactions absorb energy such as heat, light, or electricity. ☑

Electricity is often used to supply energy to endergonic reactions. Electricity supplies energy to a reaction that puts a coat of metal onto a surface. This reaction is called electroplating. Electricity also is used to supply energy to separate aluminum metal from its ore.

In the following endergonic reaction, energy from electricity keeps the reaction going.

$$2Al_2O_3(l) \xrightarrow{\text{elec.}} 4Al(l) + 3O_2(g)$$

What is an endothermic reaction?

An **endothermic** (en duh THUR mihk) **reaction** absorbs energy, usually as heat. When an endothermic reaction takes place in a beaker, it can make the beaker feel cold. Physical changes also can be described as endothermic. For example, a salt dissolving in water is an endothermic physical change.

An endothermic reaction is used to make homemade ice cream. In an ice-cream maker, salt is added to a bucket of ice and water. The salt dissolves and absorbs heat. This makes the mixture of salt and water colder. Without salt, the ice would not make the ice cream mixture cold enough to freeze.

Some reactions are extremely endothermic. When barium hydroxide ($BaOH_2$) reacts with ammonium chloride (NH_4Cl) in a beaker of water, it is so endothermic that it causes a drop of water on the outside of the beaker to freeze. Cold packs contain ammonium nitrate crystals and water. They are another example of an endothermic reaction.

How is energy conserved in chemical reactions?

You read in an earlier chapter that energy can change from one form to another, but the total amount of energy never changes. This principle is usually called the law of conservation of energy. Does this mean that the total amount of energy remains constant in chemical reactions, too? Yes. Consider the burning of methane (CH_4) in air. Methane is the major component of natural gas, and is described by the following equation. Note that energy is included as a product and recall that oxygen is in the air.

$$CH_4(g) + 2O_2(g) \rightarrow CO_2(g) + 2H_2O(g) + energy$$

During this process, some of the chemical energy of the reactants is released as thermal energy and light. However, the sum of the energy released and the chemical energy of the products is exactly equal to the chemical energy of the reactants in an exergonic chemical reaction. In other words, the chemical energy released plus the chemical energy of the products is equal to the chemical energy of the reactants. So the total amount of energy before and after the reaction remains the same. Similarly, the total amount of energy remains the same in endergonic chemical reactions. Summarizing, the law of conservation of energy applies to chemical reactions, as well as to other types of energy transformations.

Think it Over

3. Compare and Contrast What is the difference between an endergonic reaction and an endothermic reaction?

_____

_____

_____

_____

● After You Read

Mini Glossary

endergonic reaction: a reaction that absorbs energy

endothermic reaction: a reaction that absorbs energy, usually as heat

exergonic reaction: a reaction that releases energy

exothermic reaction: a reaction that releases energy, usually as heat

1. Review the terms and their definitions in the Mini Glossary. Write a sentence giving an example of the type of reaction that can cause its container to get cold or freeze water.

_____

_____

2. Complete the table. The first row gives examples of different kinds of reactions. Name the type of reaction for each example. Then describe the reaction.

Example of Reaction			
	Glow Stick	**Dynamite Exploding**	**Electroplating Metals**	**Cold Pack**
Type of Reaction	exergonic reaction		endergonic reaction	
Description of the Reaction		release energy in the form of heat		

3. <u>Mark the Text</u> What idea was the hardest for you to understand in this section? How would you explain that idea to a friend?

_____

_____

_____

Chemical Reactions

section ❹ Reaction Rates and Equilibrium

● Before You Read

What are some reasons paint dries faster on a warm day than a cool day?

_____

_____

● Read to Learn

Reaction Rates

Some chemical reactions, such as the combustion of rocket fuels, take place rapidly and release tremendous amounts of energy in a matter of seconds. Other reactions, such as the rusting of steel, proceed so slowly that you hardly notice any change from one week to the next.

How can you express the rate of a chemical reaction or the speed of any other change that occurs over a period of time? For a chemical reaction, the **reaction rate** is the rate at which reactants change into products.

Consider the synthesis reaction described by the following chemical equation:

$$2H_2(g) + O_2(g) \rightarrow 2H_2O(g) + energy$$

A chemist might choose one of several ways to state the rate of this reaction. One example is the rate at which oxygen or hydrogen is used up. Another example is the rate at which water is produced. A third example is the rate at which energy is released. And because many factors influence the rates of chemical reactions, the chemist would also state the conditions under which the reaction occurred.

What You'll Learn

- how to express reaction rates
- how catalysts and inhibitors affect reactions
- how to define equilibrium

> **Study Coach**
>
> As you read the section, make a list of the factors that affect reaction rates and note their effects.

> **FOLDABLES**
>
> ❶ **Build Vocabulary**
> Track the factors that affect chemical reaction rates. Label the foldable as shown.
>
>
>
> Temperature
> Concentration
> Surface Area
> Pressure/ Volume
> Catalysts and Inhibitors

What things affect reaction rates? You might already realize that putting things in a refrigerator slows down the rate of decomposition of foods, just like putting them over a fire speeds up the rate at which they cook, or decompose. Temperature is one factor in the reaction rate. Thermal energy speeds up the rate at which atoms and molecules in a substance move. The more thermal energy they have, the faster they move. The less energy they have, the slower they move. Speeding up particles also increases their chance of bumping into each other. Chemists use a model that controls the rate the particles move. Changing other factors such as concentration and surface area, they can predict the reaction rate. The model is called the **collision model**, and it says that atoms, ions, and molecules must collide to react.

How do temperature and concentration affect reactions? Recall that an increase in temperature increases the speed at which a particle moves. It also makes it more likely to collide with another particle. Concentration is how closely together particles are arranged. A high concentration means particles are close together because there are many particles in a small space. Particles are likely to bump into one another in a high concentration. If we heat them, it increases the likelihood that collisions will occur. The reaction rate then increases with the rate of collision. Increased temperature and increased concentration lead to increased reaction rates. ☑

How does surface area affect the rate? Recall that grinding a grain of salt into a powder increases the rate at which the salt dissolves in water. It dissolved faster because the surface area was increased and more liquid could touch the tinier grains.

Even on a hot day a wheat grain is not likely to burst into flame. However, if you crush and grind the grain into a powder and then mix it with the oxygen in the air, it can create an explosion and fire. By pulverizing the grain there are more opportunities for collisions of the powder and oxygen. This increases the reaction rate and dramatically increases the rate of combustion.

✔ **Reading Check**

1. **Explain** How does heating create more collisions?

_____

_____

_____

💡 **Think it Over**

2. **Predict** How could you keep a flammable liquid from catching fire?

_____

_____

_____

Compression   Expansion

Picture This

3. Determine What does compression and expansion do to the volume of a gas? To the concentration?

_____

_____

_____

_____

Agitation Stirring a mixture allows the reactants in it to collide more often. Rapid stirring speeds up the reaction rate. Slower stirring makes the reaction rate slower.

Pressure Increasing the pressure of gases reduces the space in which the atoms or molecules can move. That means more collisions can occur. The reaction rate goes faster when the pressure rises. The rate slows when the pressure drops.

What are catalysts?

A **catalyst** (KA tuh lust) is a substance that speeds up a chemical reaction without being permanently changed itself. A chemist might add a catalyst to a reaction that would be too slow to be useful. When a catalyst is added to a reaction, the mass of the products does not change. A catalyst lowers the energy of activation for a reaction. Collisions among reactants become more effective. The reaction rate increases.

What are inhibitors?

Sometimes, it is helpful to prevent or to slow down chemical reactions. An **inhibitor** is a substance that slows down a chemical reaction. Like catalysts, inhibitors do not change the amount of product produced in a chemical reaction. Food preservatives called BHT and BHA are inhibitors. They slow down reactions and prevent foods, such as cereals and crackers, from spoiling. ☑

☑ Reading Check

4. Define What is an inhibitor?

_____

_____

_____

Equilibrium

When you examine the general formula for a chemical reaction (reactants → products) you might think of a one-way street where vehicles may travel in one direction. Some reactions are like that. The decomposition of potassium chlorate ($KClO_3$) into potassium chloride (KCl) and oxygen (O_2) is just such a reaction.

$$2KClO_3(s) \rightarrow 2KCl(s) + 3O_2(g)$$

Other reactions can occur in both directions and are said to be reversible. A **reversible reaction** is one that can occur in forward and reverse directions like a two-way street.

When a reversible reaction's forward and reverse reactions take place at exactly the same rate, a state of balance, or equilibrium, exists. **Equilibrium** is a state in which forward and reverse reactions, or processes, proceed at equal rates. An equilibrium state is indicated with double-reaction arrows, as shown below. ☑

$$reactants \rightleftharpoons products$$

Factors affecting equilibrium When a state of equilibrium exists, the amounts of reactants and products remain constant. Sometimes stresses can speed up or slow down one of the opposing reactions. Instead of remaining constant, the amounts of reactants and products favor one of the directions of the reaction. The equilibrium becomes temporarily unbalanced. **Le Chatelier's principle** states that if a stress is applied to a system at equilibrium, the equilibrium shifts in the direction that opposes the stress. But in time, the forward and reverse reactions again reach a state of balance.

Like any reaction, equilibrium is affected by temperature, pressure, concentration, and volume as well.

● After You Read

Mini Glossary

activation energy: minimum amount of energy needed to start a reaction

catalyst: a substance that speeds up a chemical reaction without being permanently changed itself

collision model: atoms, ions, and molecules must collide to react

equilibrium: the state in which forward and reverse reactions or processes proceed at equal rates

inhibitor: a substance that slows down a chemical reaction

Le Chatelier's principle: if stress is applied to a system in equilibrium one side of the equilibrium shifts to oppose the stress

reaction rate: the rate at which reactants are changed into products

reversible reaction: a reaction that can occur in both the forward and reverse directions

1. Review the terms and their definitions in the Mini Glossary. Write a sentence giving an example of the type of reaction that requires a catalyst to proceed.

   _____

   _____

2. Describe a situation in which an equilibrium becomes stressed and eventually returns to equilibrium.

   _____

   _____

   _____

   _____

3. ◀ **Study Coach** How did listing the factors and their effects help you understand the section?

   _____

   _____

   _____

   _____

End of Section

Radioactivity and Nuclear Reactions

section ❶ The Nucleus

What You'll Learn

- what particles make up an atom and its nucleus
- how the nucleus is held together
- what radioactivity is

Study Coach

Make Flash Cards Make flash cards of important terms you read about in this section. Write term on one side of a flash card and its definition on the other side. Keep reviewing the flash cards until you know all the terms and their definitions.

✔ **Reading Check**

1. Identify What type of charge does an electron have?

_____

_____

● Before You Read

Have you ever heard of an object, such as a chair, being described as unstable? In the space below, give examples of an object that is unstable and an object that is stable.

_____

_____

_____

● Read to Learn

Describing the Nucleus

Even though you may not know it, particles are constantly hitting you. Most of these particles are from stable atoms. But some of these particles are from unstable atoms in soil, rocks, and the air. Which atoms are unstable? What particles do they emit, or give off? The answers to these questions begin with the nucleus of an atom.

What are protons and neutrons?

Recall that an atom is made up of very small particles. The particles are protons, neutrons, and electrons. The protons and neutrons are located in the nucleus at the center of an atom. Protons are positively charged particles. Neutrons have no electric charge. Neutrons are electrically neutral. Since the nucleus contains positively charged protons, it has a positive charge. Each proton has one positive electrical charge, or 1+. The total amount of positive charge in a nucleus is equal to the number of protons that the nucleus has. The number of protons in a nucleus is called its atomic number.

Atoms usually contain the same number of protons as electrons. Electrons have a negative charge. Electrons are attracted to the positively charged nucleus. This electric attraction pulls the electrons close to the nucleus. ✔

Copyright © Glencoe/McGraw-Hill, a division of The McGraw-Hill Companies, Inc.

Is the nucleus the largest part of an atom?

Protons and neutrons are packed together tightly in the nucleus of an atom. The particles in the nucleus are so close that the nucleus takes up only a tiny part of an atom. The part outside the nucleus in which the electrons are located is much larger than the nucleus. Think of an atom as a football stadium. Its nucleus would be the size of a marble. Although the nucleus takes up very little space, it contains almost all the mass of an atom.

What are isotopes?

Look at the two helium nuclei in the figure below. How are they different? They both have two protons, but the one on the right has one more neutron. All atoms of the same element have the same number of protons in the nucleus.

However, all the atoms of the same element do not always have exactly the same number of neutrons. Isotopes are atoms of the same element that have different numbers of neutrons. The atoms of all isotopes of the same element do have the same number of protons and electrons and the same chemical properties.

Helium-3   Helium-4

How is a nucleus numbered?

You can describe a nucleus using its atomic number. The atomic number of an atom is the number of protons in the nucleus. You can also describe a nucleus using its atomic mass. The nucleus contains almost all the mass of an atom, so the mass number of an atom is the number of protons and neutrons in the nucleus.

Look at the diagram of helium-3 above. Its mass number is 3 because it contains a total of three protons and neutrons. But its atomic number is 2 because it contains two protons. Look at the diagram of helium-4 above. Its mass number is 4 because it contains a total of four protons and neutrons. However, the atomic number is 2 because the helium nucleus contains two protons.

FOLDABLES

A **Find Main Ideas** Make a Foldable like the one shown below. As you read this section, write down main ideas about the strong force.

The Strong Force

Picture This

2. **Compare and Contrast** How are the two helium isotopes different?

_____

_____

The Strong Force

Particles with the same charges repel, or push away from, each other. Why don't the positively charged protons in the nucleus repel each other? Another force holds the particles of the nucleus together. The **strong force** is the force that makes protons and neutrons attract each other and stay together.

The strong force is one of the four basic forces in nature. The strong force is 100 times stronger than the electric force, but it only works when particles are close together. The figure below shows the strong force and the electric force. The two particles on the left are close, like the protons and neutrons in a nucleus, so the strong force is working. But the particles on the right are farther apart. So the strong force weakens, and the electric force takes over. So when protons are far apart, they are repelled by the electric force.

Picture This

3. **Interpret Scientific Illustrations** In the figure, circle the example showing that the strong force is greater than the electric force.

Strong force   No strong force

Electric force   Electric force

Total force   Total force

How do forces work in a small nucleus?

Not all nuclei (singular, *nucleus*) are the same size. Small nuclei have only a few protons and neutrons. The strong force holding the particles together is greater than the electric force that is pushing the particles apart. As a result, the protons and neutrons are held tightly together.

How do forces work in a large nucleus?

Even though there are more particles in a large nucleus than in a small nucleus, the strong force in a large nucleus is about the same strength as the strong force in a small nucleus. The strong force holds together particles that are closest together. In a nucleus with many protons, the electric force repels protons that are far apart.

The diagram on the next page shows how the electric force that pushes the protons apart is greater in a large nucleus than in a small nucleus. The increased repulsive force causes the particles in a large nucleus to be held together less tightly than those in a small nucleus. ☑

✔ **Reading Check**

4. **Describe** Name the force that causes protons to push away from each other.

_____

_____

Copyright © Glencoe/McGraw-Hill, a division of The McGraw-Hill Companies, Inc.

Strong force

Total force

Electric force

Strong force

Total force

Electric force

Radioactivity

When the strong force can hold a nucleus together forever, the nucleus is stable. If the strong force is not large enough to hold the nucleus together, the nucleus becomes unstable and can break apart or decay. When a nucleus decays, it emits particles and energy. **Radioactivity** is the process of a nucleus decaying and emitting particles and energy. Large nuclei are more unstable than small nuclei. All nuclei with more than 83 protons are radioactive. Some smaller nuclei are also radioactive. Even a nucleus with only one proton could be radioactive.

Who discovered radioactivity?

In 1896, Henri Becquerel made an interesting discovery. He left uranium salt in a drawer with a photographic plate. When he developed the plate, he saw an outline of the uranium salt on it. Becquerel realized that the uranium must have given off rays, or radiation, that darkened the film.

Two years after Henri Becquerel's discovery of radioactivity, there was another important discovery. Marie and Pierre Curie discovered two new elements. The elements were polonium and radium. Both new elements were radioactive. Marie and Pierre Curie wanted to get a large sample of radium so they could study it. It took them more than three years to obtain about 0.1 g of radium from several tons of the mineral pitchblende. ☑

Picture This

5. Identify Circle the nucleus that has the larger repulsive force.

✔ Reading Check

6. Recall Who discovered the elements polonium and radium?

_____

_____

● After You Read

Mini Glossary

radioactivity: the process of a nucleus decaying and emitting particles and energy

strong force: the force that makes protons and neutrons in the nucleus attract each other and stay together

1. Review the terms and their definitions in the Mini Glossary. Write a sentence on the lines below that shows your understanding of the term *radioactivity*.

_____

_____

_____

2. Complete the table below to organize information about stable and unstable nuclei.

Type of Nucleus	Comparison of Strong Force and Electric Force	Radioactive?	Example
Stable			
Unstable			

3. **Study Coach** Think about what you have learned. How did making flash cards of important terms help you learn the content?

_____

_____

_____

End of Section

Radioactivity and Nuclear Reactions

section ➋ Nuclear Decays and Reactions

● Before You Read

When you read the word *radiation*, what do you think of? Brainstorm some words or phrases that come to mind and write them on the lines below.

_____

_____

What You'll Learn
- how alpha, beta, and gamma radiation are similar and different
- what the half-life of a radioactive material is
- how radioactive dating is used

● Read to Learn

Nuclear Radiation

When an unstable nucleus decays, it breaks apart. As it decays, the nucleus emits particles and energy called nuclear radiation. The three types of nuclear radiation are alpha, beta (BAY tuh), and gamma radiation. Alpha and beta radiation are particles. Gamma radiation is an electromagnetic wave.

Alpha Particles

An **alpha particle** is made of two protons and two neutrons. During alpha radiation, the decaying nucleus emits an alpha particle. An alpha particle is the same as the nucleus of a helium atom. An alpha particle has a charge of 2+ and an atomic mass of 4. This means it has two protons and two neutrons. An alpha particle's symbol is the same as the symbol of a helium (He) atom.

Alpha particles have much more mass than beta or gamma radiation and a greater electric charge. Alpha particles can penetrate, or pass through, matter, but it makes them lose energy quickly. When alpha particles penetrate matter, they attract negatively charged electrons from the atoms that they pass. This electric force pulls electrons away from the atoms. This process causes alpha particles to lose energy and slow down. Alpha particles are heavier and move more slowly than the other two types of radiation. Alpha radiation cannot penetrate material as deeply as beta and gamma radiation.

Mark the Text

Identify the Main Idea
As you read the section, highlight the main idea in each paragraph.

FOLDABLES™

Ⓑ Compare and Contrast Make a Foldable like the one shown below. As you read this section, take notes on how alpha, beta, and gamma radiation are similar and how they are different.

Copyright © Glencoe/McGraw-Hill, a division of The McGraw-Hill Companies, Inc.

How can humans be harmed by alpha particles?

Alpha particles are heavy and move slowly. You can think of them as bowling balls moving in slow motion. They may not be able to penetrate material deeply, but they can do a lot of damage to whatever they hit. Alpha particles can be harmful if they are released inside the human body. These particles can damage cells. Damaged cells do not work properly and may cause illness and disease.

How can alpha particles help you?

Smoke detectors work by emitting alpha particles. The alpha particles collide with molecules in the air. The molecules break apart and form atoms with positive and negative charges. These charged particles flow within the smoke detector, creating an electric circuit. If smoke particles enter the smoke detector, they break this circuit. Once the circuit is broken, the smoke detector sounds an alarm.

What is transmutation?

Recall that an alpha particle is made up of two protons and two neutrons emitted from a decaying nucleus. The nucleus now has two fewer protons and two fewer neutrons than it had originally. When an atom loses protons, it becomes a different element. **Transmutation** is the process of changing one element to a different element by the decaying process.

The nucleus that emitted the alpha particle has changed. It now has two fewer protons. Its atomic number is two less than the atomic number of the original element. The new element also has two fewer neutrons. Its mass number is four less than the original element.

The figure shows a polonium nucleus. Polonium has 84 protons and a total of 210 protons and neutrons. During transmutation, the nucleus gives off an alpha particle. The nucleus now has 82 protons and a total of 206 protons and neutrons. The element with an atomic number of 82 and a mass number of 206 is lead. The polonium atom has become a lead atom.

$$^{210}_{84}\text{Po} \quad - \quad ^{4}_{2}\text{He} \quad = \quad ^{206}_{82}\text{Pb}$$

Transmutation

Copyright © Glencoe/McGraw-Hill, a division of The McGraw-Hill Companies, Inc.

Think it Over

1. **Compare** Alpha particles can be helpful and harmful to humans. List one way they can help and one way they can harm.

   Help:_____

   Harm:_____

Picture This

2. **Identify** What is the symbol for lead?

   _____

Beta Particles

A second type of radioactive decay is beta radiation. Sometimes in an unstable nucleus, a neutron decays into a proton and emits an electron. The electron leaves the nucleus. This electron is a beta particle. A **beta particle** is an electron that a neutron emits when it decays into a proton. The symbol for a beta particle is $_{-1}^{0}e$. ☑

How does transmutation relate to beta particles?

An atom that loses a beta particle undergoes transmutation. It changes into a different element. The figure below shows an iodine nucleus giving off a beta particle. Before the nucleus decays, it has 53 protons and a total of 131 protons and neutrons. One of the neutrons becomes a proton by giving off a beta particle. Now, the number of protons increases by one and the number of neutrons decreases by one. The new element, xenon, has 54 protons. Notice that the mass number has not changed. There is still a total of 131 protons and neutrons in the nucleus.

$$_{53}^{131}I \longrightarrow _{-1}^{0}e + _{54}^{131}Xe$$

How can beta particles harm you?

Beta particles move much faster than alpha particles because they are smaller and lighter. Beta particles can also penetrate deeper into the material they hit. A beta particle can pass through a sheet of paper. A sheet of aluminum foil will stop a beta particle. Beta particles, like alpha particles, can damage cells if they are released inside the human body.

Gamma Rays

The third type of radiation is gamma radiation. Gamma radiation is not emitted as particles, like alpha and beta radiation. Gamma radiation is emitted as electromagnetic waves called gamma rays. **Gamma rays** are electromagnetic waves with the highest frequencies and the shortest wavelengths in the electromagnetic spectrum. The symbol for a gamma ray is γ, which is the Greek letter gamma.

✔ **Reading Check**

3. Classify Which of the following is emitted when a neutron decays into a proton? Circle your answer.

a. alpha particle

b. beta particle

c. gamma rays

Picture This

4. Identify In the figure, circle the mass numbers and underline the atomic numbers.

What are the properties of gamma rays?

Gamma rays have no mass and no charge. They travel at the speed of light. A nucleus usually emits gamma rays when an alpha or beta particle is created. Gamma rays can pass easily through a sheet of paper or aluminum foil because they travel very fast and have no mass. It takes thick blocks of a material such as concrete or lead to stop gamma rays. Gamma rays cause less damage to cells inside the human body than alpha and beta particles.

Nuclear Fission

In the 1930s, the physicist Enrico Fermi tried bombarding uranium (U) nuclei with rapidly moving neutrons. He incorrectly thought the neutrons would combine with the nuclei and form larger, heavier nuclei. In 1938, Otto Hahn and Fritz Strassmann found that when a neutron hits a uranium-235 (U-235) nucleus, the nucleus splits apart into smaller nuclei.

In 1939, Lise Meitner theorized that when a neutron hits a uranium-235 nucleus, the nucleus becomes so unstable that it splits into two smaller nuclei. The process of splitting a nucleus into smaller nuclei is called **nuclear fission**. The word *fission* means "to divide."

What nuclei can split during nuclear fission?

Only large nuclei, such as the nuclei of uranium and plutonium, can split apart during nuclear fission.

Nuclear fission begins with a neutron hitting a U-235 nucleus. This produces a nucleus of U-236. U-236 is so unstable that it immediately splits into a barium nucleus and a krypton nucleus. The process also produces several neutrons. The total mass of the products of a fission reaction is a little less than the total mass of the original nucleus and the neutron. This small amount of missing mass becomes another product of nuclear fission. The missing mass is changed into a large amount of energy. In the figure on the next page, the wavy lines stand for the energy produced through fission. ✓

5. Identify the products of nuclear fission of uranium 235.

_____

_____

_____

Nuclear Fission Reaction of U-235

$^{91}_{36}$Kr

n

$^{235}_{92}$U

$^{236}_{92}$U
(Unstable nucleus)

$^{142}_{56}$Ba

n

n

n

+ energy

Picture This

6. **Calculate** What is the sum of the mass numbers for Kr and Ba? Why does this not equal the mass number of U-236?

_____

_____

_____

How are mass and energy related?

The diagram above shows that in the process of nuclear fission, some mass is lost. This mass is changed into energy. Albert Einstein proposed that mass and energy are related. He said that mass can be changed into energy and energy can be changed into mass. This is Einstein's special theory of relativity. The theory says that energy in joules is equal to mass in kilograms multiplied by the speed of light squared. The mass-energy equation expresses this theory.

$$\textbf{Energy (joules)} = \textbf{mass (kg)} \times [\textbf{speed of light (m/s)}]^2$$
$$E = mc^2$$

Einstein's theory of relativity states small amount of mass can be changed into a huge amount of energy. For example, if one gram of mass is changed into energy, it releases about 100 trillion joules of energy.

How does a chain reaction occur during nuclear fission?

In a nuclear fission reaction, neutrons are released. These neutrons can hit other nuclei. Then the nuclei split and emit more neutrons. These neutrons hit other nuclei, and the fission reaction repeats. A **chain reaction** is a series of fission reactions caused by neutrons released in each reaction.

If a chain reaction is uncontrolled, a huge amount of energy is released in a short amount of time. However, chain reactions can be controlled. Adding materials that absorb the free neutrons is one way to control chain reactions. These materials keep the neutrons from hitting other nuclei and continuing the reaction.

Not all fission reactions cause a chain reaction. A chain reaction cannot occur if there is less than critical mass to provide enough free neutrons. The critical mass is the amount of starting material that makes each fission reaction produce at least one more fission reaction.

Think it Over

7. **Illustrate** Draw a diagram of a chain reaction in the space below.

Nuclear Fusion

You have learned that nuclear fission releases great amounts of energy. When one nucleus of uranium-235 splits, it releases about 30 million times more energy than when one molecule of dynamite explodes. Another type of nuclear reaction, called nuclear fusion, releases even more energy than nuclear fission. In a **nuclear fusion** reaction, two small, light nuclei combine to form one large, heavy nucleus. Fusion combines atomic nuclei, and fission splits nuclei apart.

How are temperature and fusion related?

In a fusion reaction, two nuclei combine into one. How can two nuclei get close enough to combine? The nuclei have to be moving very fast. Nuclei are positively charged. Positively charged objects repel each other. If nuclei move fast enough, their kinetic energy overcomes the electric force pushing them apart. Then the two nuclei get close enough to combine. ☑

A particle's kinetic energy increases as temperature increases. The temperature must be millions of degrees Celsius for nuclei to move fast enough to combine in a fusion reaction. These high temperatures exist on the Sun and other stars.

How is the Sun's energy produced?

The Sun is made mostly of hydrogen. It produces energy by the fusion of hydrogen nuclei. The figure shows one step of many in the fusion process. A proton (H-1) and a hydrogen isotope (H-2) combine to form an isotope of helium (He-3). In the final step, two He-3 atoms fuse to form one He-4 atom and two protons. In this process, the extra mass changes into a huge amount of energy. The thermal energy and light Earth receives from the Sun comes from this process.

About one percent of the Sun's mass has so far been changed into energy. It is estimated that the Sun has enough hydrogen for fusion reactions to continue for another 5 billion years.

✔ Reading Check

8. Explain how two nuclei can overcome the electronic force pushing them apart.

_____

_____

_____

Picture This

9. Label the proton, the hydrogen isotope, and the helium isotope in the figure.

H-1

He-3 +

H-2

● After You Read

Mini Glossary

alpha particle: particle made of two protons and two neutrons that are emitted from a decaying atomic nucleus

beta particle: an electron that a neutron emits when it decays into a proton

chain reaction: series of repeated fission reactions caused by neutron releases from each reaction

transmutation: the process of changing one element to a different element by the decaying process

1. Review the terms and definitions in the Mini Glossary above. Choose two of the terms that are related and write a sentence using both terms.

_____

_____

2. Complete the table below to organize what you learned about different types of nuclear radiation. Some of the table is filled in to help you start.

	Alpha Radiation	Beta Radiation	Gamma Radiation
Symbol			γ
How it forms	given off by the decaying nucleus		
What it's made of		an electron	
Speed it travels			moves at the speed of light
Mass			has no mass

End of Section

Radioactivity and Nuclear Reactions

section ❸ Radiation Technology and Applications

What You'll Learn

- how a Geiger counter measures radiation
- why you are always surrounded by some radiation
- how radiation can be useful
- what a material's half-life is

Mark the Text

Locate Information

Underline every heading in the section that asks a question. Then highlight the answers to the questions as you find them.

FOLDABLES

Ⓒ Make a Checklist

Make a foldable like the one shown below. Write each paragraph head on a line and check it off as you read it. Write notes under each head.

Radiation Checklist:
☐
☐
☐

● Before You Read

Many devices warn you about something that could cause a problem or be dangerous. For example, a smoke detector makes a loud noise to let you know there may be a fire. On the lines below, list three devices and their warning signals.

_____

_____

_____

● Read to Learn

Radiation Detectors

You can't see or feel alpha particles, beta particles, or gamma rays. You must use special instruments to tell if they are present. These instruments detect radioactivity. Some radioactive particles have an electric charge. The charged particles form ions in the matter they pass through. Radiation detectors are instruments that detect newly formed ions.

How does a Geiger counter measure radiation?

A Geiger counter is a device that measures radiation by producing an electric current when it detects a charged particle. Inside a Geiger counter is a negatively charged copper tube. A positively charged wire runs through the tube. The tube is filled with gas at a low pressure. When radiation enters the tube at one end, it knocks electrons off other gas atoms. The process continues and produces an electron avalanche. The positively charged wire in the tube attracts the free electrons. When a large number of electrons touch the wire, a short, strong current is produced in the wire. An amplifier strengthens the current, creating a clicking sound or a flashing light. The number of clicks or flashes of light that occur each second tells how strong the radiation is.

Electrons
Ionized gas atom
Radioactive particle path
Window
Amplifier and counter
Voltage source
e
e⁻

Background Radiation

The air, the ground, and even the walls of your house give off radiation. This type of radiation is background radiation. Background radiation is found in small amounts. Radioactive isotopes that occur in nature emit background radiation. Rocks, soil, and air contain these isotopes. Bricks, wood, and stone also contain small amounts of radioactive isotopes. Even the food, water, and air used by animals and plants have them. As a result, all animals and plants have small amounts of these isotopes. ☑

Where does background radiation come from?

Background radiation comes from several sources. The circle graph below shows the sources of background radiation received on average by a person living in the United States. The largest source comes from the decay of radon gas. Decay of uranium-238 in soil and rocks produces radon gas. Radon gas can move from the soil and rocks into houses and basements.

Sources of Background Radiation

Rocks and soil 8%
Cosmic rays 8%
X rays 11%
Inside the body 11%
Other 7%
Radon 55%

Some background radiation comes from high-speed nuclei that hit Earth's atmosphere. These nuclei are called cosmic rays. When cosmic rays hit the atmosphere, they give off alpha, beta, and gamma radiation. Because the atmosphere absorbs most of this radiation, it does not reach Earth's surface. At very high altitudes, the atmosphere is thinner than it is at Earth's surface. There is less atmosphere to absorb this radiation. As a result, background radiation from cosmic rays is greater at higher altitudes.

✔ Reading Check

2. **Explain** What emits background radiation?

_____

_____

Applying Math

3. **Use a Graph** List the three largest sources of background radiation in the United States.

_____

_____

_____

Using Nuclear Reactions in Medicine

If you wanted to meet a friend at the mall, knowing what each other was wearing would make finding them easier. Doctors can find a certain molecule in a group of molecules by making it stand out. A radioactive molecule stands out by the radiation it gives off.

A **tracer** is a radioactive isotope used to find or track a molecule. In humans a non-functioning thyroid gland causes many complications. A properly functioning thyroid uses and stores iodine, so an iodine tracer would be useful in diagnosing thyroid problems. Injecting radioactive iodine 131 as a tracer tests the function of the thyroid gland. A strong radioactive reading in the thyroid means it is functioning properly. Other radioactive isotopes can be injected to kill cancerous tumors.

Half-Life

If an element is radioactive, how can you tell when it will decay and give off radiation? Not all radioactive elements decay in the same amount of time. Some decay in less than a second and others take millions of years. The **half-life** of a radioactive isotope is the time it takes for half of the nuclei in a sample of the isotope to decay. The nucleus left afterwards is called the daughter nucleus.

Radioactive Decay and Half-Lives

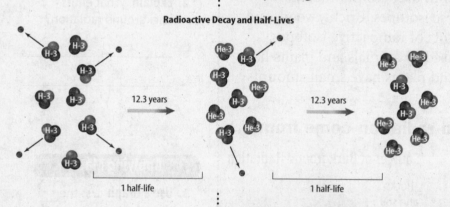

What is radioactive dating?

The half-life of an isotope can be useful. If scientists want to know how old a rock or fossil is, they measure the amount of radioactive isotope is in the sample. Then they measure the amount of daughter nuclei that is in the sample and calculate the number of half-lives of the isotope that have passed to reach this ratio. This is usually the age of the object.

Different isotopes are used to measure different things. Living things incorporate carbon into their bodies as they eat and grow. Some of that carbon is radioactive carbon-14. At its death it has all the carbon it will ever have. Measuring the ratio of carbon-12 to radioactive carbon-14 allows us to calculate an age for the organism. Because carbon-14 decays so rapidly, ages can only be determined up to about 50,000 years. Objects older than 50,000 years do not contain enough carbon-14 to measure.

Think it Over

4. **Determine** How many H-3 and He-3 atoms will there be after a third half-life of 12.3 years passes?

_____

_____

_____

● After You Read

Mini Glossary

half-life: the amount of time it takes for half the nuclei in a sample of an isotope to decay

tracer: a radioactive isotope that doctors use to locate molecules in an organism

1. Review the terms and definitions in the Mini Glossary. Write a sentence about the device used to locate molecules.

   _____

   _____

2. On the lines below, tell how a Geiger counter might be used to measure the radioactive iodine used to check a person's thyroid. Tell what a strong and a weak radiation signal means.

   _____

   _____

   _____

   _____

   _____

3. **Mark the Text** As you read this section, you highlighted the answers to questions in the text. Was it difficult to find the answers to any questions? Explain.

   _____

   _____

End of Section

chapter 21 Solutions

section ❶ How Solutions Form

What You'll Learn

- three types of solutions
- how things dissolve
- the rate solids and gases dissolve

Mark the Text

Locate Information As you read this section, highlight the factors that affect how quickly substances dissolve.

FOLDABLES

Ⓐ Draw and Label Make a half-sheet Foldable like the one below. As you read, draw three diagrams showing solute and solvent molecules before dissolving, during dissolving, and after dissolving. Label your drawings.

Dissolving:

● Before You Read

Have you ever made a drink from a powdered mix? What happened to the powder?

_____

● Read to Learn

What is a solution?

Many people like to watch hummingbirds and put up hummingbird feeders in their yards. They fill the feeders with a red liquid made of water, sugar, and red food coloring. The sweet, colored liquid attracts hummingbirds. To make this food, you add sugar to water and stir. When you stir, the sugar crystals disappear. Next, you add a few drops of red food coloring and stir again. The red color spreads evenly. Why does this happen?

The red sugar-water is a solution. A **solution** is mixture that has the same ingredients, color, density, and even taste mixed evenly throughout. You cannot see the sugar crystals in the solution because they have broken up into molecules. The food coloring also breaks up into molecules. The sugar molecules and the food coloring molecules mix evenly among the water molecules throughout the solution. The figure shows what a sugar-water solution would look like if you could see the separate molecules.

Liquid Solution

Solutes and Solvents

To describe a solution, you could say that one substance dissolves in another. In a solution, the **solute** is the substance that dissolves. The **solvent** is the substance that is doing the dissolving. In sugar-water, the sugar is the solute and the water is the solvent.

What can dissolve in a liquid?

In a solution made with a liquid and a solid, the solid is the solute and the liquid is the solvent. In salt water, salt is the solute and water is the solvent.

Some solutions are made from a gas dissolved in a liquid. In carbonated soda, carbon dioxide gas is dissolved in water. The gas is the solute and the liquid is the solvent.

Other solutions have a liquid dissolved in another liquid, such as the liquid food coloring in the hummingbird food. The solvent is usually the liquid present in the larger amount.

Are there solutions that do not contain a liquid?

Solutions also can be mixtures of gases or even mixtures of solids. The air you breathe is a solution. Air is a solution of 78 percent nitrogen, 20 percent oxygen, and small amounts of other gases. Look at the figures below. The figure on the left shows different gas molecules in a gas solution.

Some jewelry is made of mixtures of metals. Sterling silver is a mixture of silver and copper. Mixtures of metals are solid solutions. In a solid solution, one metal is the solute and the other metal is the solvent. The two metals are melted together. Solid solutions are known as alloys. An **alloy** is a mixture of elements that have metallic properties. Some musical instruments are made of an alloy called brass. Brass is a solid solution of copper and tin. The figure on the right shows atoms of copper and tin in a solid solution brass. Coins are alloys.

Gas

Solid

How Substances Dissolve

Fruit drinks and sports drinks are examples of solutions made by dissolving solids in liquids. Both contain sugar and other substances that add color and flavor to the drink. You know that sugar dissolves in water, but how does it happen?

A solid starts to dissolve at its surface. To understand how a water solution forms, you need to remember two things you have learned about water molecules. First, water molecules, like all particles, are always moving. Second, water molecules are polar. Polar means that they have a positive end and a negative end. Sugar molecules are also polar. ☑

Picture This

1. **Observe** Look at the spacing of the atoms in the two solutions on this page and the liquid solution on the first page. Which is denser, a solid solution, a gaseous solution, or a liquid solution?

_____

✔ **Reading Check**

2. **Explain** Where does a solid start to dissolve?

_____

_____

How does a solid dissolve in a liquid?

The figures below show how sugar dissolves in water. In the figure on the left, water molecules move toward a sugar crystal. The positive ends of sugar molecules attract the negative ends of water molecules.

In the middle figure, water molecules pull sugar molecules into solution. This process continues as layer after layer of sugar molecules move away from the crystal. Finally, the crystal dissolves completely.

The water molecules and sugar molecules spread out and mix evenly. They are now a homogeneous solution. The figure on the right shows sugar molecules surrounded by water molecules in a sugar-water solution. This process happens whenever a liquid solvent dissolves a solid solute.

How do liquid and gas solutions form?

Liquids and gases also can form solutions. Liquid and gas particles form solutions in a way similar to that of sugar and water. But, the process is more complex. Liquid particles and gas particles move much faster than solid particles. The movement separates the solute particles and mixes them evenly in the solvent. The result is a homogeneous solution.

How do solids dissolve in other solids?

You have learned that all particles move. Particles in a solid do not move enough to spread out and mix evenly with other solids. Solids must be melted into liquids and then mixed together. In the liquid state, the atoms can move more freely. They spread out and form a homogeneous solution. The atoms stay in solution after cooling. ☑

☑ **Reading Check**

4. **Explain** what must be done to solids so they can be mixed together.

_____

_____

Rate of Dissolving

Sometimes a solute dissolves quickly into a solvent. At other times, it dissolves more slowly. There are some things you can do to make a solute dissolve faster. You can stir the solution or heat the solution. If the solute is a solid, you can break it into smaller pieces.

How does stirring speed up dissolving?

Stirring a solution speeds up the dissolving process by making the solvent and solute particles move faster. More solvent particles come into contact with more solute particles. The solid solute dissolves more quickly.

How does breaking up a solid solute speed up dissolving?

Suppose you put a piece of rock candy in your drink to sweeten it. You may have to wait a long time for the candy to dissolve. Now, suppose you crush the rock candy into a powder before adding it to your drink. The small pieces of candy dissolve much more quickly than the chunk of candy. Why?

When you break a solid solute into smaller pieces, you increase its surface area. Remember that the dissolving process takes place at the surface of the solute. More surface area means that more solute comes in contact with the solvent. When the surface area of the solute increases, the solute dissolves more quickly. ☑

How does heating speed up dissolving?

When you make hot chocolate from a mix, you mix the powder into a hot liquid. The sugar in the mix dissolves faster in a hot liquid solvent. Solvent particles move faster when the temperature of the solvent increases. Fast-moving solvent particles have more chances to come in contact with solute particles. The more often they come in contact, the faster the solute particles break loose and dissolve.

Can you combine these methods?

If you use more than one of these methods at the same time, you can speed up the dissolving process even more. Suppose you place a sugar cube in cold water. You know the sugar will dissolve eventually. If you heat the water, the sugar will dissolve at a faster rate. If you heat the water and stir the solution, you increase that rate even more. Finally, if you crush the sugar cube, heat the water, and stir the solution, the sugar will dissolve at the fastest rate. The rate of dissolving increases with each additional method you use.

✔ **Reading Check**

5. **Determine** Why does breaking up a solid solute into smaller pieces speed up the dissolving process?

_____

_____

_____

💡 **Think it Over**

6. **Infer** What do both stirring and heating do that increases the rate that particles dissolve?

_____

_____

_____

❂ After You Read

Mini Glossary

alloy: a mixture of elements with metallic properties

solute: the substance that is dissolved in a solution

solvent: the substance that does the dissolving in a solution

1. Review the terms and definitions in the Mini Glossary. Select one term and write a definition of the term in your own words.

_____

2. Complete the outline to help you organize what you learned about solutions.

> **Solutions**
>
> I. Solutions
>
>   A. A solution is _____
>
>   _____
>
>   B. An example of a solution made with a liquid and a solid is
>
>   _____
>
>   C. An example of a solution made with a liquid and a gas is _____
>
>   D. An example of a solution made with two solids is _____
>
> II. Solutes and solvents
>
>   A. A solute is _____
>
>   B. A solvent is _____
>
> III. How substances dissolve
>
>   A. Water molecules approach the solid solute.
>
>   B. _____
>
>   C. _____
>
> IV. Rate of dissolving is affected by
>
>   A. _____
>
>   B. _____
>
>   C. _____

End of Section

chapter 21 Solutions

section ② Concentration and Solubility

● Before You Read

The labels of some containers of fruit juice say "Not from concentrate." What do you think this means?

_____

_____

What You'll Learn

- what solubility is
- about the concentration of solutions
- three types of solutions
- factors that affect gas solutions

● Read to Learn

How much can dissolve?

Suppose you like your lemonade very sweet. You add a teaspoon of sugar to it and stir. The sugar dissolves. If you keep adding sugar to the lemonade, you will reach a point when no more sugar will dissolve in it. The excess sugar crystals fall to the bottom of the glass. They do not go into the solution.

You have reached the solubility of sugar in a given amount of water. <u>Solubility</u> (sahl yuh BIH luh tee) is the greatest amount of solute that can dissolve in a specific amount of solvent at a given temperature.

Are the solubilities of all substances the same?

You can dissolve more than 32 g of salt in 113 g of water at 25°C. But, you can only dissolve about 12 g of baking soda in 113 g water. Salt and baking soda have different solubilities in water. The difference in the solubilities of solutes depends on the nature of the solute and the nature of the solvent.

Concentration

Suppose you and a friend are making lemonade. You add one teaspoon of lemon juice to a glass of water. Your friend adds four teaspoons of lemon juice to the same amount of water. Your friend's glass has more flavor. It has a large amount of solute dissolved in the solvent. It is concentrated. Yours is dilute because it has less solute. <u>Concentration</u> of a solution measures the amount of solute dissolved in a given amount of solvent.

Mark the Text

Check for Understanding

As you read this section, highlight any sentences that you read more than once. After you finish the section, go back and read the highlighted sentences again.

FOLDABLES™

B **Classify** Make a Foldable like the one below. As you read this section, take notes about the different types of solutions.

Saturated

Unsaturated

Supersaturated

1. **Identify** Which of the concentrated drink solution ingredients is the solute? (Circle your answer.)

a. 10 mL

b. 90 mL

c. 100 mL

How much solute is in a concentrated solution?

Concentrated and dilute are not precise terms. Concentration of solutions can be described precisely, though. One way is to state the concentration as a percentage by volume of the solute. For example, the label on a bottle of orange-flavored drink states that the drink contains 10 percent fruit juice. This means that in 100 mL of the drink, there are 10 mL of juice. Or, to make 100 mL of a 10 percent solution of orange drink, the manufacturer added 10 mL of juice to 90 mL of water. To be sure you are getting the highest concentration of juice, choose a drink that is 100 percent juice.

Types of Solutions

Each solute has its own <u>solubility</u>—the maximum amount of solute that can be dissolved in a given amount of solvent. Solutions are defined then as saturated, unsaturated, or supersaturated.

What is a saturated solution?

The table lists the solubilities of some compounds in 100 g of water at different temperatures. If you add 35 g of copper(II) sulfate, $CuSO_4$, to 100 g of water at 20°C, only 32 g will dissolve. The solution is saturated. A <u>saturated solution</u> is a solution that contains all the solute it can hold at a given temperature. If you increase the temperature of the mixture, more copper(II) sulfate can dissolve. As shown in the table, the solubility of solid solutes increases as the temperature of the liquid solvent increases.

Applying Math

2. **Use a Table** How many grams of sugar are needed to make a saturated solution of sugar in 100 g of water at 20°C?

Solubility of Compounds in g/100 g of Water			
Compound	**0°C**	**20°C**	**100°C**
Copper(II) sulfate	23.1	32.0	114
Potassium bromide	53.6	65.3	104
Potassium chloride	28.0	34.0	56.3
Potassium nitrate	13.9	31.6	245
Sodium chlorate	79.6	95.9	204
Sodium chloride	35.7	35.9	39.2
Sucrose (sugar)	179.2	203.9	487.2

What is a solubility curve?

The graph shows some of the information from the table on the previous page. Each line on the graph is a solubility curve for a substance. You can use this to find how much solute dissolves in 100 g of water at a given temperature. To find how much sodium chloride, NaCl, dissolves in 100 g of water at 50°C, find 50°C on the x-axis. Trace a line upward from 50°C to the curve for NaCl. Read the amount shown on the y-axis at that point. About 35 g of NaCl dissolves in 100 g of water at 50°C. As the temperature of the water increases, more solute can dissolve. This is true for most liquid solvents and solid solutes. ☑

What is an unsaturated solution?

You learned that 32 g of copper(II) sulfate forms a saturated solution with 100 g of water at 20°C. What if a solution has less than 32 g of copper(II) sulfate? Then it is an **unsaturated solution**, a solution that can dissolve more solute at a given temperature.

The term *unsaturated solution* is not precise. You know exactly how much copper(II) sulfate makes a saturated solution in 100 g of water at 20°C. An unsaturated solution can have any amount of copper(II) sulfate less than 32 g in 100 g of water at 20°C.

How can a solution be supersaturated?

Suppose you make a saturated solution of potassium nitrate (KNO_3) with 100 g of water at 100°C. You add 245 g of (KNO_3) to the water, just as the solubility table shows. You then let the solution cool to 20°C. What happens? Some of the (KNO_3) solute comes out of solution and falls to the bottom of the container. You can see from the table that at 20°C only 31.6 g of (KNO_3) dissolves in 100 g of water. At the lower temperature, the solvent cannot hold as much solute.

Copyright © © Glencoe/McGraw-Hill, a division of The McGraw-Hill Companies, Inc.

Most saturated solutions behave the same way as the potassium nitrate solution when cooled. But some solutions can become supersaturated. A **supersaturated solution** is a solution that has more solute than a saturated solution at the same temperature. For example, if you cool a saturated solution of sodium acetate from 100°C to 20°C, no solute comes out of the solution. This solution is supersaturated. A supersaturated solution is unstable. If a crystal of sodium acetate is dropped into this solution, crystals begin to form. The extra sodium acetate comes out of solution.

When do solutions give off energy?

The supersaturated sodium acetate solution becomes hot as sodium acetate crystallizes. New bonds form between the sodium acetate ions and water molecules. Sometimes when bonds form, energy is given off in the form of heat. Some heat packs are filled with a supersaturated solution that gives off heat as the solute crystallizes.

Some solutes take energy from their surroundings to dissolve. As a result, the temperature of the solution is reduced. Ammonium nitrate is an example. A cold pack has inner bags of water and ammonium nitrate. A solution forms when the inner bags are broken. The ammonium nitrate draws energy from the water as the solution forms. The water temperature drops and the pack feels cool. ☑

Solubility of Gases

Soda is a solution of carbon dioxide gas dissolved in flavored water. When you shake an open bottle of soda, it bubbles. Shaking or stirring a solution of a gas in a liquid allows more gas molecules to reach the surface of the liquid, where they escape from the liquid into the air.

How do pressure and temperature affect a gas dissolved in a liquid?

Increasing the pressure of a gas over a liquid forces more gas to dissolve in the liquid. Soda is bottled under a large amount of pressure. The pressure increases the amount of gas that can dissolve in the soda and also keeps the carbon dioxide gas in solution. When you open a bottle of soda, the pressure is released and bubbles of gas come out of solution.

Cooling a liquid increases the amount of gas that will dissolve in it, the opposite of dissolving amounts of most solids in a liquid. Less gas is able to dissolve in warm liquids. This is why warm soda bubbles up more than cold soda.

✔ **Reading Check**

6. **Apply** What is happening when a solution gets colder?

_____

_____

_____

💡 **Think it Over**

7. **Draw Conclusions** How could you make the bubbles come out of solution before you drink a soda?

_____

_____

_____

● After You Read

Mini Glossary

concentration: how much solute is present in a solution compared to the amount of solvent

saturated solution: a solution that contains all the solute it can hold at a given temperature

solubility: the greatest amount of solute that can dissolve in a specific amount of solvent at a given temperature

supersaturated solution: a solution that has more solute than a saturated solution at the same temperature

unsaturated solution: a solution that can dissolve more solute at a given temperature

1. Review the terms and definitions in the Mini Glossary. Write a sentence using one of the terms that describes a type of solution.

_____

_____

2. Complete the table below to help you organize the information you learned about solubility and types of solutions.

Type of Solution	How is it made?	Does it depend on temperature?	Does it depend on the amount of solute?
Saturated			
Unsaturated			
Supersaturated			

3. ◀ **Mark the Text** As you read this section, you highlighted the sentences that you read more than once. How could you use this strategy if you were studying with a friend?

_____

_____

End of Section

Solutions

section ❸ Particles in Solution

What You'll Learn
- how some solutes form positively or negatively charged particles
- how some solutions conduct electricity
- how antifreeze works

Mark the Text

Locate Information
Highlight every heading in this section that asks a question. Then highlight each answer as you find it.

FOLDABLES™

C **Summarize** Make a Foldable like the one below. As you read the section, write down the most important ideas under each heading.

| Particles With a Charge: | Ionic Solutions: | Effects of Solute Particles: |

● Before You Read

Why do you think many electrical appliances have labels warning the user not to use the appliance near water?

_____

_____

● Read to Learn

Particles with a Charge

There are charged particles in your body that conduct electricity. A particle with a charge is an **ion.** Ions are in the fluids that are in and around all the cells in your body. You could not live without ions. Some ions help nerve cells send messages to other parts of your body. Nerves control how your muscles move each time you blink your eyes or wave your hand.

An **electrolyte** is a compound that produces solutions of ions that conduct electricity in water. Strong electrolytes dissolve completely into ions in solution. Strong electrolytes conduct a strong electric current, because they contain only ions. Sodium chloride is an example of a strong electrolyte.

Weak electrolytes stay mainly as molecules when they dissolve in water. Weak electrolytes produce only a few ions and conduct current weakly. Acetic acid in vinegar is an example of a weak electrolyte.

Nonelectrolytes are substances that do not form ions in water and cannot conduct electricity. Examples of nonelectrolytes are organic molecules like sucrose (sugar) and ethyl alcohol.

How do ionic solutions form?

Ionization Ionic solutions form in two ways. Some electrolytes are molecules made up of neutral atoms. To form ions, these molecules must be broken apart in a way that the atoms take on a charge. This process of forming ions is **ionization.**

HCl + H₂O → H₃O⁺ + Cl⁻

Picture This

1. **Use a Scientific Illustration** In the figure, circle the positive ion.

Hydrogen chloride is an electrolyte. The above figure shows how hydrogen chloride molecules that are made up of neutral atoms go through the process of ionization when they mix with water molecules. Recall that hydrogen chloride and water are both polar molecules. Water molecules surround the hydrogen chloride molecules and pull them apart into positive hydrogen ions and negative chloride ions, Cl^-. A symbol for a hydrogen ion in water is H_3O^+.

Dissociation The second way that ionic solutions form is by the separation of ionic compounds. Some electrolytes, such as sodium chloride, are ionic compounds. Ionic compounds already contain ions that are attracted into solution by the surrounding water molecules. The process in which an ionic solid separates into its positive and negative ions is **dissociation**.

The figure below shows what happens when a crystal of sodium chloride goes through the process of dissociation in water. Water molecules pull the ions from the surface of the crystal. Water molecules surround and separate the sodium ions and chloride ions. The positive ends of the water molecules are attracted to the negative chloride ions. The negative ends of the water molecules are attracted to the positive sodium ions. The sodium ions and chloride ions dissociate from each another. Because the sodium and chloride ions can move freely through the solution, they can conduct an electric current.

Think it Over

2. **Contrast** What is the difference between ionization and dissociation?

_____

_____

_____

_____

Na⁺ ion   Cl⁻ ion

Water molecules

Sodium and Chloride Ions Mixed with Water

Picture This

3. **Identify** In the figure, circle the sodium chloride crystal.

Effects of Solute Particles

You have learned about many different solutes—polar and nonpolar, electrolyte and nonelectrolyte. All solute particles can affect the physical properties of a solvent, such as its freezing point and its boiling point. How a solute affects the freezing point or the boiling point of a solvent depends on the number of solute particles in solution. Some of these effects are useful. Adding antifreeze to the water in a car radiator lowers the freezing point of the radiator fluid.

How does antifreeze lower the freezing point?

As a substance freezes, its particles arrange themselves in an orderly pattern. When a solute is added, the solute particles interfere with this pattern, making it harder for the solvent to freeze. To overcome the interference, a lower temperature is needed to freeze the solvent.

The figure shows how water molecules in pure water can move easily from the liquid state to the solid state. Pure water freezes at 0°C. When a solute is added to the water, the solute particles interfere with water molecules as they try to form a crystalline pattern. In a car radiator, antifreeze molecules added to water block the formation of ice crystals. When enough solute particles are present, water cannot freeze at 0°C. The temperature must be lower to overcome the interference of solute particles.

Liquid state

Pure water

Solid state

$T_{f \text{ (water)}}$ 0°C

Liquid state

$T_{f \text{ (solution)}}$ less than 0°C

Aqueous solution

Solid state

Solute

💡 **Think it Over**

4. **Explain** how antifreeze keeps water in a car radiator from freezing at 0°C.

_____

_____

_____

_____

Picture This

5. **Illustrate** In the figure, highlight the particles that keep the water from freezing at its normal temperature.

Why can some animals live in cold climates?

Some animals that live in very cold climates have their own kind of antifreeze. Caribou have substances in their bodies that keep their legs from freezing when the temperature falls below 0°C. Fish that live in very cold water also have a natural kind of antifreeze. The antifreeze keeps ice crystals from forming in their tissues. Many insects have a similar antifreeze chemical that protects them from freezing temperatures.

How can the boiling point of water be raised?

You may be surprised to learn that antifreeze raises the boiling point of water. Solute particles interfere with the evaporation of solvent particles. More energy is needed for the solvent particles to escape from the liquid surfaces. The more solute particles in the solution, the higher the boiling point of the solution will be.

How does antifreeze work in a car radiator?

In the figure below, the beaker on the left represents water in a car radiator without antifreeze. As pure water is heated, water molecules move easily to the surface and vaporize. The number of molecules that vaporize depends on the temperature of the water. Water molecules move faster as the temperature rises and more molecules can vaporize. When the pressure of the water vapor equals the pressure of the atmosphere, the water boils.

The beaker on the right shows what happens when a solute, such as antifreeze, is added to the water. The solute particles are distributed evenly throughout the solution, including the surface. Solute particles block part of the surface, so fewer water molecules can reach the surface and vaporize. The solution cannot boil because the vapor pressure of the solution is lower than the vapor pressure of the solvent. Energy must be added to overcome the interference and raise the vapor pressure of the solution to make it boil. The added energy means the solution boils at a temperature higher than the boiling point of the pure water.

Think it Over

6. **Draw Conclusions** What substance would you add to a car radiator during the hot summer months to prevent the radiator from boiling over?

_____

Picture This

7. **Draw** arrows above the beakers in both figures to show the direction of the atmospheric pressure on the water.

● After You Read

Mini Glossary

dissociation: the process in which an ionic solid separates into its positive and negative ions

electrolyte: a compound that produces solutions of ions that conduct electricity in water

ionization: the process of forming ions

nonelectrolyte: substances that do not form ions in water and cannot conduct electricity

1. Review the terms and their definitions in the Mini Glossary. Write a sentence using the terms *ionization* and *electrolyte* correctly.

_____

_____

2. Write the letter of the term in Column 2 that matches the description or example in Column 1.

Column 1

_____ **1.** The process used by sodium chloride to break apart into ions

_____ **2.** Sodium chloride

_____ **3.** The process used by hydrogen chloride to break apart into ions

Column 2

a. electrolyte

b. ionization

c. dissociation

3. **◀ Mark the Text** You highlighted every heading in this section that asks a question and then highlighted each answer as you found it. How would this strategy help you study for a test?

_____

_____

End of Section

chapter 21 Solutions

section ❹ Dissolving Without Water

● Before You Read

You probably have gotten stains on your clothes that didn't rinse out with just water. Write about a stain that you could not remove with water.

_____

_____

● Read to Learn

When Water Won't Work

Water often is called the universal solvent because it dissolves so many things. But, there are some things that water cannot dissolve, such as a salad dressing with vinegar and oil. The water in vinegar cannot dissolve the oil. Why not?

Water molecules have positive and negative ends. The charged areas of water molecules help water dissolve polar solutes. Some substances are nonpolar. A **nonpolar** material does not have positive and negative areas. Nonpolar substances are not attracted to polar substances, including water. Most nonpolar substances do not dissolve in water.

How do nonpolar solutes behave?

Think again about the example of a vinegar-and-oil salad dressing. If you do not mix the dressing, it separates into two layers. The bottom layer is vinegar, which is a solution of acetic acid in water. The top layer is salad oil.

Most salad oils are made of large molecules of carbon and hydrogen atoms called hydrocarbons. Hydrocarbon molecules share the electrons in their bonds in a nearly equal way. So, hydrocarbon molecules do not have separate positive and negative ends. The nonpolar oil molecules are not attracted to the polar water molecules in the vinegar solution. The water molecules in the salad dressing cannot dissolve the nonpolar oil molecules. That is why you have to shake a vinegar-and-oil dressing to mix it before using it.

Copyright © Glencoe/McGraw-Hill, a division of The McGraw-Hill Companies, Inc.

What You'll Learn

- what solutes do not dissolve in water
- how polar and nonpolar solvents work in water
- how to choose the right solvent for cleaning

> **Study Coach**

Create a Quiz As you read this section, write five questions that could be used on a quiz. Write different types of questions, such as multiple choice and fill-in-the-blank. Be sure to answer the questions after you write them.

FOLDABLES™

❶ Classify Make a Foldable like the one below. As you read this section, write down the different nonpolar solvents and what they can be used for.

> Nonpolar:

Why are alcohols special?

Molecules of some substances have a polar end and a nonpolar end. These substances can form solutions with polar and nonpolar solutes. Alcohols are examples of this kind of substance. The polar end of an alcohol molecule dissolves polar solutes. The nonpolar end dissolves nonpolar solutes.

The figure shows a molecule of the alcohol ethanol and its formula. The –OH group is polar. The other part of the molecule, $-C_2H_5$, is nonpolar. Ethanol can dissolve iodine, a nonpolar substance. It also can dissolve water, a polar substance.

Picture This

1. **Determine** Circle the nonpolar part of the ethanol molecule.

Useful Nonpolar Molecules

You may have nonpolar solvents around your house. Mineral oil is a useful nonpolar solvent. It can be used to remove candle wax from candleholders. Mineral oil can also help remove bubble gum from some surfaces. Mineral oil, candle wax, and bubble gum are all nonpolar substances.

Oil-based paints contain pigments dissolved in oils. Oils are nonpolar molecules. To remove wet paint or to make it thinner, you must use a nonpolar solvent, such as turpentine.

Another useful nonpolar substance is the gasoline used in cars and lawnmowers. Gasoline is a solution of different hydrocarbons. Remember that hydrocarbons are nonpolar substances.

Dry cleaners use nonpolar solvents to remove oil and grease stains. The word *dry* means that no water is used in the process. Molecules of a nonpolar solute can slip easily among molecules of a nonpolar solvent. Dry cleaning can remove stains from clothes that cannot be removed when the clothes are washed in water.

"Like dissolves like" is a useful rule to remember. It means that polar solvents dissolve polar solutes and nonpolar solvents dissolve nonpolar solutes.

Think it Over

2. **Interpret** Explain in your own words what the phrase "like dissolves like" means.

_____

_____

_____

When are nonpolar solvents not helpful?

Nonpolar solvents can be dangerous when they are not used carefully. Many nonpolar solvents are flammable, which means that they burn easily. Some nonpolar solvents are toxic, which means they can harm you if they touch your skin or if you breathe their vapors. Never use nonpolar solvents in a closed room. These solvents evaporate more readily than water. You must always have fresh air coming into the room so you do not breathe in the vapors of a nonpolar solvent.

How does soap work?

You have natural oils on your skin and hair. These oils keep your skin and hair from drying out. They also attract and hold dirt. Dirt forms a nonpolar mixture when it combines with oils. Water won't wash this nonpolar mixture away, because water is a polar solvent. You need a substance with both polar and nonpolar properties. You need to use soap.

Look at the soap molecule in the figure above. Soaps start out as large fatty acid molecules. Fatty acids have long hydrocarbon ends that are nonpolar. A carboxylic acid group, –COOH, is at the other end.

To make soap, the hydrogen atom of the carboxylic acid group is removed. Without the hydrogen atom, the end has a negative charge. The negative end attracts a positive ion of sodium or potassium. The finished soap molecule has a nonpolar end and an ionic end. The figure below shows how soap helps oil and water combine. The ionic end of a soap molecule dissolves in water. The nonpolar end dissolves in oily dirt. Together the two ends of a soap molecule remove dirt so it can be rinsed away.

Copyright © Glencoe/McGraw-Hill, a division of The McGraw-Hill Companies, Inc.

Picture This

3. Illustrate Circle the part of the soap molecule that is ionic.

_____

_____

_____

Picture This

4. Explain Why can't you wash away oily dirt with only water?

_____

_____

_____

5. Identify Name three foods that are good sources of vitamin A.

_____

_____

_____

6. Determine Name a vitamin that is made of polar molecules.

_____

💡 **Think it Over**

7. Describe What is the difference between fat-soluble vitamins and water-soluble vitamins?

_____

_____

_____

Polarity and Vitamins

You know that your body needs vitamins to stay healthy. Vitamin A is an important vitamin. Good sources of Vitamin A include liver, lettuce, cheese, eggs, carrots, sweet potatoes, and milk. ✔

Nonpolar Vitamins Vitamin A is nonpolar and can dissolve in fat in your body because fat is also a nonpolar substance. Fat also dissolves vitamin D, vitamin E, and vitamin K. Vitamins that dissolve in fat are called fat-soluble vitamins. Fat and fat-soluble vitamins do not wash away with the water that is in your body. The vitamins stay in your tissues. If you take too much of a fat-soluble vitamin, you can have a high concentration of the vitamin in your body. Some fat-soluble vitamins can be very harmful at high concentrations. You should not take large amounts of fat-soluble vitamins unless your doctor tells you to.

Polar Vitamins Other vitamins, such as vitamins B and C, are polar. You can see the structure of vitamin C in the figure to the right. Notice it has many carbon-to-carbon bonds. This might make you think vitamin C is nonpolar. But vitamin C also has oxygen-to-hydrogen bonds that resemble the bonds in water. These oxygen-to-hydrogen bonds make vitamin C a polar molecule. ✔

Polar vitamins dissolve in the water that is in your body. Vitamins that dissolve in water are called water-soluble vitamins. Any extra amounts of water-soluble vitamins in your body wash away. Because of this, you must continue to replace the water-soluble vitamins that your body uses. You can do this either by eating foods that contain these vitamins or by taking vitamin supplements. Foods that are good sources of vitamin C include orange juice, green peppers, cantaloupe, and broccoli.

In general, it is a good idea to get most of your vitamins from eating a healthful diet. That way you will not risk a harmful overdose.

● After You Read

Mini Glossary

nonpolar: a material that does not have a positive area and a
negative area

1. Review the term and its definition in the Mini Glossary. Write a sentence describing how a nonpolar liquid behaves when added to water.

   _____

   _____

2. Write the name of each substance in the box that contains the kind of solvent that can dissolve it.

   vitamin C   bubble gum   salad oil   candle wax   vinegar

Nonpolar solvent	**Polar solvent**

3. **Study Coach** You created a quiz based on what you have learned in this section, then answered the questions. What kind of question best helped you learn the material?

   _____

   _____

   _____

End of Section

Acids, Bases, and Salts

section ❶ Acids and Bases

What You'll Learn

- how acids and bases are similar and different
- formulas and uses of common acids and bases

Mark the Text

Underline As you read, underline any words or sentences you think might be important to remember. When you finish reading, look back at what you underlined to make sure you understand it.

FOLDABLES

A **Compare and Contrast**
Make the following Foldable to help you understand acids and bases.

Describe	Properties	Common Examples
Acids		
Bases		

● Before You Read

Think of a food or drink you like that tastes sour. What do you think gives it a sour taste?

● Read to Learn

Acids

What do you think about when you hear the word *acid*? Do you think of a substance that can burn your skin or put a hole in metal? Many sour foods contain acids. Some acids are dangerous to handle. Others are safe, and some are good to eat.

What are the properties of acids?

An **acid** is a substance that produces hydrogen ions, H^+, in a water solution. When an acid dissolves in water, some of the hydrogen atoms are released as hydrogen ions, H^+. It is the ability to produce these H^+ ions that gives acids their characteristic properties. When an acid dissolves in water, H^+ ions interact with water molecules to form hydronium ions. A **hydronium ion** (hi DROH nee um • I ahn), H_3O^+, is a combination of an H^+ ion and a water molecule.

Acids have some common properties. All acids taste sour. But you should never taste a substance to see if it is acidic. Some acids can burn you. Acids are corrosive, which means they seem to eat away some metals. When an acid reacts with a metal, hydrogen gas and metallic compounds form. Acids also cause indicators to change color. An **indicator** is an organic compound that changes color in acid and base. Litmus paper is an indicator that turns red in acids.

What are some common acids?

Many foods contain acids. Citrus fruits contain citric acid. Lactic acid is found in yogurt and buttermilk. Pickled foods contain vinegar, also known as acetic acid. Your stomach uses hydrochloric acid to help it digest food.

Common Acids and Their Uses

Name, Formula	Use (example)	Other Information
Acetic acid, CH_3COOH	Food preservation and preparation	When in solution with water, it is known as vinegar.
Acetylsalicylic acid, $HOOC-C_6H_4-OOCCH_3$	Pain relief, fever relief, inflammation reduction	Known as aspirin
Ascorbic acid, $H_2C_6H_6O_6$	Antioxidant, vitamin	Called vitamin C
Carbonic acid, H_2CO_3	Carbonated drinks	Involved in cave, stalactite, and stalagmite formation and acid rain
Hydrochloric acid, HCl	Digestion as gastric juice in stomach, to clean steel in a process called pickling	Commonly called muriatic acid
Nitric acid, HNO_3	To make fertilizers	Colorless, yet yellows when exposed to light
Phosphoric acid, H_3PO_4	To make detergents, fertilizers and soft drinks	Slightly sour but pleasant taste, detergents containing phosphates cause water pollution
Sulfuric acid, H_2SO_4	Car batteries, to manufacture fertilizers and other chemicals	Dehydrating agent, causes burns by removing water from cells

Some common acids and their uses are listed in the table. Many of these acids are important in making products, such as fertilizer. Remember that many acids can burn your skin.

Bases

A **base** is any substance that forms **hydroxide ions,** OH^-, in a water solution. A base is also any substance that accepts H^+ ions from acids.

Unlike acids, not many foods are bases. Egg whites and baking powder are two foods that are basic. Some medicines, such as antacids, are basic. A common base is soap. A characteristic of bases is that they feel slippery, like soapy water. Many cleaning products contain bases. Bases are important in industry, also. For example, sodium hydroxide is a base that is used to separate cellulose fibers from wood pulp to make paper.

Picture This

1. **Identify** Look at the acids listed in the table above. What is the first element in the chemical formula of most acids?

What are the properties of bases?

Bases are the opposites of acids. While bases and acids do share some features, bases also have their own properties. When they are not dissolved in water, many bases are solids in the form of crystals. In solution, bases feel slippery and taste bitter. Like strong acids, strong bases are corrosive. Bases can burn you. Never taste or touch a substance to see if it is basic. Bases also cause indicators to change color. Litmus paper is an indicator that turns blue in bases.

What are some common bases?

The table below lists some common bases and their uses. You may have used many common bases found in cleaning products and not even known it.

Picture This

2. **Identify** Look at the bases listed in the table. What molecule do most of the bases have in their chemical formulas?

_____

Common Bases and Their Uses		
Name, Formula	**Uses**	**Other Information**
Aluminum hydroxide, $Al(OH)_3$	Color-fast fabrics, antacid, water purification	Sticky gel that collects suspended clay and dirt particles on its surface
Calcium hydroxide, $Ca(OH)_2$	Leather-making, mortar and plaster, lessen acidity of soil	Called caustic lime
Magnesium hydroxide, $Mg(OH)_2$	Laxative, antacid	Called milk of magnesia when in water
Sodium hydroxide, $NaOH$	To make soap, oven cleaner, drain cleaner, textiles, and paper	Called lye and caustic soda; generates heat (exothermic) when combined with water, reacts with metals to form hydrogen
Ammonia, NH_3	Cleaners, fertilizer, to make rayon and nylon	Irritating odor that is damaging to nasal passages and lungs

☑ **Reading Check**

3. **Explain** Why is water the main solvent for acidic and basic solutions?

_____

_____

_____

_____

Solutions of Acids and Bases

Many products that contain acids and bases are solutions. Water is the main solvent for these solutions because water molecules have polarity. Remember, polarity means a molecule has a slight positive charge on one end and a slight negative charge on the other end. ☑

What happens when acids dissolve in water?

Remember that an acid produces hydrogen ions (H^+) in water. When an acid dissolves in water, the negative ends of nearby water molecules attract the positive hydrogen in the acid. The acid separates into ions, which is called dissociation. What is left after dissociation are negative ions and positive H^+ ions. The H^+ ions combine with water molecules to form hydronium ions (H_3O^+). Therefore, an acid compound produces hydronium ions when dissolved in water. The figure shows the dissociation of hydrogen chloride (HCl) into hydronium and chloride ions.

Picture This

4. **Identify** When HCl dissociates, what ions does it produce?

_____

_____

What happens when bases dissolve in water?

Bases form hydroxide ions (OH^-) in water. In the table of bases on the previous page, you can see that most bases have –OH in their formulas. When bases dissolve in water, the positive ends of nearby water molecules attract the OH^- ions in the base. The base dissociates. What is left after dissociation are positive ions and negative OH^- ions. Unlike acid dissociation, the OH^- ions do not combine with water molecules. The figure shows the dissociation of sodium hydroxide (NaOH).

$$NaOH + H_2O \rightarrow Na^+ + OH^- + H_2O$$

Picture This

5. **Identify** When NaOH dissociates, it produces a molecule of water, H_2O and two ions. What ions does it produce?

_____

_____

How is ammonia different from other bases?

Ammonia, NH_3, is a base that does not contain –OH. In water, ammonia actually dissociates water molecules. An ammonia molecule attracts a hydrogen ion from a water molecule to form an ammonium ion, NH_4^+. The rest of the water molecule is a hydroxide ion, OH^-. Ammonia is found in many household cleaners. You should never use products containing ammonia with other cleaners that contain chlorine (sodium hypochlorite), such as bathroom bowl cleaners and bleach. Ammonia reacts with sodium hypochlorite and produces toxic gases that can severely damage lung tissue and cause death.

● After You Read

Mini Glossary

acid: a substance that produces hydrogen ions, H$^+$, in a water solution

base: any substance that forms hydroxide ions, OH$^-$, in a water solution or a substance that accepts H$^+$ ions from acids

hydronium ion: a combination of a water molecule and a hydrogen ion

hydroxide ion: OH$^-$ ion formed when bases dissolve in water

indicator: an organic compound that changes color in acid and base

1. Review the terms and their definitions in the Mini Glossary. Write a sentence using the term for the substance that makes many foods taste sour.

_____

_____

2. Fill in the Venn diagram with properties of acids and bases. Be sure to put any properties that acids and bases have in common in the part where the ovals overlap.

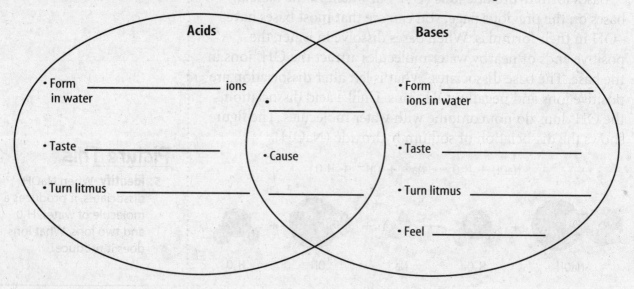

Acids

- Form _____ ions in water

- Taste _____

- Turn litmus _____

- Cause _____

Bases

- Form _____ ions in water

- Taste _____

- Turn litmus _____

- Feel _____

3. ◄ **Mark the Text** You underlined words and sentences in this section that you thought would be important to remember. How did this help you learn about the topics in the section?

_____

_____

_____

End of Section

Acids, Bases, and Salts

section ⊖ Strength of Acids and Bases

⬤ Before You Read

Describe the difference between a food that is very sour and a food that is just a little sour.

_____

_____

Copyright © Glencoe/McGraw-Hill, a division of The McGraw-Hill Companies, Inc.

⬤ Read to Learn

Strong and Weak Acids and Bases

Some acids, like the sulfuric acid in car batteries, can burn your skin. Some acids are stronger than others. The strength of an acid or a base depends on how many acid molecules dissociate into ions in water. A **strong acid** is one in which almost all the acid molecules dissociate in water.

Hydrochloric acid (HCl), nitric acid (HNO$_3$), and sulfuric acid (H$_2$SO$_4$), are examples of strong acids. A **weak acid** is one in which only a small number of the acid molecules dissociate in water. Acetic acid (CH$_3$COOH) and carbonic acid (H$_2$CO$_3$) are weak acids.

Ions in solution can conduct electric current. Acids and bases can carry an electric current because they dissociate into ions. The more ions in a solution, the more electric current the solution can conduct. Strong acids have many ions in solution, so they conduct current well. Weak acids have few ions in solution and do not conduct current as well. ☑

How do you write chemical equations for acid dissociation?

An equation for the dissociation of a strong acid uses a single arrow that points toward the ions that form.

$$HCl(g) + H_2O(l) \rightarrow H_3O^+(aq) + Cl^-(aq)$$

What You'll Learn

- what makes acids and bases strong or weak
- how strength and concentration are alike and different

▸ Study Coach

Identifying the Main Point Write down the main point of each paragraph as you read this section. After you read, look over the main points to make sure you understand them.

☑ **Reading Check**

1. **Identify** Which conducts an electric current best, a strong acid or a weak one?

_____

B Compare and Contrast
Make the following Foldables to understand how the strength and concentration of acids and bases are the same and different.

Strength: Concentration:

Weak Acids In weak acids, like acetic acid, only some of the acid dissociates. An equation for the dissociation of a weak acid uses double arrows pointing in opposite directions.

$$CH_3COOH(l) + H_2O(l) \rightleftharpoons H_3O^+(aq) + CH_3COO^-(aq)$$

What are strong and weak bases?

A **strong base** is one that dissociates completely in solution. Sodium hydroxide (NaOH) is an example of a strong base. An equation for the dissociation of a strong base uses a single arrow that points toward the ions that form.

Strong base: $NaOH(s) \rightarrow Na^+(aq) + OH^-(aq)$

A **weak base** is one that does not dissociate completely in solution. Ammonia (NH_3) is a weak base. An equation for the dissociation of a weak base uses double arrows pointing in opposite directions.

Weak base: $NH_3(aq) + H_2O \rightleftharpoons NH_4^+(aq) + OH^-(aq)$

How are strength and concentration described?

The words *strong* and *weak* refer to how easy it is for the acid or base to dissociate in solution. Strong acids and bases dissociate completely. Weak acids and bases dissociate only partially.

The words *dilute* and *concentrated* tell how much acid or base is dissolved in the solution. A dilute solution means there is a small amount of acid or base in the solution. A concentrated solution means there is a large amount of acid or base in the solution.

You can have dilute solutions of strong acids and bases. You also can have concentrated solutions of weak acids and bases. The figure on the left does not have as many particles in the solution. But, all of the particles are ions. This is a dilute solution of a strong acid. The next figure has more particles in solution, but, not all of the particles are ions. Some of them are still acetic acid. This is a concentrated solution of a weak acid.

Picture This

2. **Describe** In the figure, the solution of acetic acid is more concentrated than the solution of hydrochloric acid. Does this mean acetic acid is stronger than hydrochloric acid? Explain.

_____

_____

_____

_____

This is a dilute solution of HCl.

This is a concentrated solution of acetic acid.

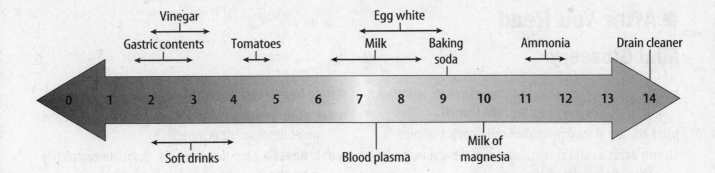

pH of a Solution

If you have a swimming pool or keep tropical fish, you know that the pH of the water must be kept at certain levels. The **pH** of a solution is a measure of the concentration of H^+ ions in the solution. The pH of a solution is measured on a scale ranging from 0 to 14. The greater the concentration of H^+ ions is, the lower the pH and the more acidic the solution.

The figure shows a pH scale and the pH of some common acid and base substances. Solutions with a pH lower than 7 are acidic. Solutions with a pH greater than 7 are basic. The greater the pH value, the more basic the solution is. A solution with a pH of 7 is called neutral. In a neutral solution, concentrations of H^+ and OH^- ions are equal. Pure water at 25°C has a pH of 7.

You can measure pH with universal indicator paper. The paper changes color when H_3O^+ or OH^- ions are present in solution. The color of the paper is then compared to colors in a chart to find the pH. You also can find pH with an electronic pH meter. The meter has electrodes that determine the pH when they are placed in a solution.

Why is pH of blood important?

The pH of your blood must remain between 7.0 and 7.8. If your blood pH goes outside of this range, enzymes do not work. Enzymes are protein molecules that act as catalysts for many of your body's reactions. Why doesn't your blood pH change when you eat acidic foods? Your blood contains compounds called buffers that allow small amounts of acids or bases to be absorbed without changing the pH. ☑

Buffers are solutions containing ions that react with additional acids or bases to decrease their effects on pH. The buffers help your blood stay at an almost constant pH of 7.4. One buffer system in the blood involves bicarbonate ions, HCO_3^-.

Copyright © Glencoe/McGraw-Hill, a division of The McGraw-Hill Companies, Inc.

3. Use a Scale Which is more acidic, a pH of 9 or a pH of 3?

_____

FOLDABLES

C Organize Information
Make the following Foldable to help you organize information about pH. Write important information about acids under the first tab and important information about bases under the second tab.

✓ **Reading Check**

4. Explain What are the compounds called that allow you to eat acidic foods without causing a pH change in your blood?

_____

● After You Read

Mini Glossary

buffer: a solution containing ions that react with additional acids or bases to decrease their effects on pH

pH: a measure of the concentration of H^+ ions in a solution

strong acid: an acid in which almost all the acid molecules dissociate in water

strong base: a base that dissociates completely in solution

weak acid: an acid in which only a small number of the acid molecules dissociate in water

weak base: a base that does not dissociate completely in solution

1. Review the terms and their definitions in the Mini Glossary. Choose two related terms and use them together in a sentence.

_____

_____

2. Complete the table to describe acids and bases.

Understanding Acids and Bases	
Property	**Description**
Strength	• Strong acids and bases almost completely dissociate in solution. • Weak acids and bases _____ _____.
Concentration	• Concentrated acids and bases _____ _____ • Dilute acids and bases have few particles in solution.
pH	• Acids have a pH lower than 7. • Bases have a pH _____

3. **Study Coach** As you read this section, you wrote down the main points of the section. How did these help you understand the section better?

_____

_____

End of Section

Acids, Bases, and Salts

section ❸ Salts

● Before You Read

What does the word *salt* mean to you? How do you use salt?

_____

_____

● Read to Learn

Neutralization

<u>Neutralization</u> is a chemical reaction between an acid and a base that happens in a water solution. For example, NaOH neutralizes HCl. Hydronium ions from the acid combine with hydroxide ions from the base, producing neutral water.

$$H_3O^+(aq) + OH^- \rightarrow 2H_2O(l)$$

Antacids are medicines that contain bases or other compounds that neutralize the HCl in your stomach. One of these antacids is sodium bicarbonate—NaHCO$_3$.

$$HCl(aq) + NaHCO_3(s) \rightarrow NaCl(aq) + CO_2(g) + H_2O(l)$$

How is salt formed?

HCl, shown above, is being neutralized by NaOH. But only half of the ions are shown. The other ions react to form a salt. A <u>salt</u> is a compound formed when the negative ions from an acid combine with the positive ions from a base. When HCl reacts with NaOH, the salt formed is sodium chloride.

$$Na^+(aq) + Cl^-(aq) \rightarrow NaCl(aq)$$

What You'll Learn
- identify a neutralization reaction
- about salt and how it forms
- about soaps and detergents

Study Coach

Make Flash Cards As you read the section, look for the main points in each paragraph. Write each main point on a note card. Use the note cards as flash cards to help you remember the main points.

FOLDABLES™

ⓓ Organize Information
Make the following Foldable to help you organize information about how salts form.

How do acid-base reaction equations look?

This general equation shows acid-base reactions in water:

$$\text{acid} + \text{base} \rightarrow \text{salt} + \text{water}$$

The equation for the reaction between HCl and NaOH is:

$$HCl(aq) + NaOH(aq) \rightarrow NaCl(aq) + H_2O(l)$$

Salts

Salt is necessary for many organisms. Some common salts and their uses are shown in the table. Most salts are made up of a positive metal ion and an ion with a negative charge, such as Cl^- or CO_3^{2-}. Ammonium salts contain the ammonium ion, NH_4^+, instead of a metal.

Picture This

1. Identify Which salt is used to make fertilizers?

_____

_____

Some Common Salts and Their Uses		
Name, Formula	**Common Name**	**Uses**
Sodium chloride, NaCl	Salt	Food, manufacture of chemicals
Sodium hydrogen carbonate, $NaHCO_3$	Sodium bicarbonate Baking soda	Food, antacids
Potassium nitrate, KNO_3	Saltpeter	Fertilizers
Ammonium chloride, NH_4Cl	Sal ammoniac	Dry-cell batteries

Titration

You can find the concentration of an acid or base by titration. In **titration** (ti TRAY shun), a solution of known concentration is used to find the concentration of another solution.

To do a titration on an acid solution, measure the volume of the solution of unknown concentration. Add a few drops of an indicator to the solution. Phenolphthalein (fee nul THAY leen) is an indicator that has no color in an acid but turns pink in a base.

Since the unknown solution is an acid, slowly and carefully add a base solution of known concentration to the acid-and-indicator mixture. Toward the end of the titration, add base solution drop by drop until one drop turns the solution pink and the color stays. The completed titration is the end point. The acid is neutralized. Use the volume of base that was added to calculate the concentration of the acid solution.

 Reading Check

2. Explain In a titration, what does it mean when the color changes in the unknown solution?

_____

_____

_____

What are indicators?

Many natural substances are acid–base indicators. The indicator litmus comes from a combination of a fungus and an algae called lichen. The flowers of the hydrangea plant are indicators. The flowers are blue when the pH of the soil is acidic. The flowers are pink when the pH of the soil is basic. Red cabbage juice is also an indicator. The juice is deep red at pH 1, lavender at pH 7, and yellow-green at pH 10. ☑

Soaps and Detergents

At the supermarket, you will see many different kinds of soaps and detergents. Some are liquids and some are solids.

What are soaps?

<u>Soaps</u> are organic salts. They have a nonpolar organic chain of carbon atoms on one end of the molecule. On the other end, they have either a sodium or potassium salt of a carboxylic (kar bahk SIHL ihk) acid—COOH group. The figure below shows a soap molecule.

COO Na
Ionic head

Soaps clean well for two reasons. Look at figure on the left below. The long, nonpolar, hydrocarbon tail mixes with oils and dirt so they can be removed easily. The polar head attracts water molecules. The figure on the right shows how the attraction between the soap and water molecules helps to wash away the dirt and oil linked with the soap.

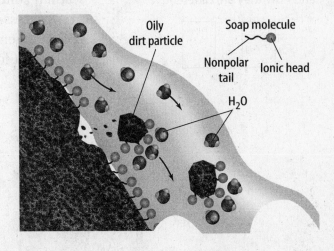

H₂O
Soap
Oil
Soap
Oil
H₂O

Oily dirt particle
Soap molecule
Nonpolar tail
Ionic head
H₂O

✓ Reading Check

3. **List** Name two acid-base indicators that exist in nature. p-

_____

_____

Picture This

4. **Describe** Which part of a soap molecule attracts dirt and oil? Which part attracts water molecules?

_____

_____

A long-chain fatty acid is reacted with sodium or potassium hydroxide to make simple soaps. The fatty acids used to make soaps are natural oils, such as canola, palm, and coconut. A problem with soaps is that tap water often contains metal ions such as calcium, magnesium, and iron. When this occurs, it is known as hard water. These metal ions can react with soap, replacing the sodium or potassium ions. This replacement makes soap insoluble in water. The ions separate out of solution in the form of soap scum. ☑

What are detergents?

Detergents are synthetic products that are made from petroleum molecules, instead of from natural fatty acids like soaps. Similar to soaps, they have long hydrocarbon chains. But instead of a carboxylic acid group ($-COOH$) at the end, they may contain a sulfonic acid group ($-S(=O)_2-OH$). These acids form salts that dissolve more easily in hard water, reducing the problem of soap scum.

Detergents can also be used in cold water. Most contain additional ingredients called builders and surfactants to enhance the ability to clean and to produce suds. Surfactants draw water to the problem area by reducing the surface tension of water, which makes it flow easier.

When can detergents be a problem?

Despite solving the problem of cleaning in hard water, detergents are not the complete solution to our needs. Some detergents contained phosphates but are no longer produced because they cause water pollution. The phosphates act like fertilizer and create algal blooms. Algae growing rapidly in these blooms, use up oxygen and minerals in a stream which in turn kills fish and other aquatic organisms that need that oxygen.

Certain sulfonic acid detergents also present problems in the form of excess foaming in water treatment plants and streams. These detergents do not break down easily and remain in the environment for long periods of time.

Copyright © Glencoe/McGraw-Hill, a division of The McGraw-Hill Companies, Inc.

✔ **Reading Check**

5. **Define** What is hard water?

_____

_____

FOLDABLES™

E Compare and Contrast
Make the following Foldables to help you understand how soaps, detergents, and esters are similar and how they are different.

Soaps/Detergents: Esters:

● After You Read

Mini Glossary

neutralization: a chemical reaction between an acid and a base that happens in a water solution

salt: a compound formed when the negative ions from an acid combine with the positive ions from a base

soap: an organic salt

titration: a process in which a solution of known concentration is used to find the concentration of another solution

1. Review the terms and their definitions in the Mini Glossary. Write a sentence about the types of solutions used with titration.

_____

_____

2. The graphic below shows the set-up for a titration. The beaker contains a known amount an acid of unknown concentration. In the boxes, write what must be added to the beaker and the pipette to do a titration. Then, describe the procedure.

Added:

_____

_____

Procedure:

_____

_____

_____

_____

Added:

_____

_____

Acid-unknown concentration

End of Section

Organic Compounds

section ❶ Simple Organic Compounds

What You'll Learn

- about organic and inorganic carbon compounds
- difference between saturated and unsaturated hydrocarbons
- identify isomers

Study Coach

Make Flash Cards After you read each page, write down two questions you think your teacher might ask on a test. Write each question on a flash card. Write the answer on the other side. Go over the questions until you know the answers.

FOLDABLES

Ⓐ Make a Venn Diagram Make a Foldable like the one below. List the characteristics of organic and inorganic compounds and of those they share.

Organic Compounds

Both

Inorganic Compounds

● Before You Read

Many people think organic means that something doesn't contain chemicals. Write what you think organic means. Give an example of something you think is organic.

_____

_____

● Read to Learn

Organic Compounds

What do you have in common with your shoes, your sunglasses, and your backpack? Those items are made of compounds that contain the element carbon—and so are you. Most compounds that contain the element carbon are **organic compounds**.

Before 1830, scientists thought organic compounds were made only by living organisms. That's how they got their name. Then, scientists began making organic compounds in laboratories.

More than 90 percent of carbon compounds are organic compounds. Other carbon compounds, like carbon dioxide and carbonates, are inorganic compounds.

Why does carbon form so many organic compounds?

An atom of carbon has four electrons in its outer energy level. It can form one covalent bond with each of these electrons. A covalent bond forms when two atoms share a pair of electrons. A carbon atom can form bonds with other carbon atoms or with atoms of other elements. There are many carbon compounds because carbon can form so many bonds. Some carbon compounds are small, like the ones used as fuels. Other carbon compounds are complex, like the ones found in medicines and plastics. But the structure of the carbon atom is not the only reason that carbon can form so many compounds.

Straight Chain

H H H H H H H
| | | | | | |
H—C—C—C—C—C—C—C—H
| | | | | | |
H H H H H H H

Branched Chain

H
|
H—C—H
|
H   H
| |
C=C—C=C
| | |
H   H   H
H

Cyclic Chain

(vanillin structure with OH, ring carbons, O—C—H and H—C=O groups)

How can carbon atoms arrange themselves?

Another reason carbon atoms can form so many organic compounds is that carbon atoms can bond together in different ways—chains, branched chains, and rings. Look at the figure above. The first structure shows carbon atoms bonded together in a straight chain. This compound is heptane. Heptane is an organic compound in gasoline. In the second structure, the carbon atoms bond together in a branched chain. This compound is isoprene, an organic compound in natural rubber. Look at the third structure. See how some of the carbon atoms form a ring? This ring is called a cyclic chain. The compound is vanillin, which is an organic compound found in vanilla flavoring. Carbon also can form single, double, or triple covalent bonds.

Hydrocarbons

There are many organic compounds that contain only carbon and hydrogen. A compound that is made of only carbon and hydrogen atoms is called a **hydrocarbon**. A furnace, stove, or water heater may burn natural gas. Natural gas contains the hydrocarbon methane. The chemical formula for methane is CH_4.

The figure shows two other ways to represent the formula of a compound. The first is a space-filling model of a methane molecule. It shows the relative size and arrangement of the

H
|
H—C—H
|
H

atoms in the molecule. The second item in the figure is a structural formula for a methane molecule. Structural formulas use lines to show the bonds between atoms. Each line represents a covalent bond. The structural formula of methane shows one carbon atom bonded to four hydrogen atoms. Chemists usually use chemical formulas or structural formulas when they write about chemical reactions.

Picture This

1. **Conclude** Why is the structure of isoprene called a branched chain?

_____

_____

_____

Picture This

2. **Interpret a Scientific Illustration** Circle the hydrogen atoms in the space-filling model and in the structural formula of methane.

Applying Math

3. **Finding Patterns** The chemical formula for pentane is C_5H_{12}. Using the pattern in the table, draw the structural formula for pentane.

Single Bonds

To understand organic compounds you must understand how carbon atoms form bonds. They can form single, double, or triple covalent bonds. Hydrocarbons with only single-bonded carbon atoms are called **saturated hydrocarbons.** The compound is saturated because each carbon atom is bonded to as many hydrogen atoms as possible.

Some Hydrocarbons				
Name	Methane	Ethane	Propane	Butane
Chemical Formula	CH_4	C_2H_6	C_3H_8	C_4H_{10}
Structural Formula	H | H—C—H | H	H H | | H—C—C—H | | H H	H H H | | | H—C—C—C—H | | | H H H	H H H H | | | | H—C—C—C—C—H | | | | H H H H

The table lists four saturated hydrocarbons. Look at the structural formulas. You can see how each carbon atom looks like a link in a straight chain connected by single covalent bonds.

Boiling Points of Hydrocarbons The graph below shows the boiling points of five saturated hydrocarbons. Methane has one carbon atom and a boiling point of about –160°C. Ethane has two carbon atoms and a boiling point of about –90°C. Propane has two carbon atoms. What is propane's boiling point? Do you see a pattern? The boiling points of saturated hydrocarbons increase as the number of carbon atoms in the chain increases.

Applying Math

4. **Use Graphs** What is the approximate boiling point of propane?

Boiling Points of Hydrocarbons

What are isomers?

<u>Isomers</u> are compounds that have the same chemical formula, but have different molecular structures and shapes. Thousands of hydrocarbons are isomers. Butane and isobutane are two of them. ☑

If you have cooked on a camping stove, you may have used the hydrocarbon butane. The chemical formula of butane is C_4H_{10}. Another hydrocarbon called isobutane also has the chemical formula C_4H_{10}. How can two different molecules have the same chemical formula but different names?

Look at the structural figures below. The first one shows a molecule of butane. In butane, the carbon atoms form a straight chain. But in isobutane, the second structural figure, the chain of carbon atoms is branched. The arrangement of carbon atoms changes the shape of each molecule. The arrangement of carbon atoms also often affects each molecule's physical properties. Butane and isobutane are isomers.

Butane
C_4H_{10}

Isobutane
C_4H_{10}

Look at the table below. It lists properties for butane and isobutane. The melting point and boiling point of isobutane are lower than the melting point and boiling point of butane. Generally, the more branches an isomer has, the lower its melting point and boiling point are.

Properties of Butane Isomers

Property	Butane	Isobutane
Description	Colorless gas	Colorless gas
Density	0.60 kg/L	0.603 kg/L
Melting point	−135°C	−145°C
Boiling point	−0.5°C	−10.2°C

✔ Reading Check

5. Identify Name two isomers.

_____

_____

Picture This

6. Illustrate Circle the isomer of butane that is in the form of a straight chain.

Applying Math

7. Use Negative Numbers Is the boiling point of isobutane higher or lower than the boiling point of butane? By how much?

Picture This

8. Draw and Label In the space below, draw a picture of your right hand and its mirror image. Label one "right hand" and the other "mirror image."

✔ **Reading Check**

9. Apply Fill in the blanks to complete the sentence: Unsaturated hydrocarbons are compounds that have at least one (a.) _____ bond or (b.) _____ bond.

(a.) _____

(b.) _____

Are there other kinds of isomers?

There are many other kinds of isomers. Not all isomers are in the form of a straight chain molecule and a branched chain molecule. Some isomers differ only slightly in the way their atoms are arranged. Some isomers form what are called right-handed and left-handed molecules.

Try this: Hold your right hand in front of a mirror. What do you see? You see an image that looks like it is your left hand. This is a mirror image. The same idea describes right-handed and left-handed molecules. They are mirror images of each other. Everything else about their structure is the same. Right-handed and left-handed isomers have nearly identical physical and chemical properties.

Multiple Bonds

Hydrocarbons can also have multiple bonds. The last three letters in the name of a hydrocarbon tell what type of bond is in the molecule. Compounds that end in –ane have only single bonds between the carbons. Compounds that end in –ene have at least one double bond. Those that end in –yne have at least one triple bond. Hydrocarbons with one double or triple bond are called an **unsaturated hydrocarbons.**

A gas that is given off by ripening fruit is ethylene, a hydrocarbon known as ethene (eh THEEN). The chemical formula for ethene is C_2H_4. It has one double bond as shown in the figure. The figure also shows the structural formula for ethyne (eh THIHN). The ethyne molecule has a triple bond. ✔

Carbon Rings

Some carbon chains can join to form rings. Cyclopropane is a circle of four carbon atoms. In the figure above there is a special form of ringed carbon atoms called a benzene ring. **Benzene** forms a cyclic six carbon ring that has alternating single and double bonds. All of the electrons of these six carbons are shared equally making this structure very strong and stable. It has its own special symbol—a circle inside a hexagon.

Some organic compounds contain more than one ring structure. Naphthalene has two rings fused together. Others may have three or more rings. These make up compounds with distinctive smells and odors.

● After You Read

Mini Glossary

benzene: a cyclic hydrocarbon whose carbon atoms are joined with alternating single and double bonds

hydrocarbon: a compound that is made of only carbon and hydrogen atoms

isomers: compounds that have the same chemical formula but have different molecular structures and shapes

organic compound: a compound that contains the element carbon

saturated hydrocarbon: a hydrocarbon made of only single-bonded carbon atoms

unsaturated hydrocarbon: a hydrocarbon that has at least one double bond or triple bond

1. Review the terms and their definitions in the Mini Glossary. Write a sentence using a term that could apply to compounds with only single bonds.

   _____

   _____

2. Complete the flow chart by filling in the type of compound and its characteristics.

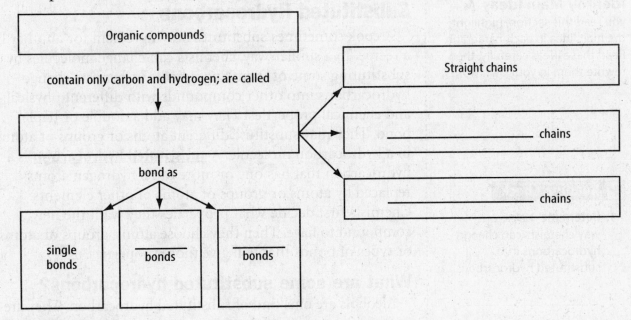

3. **Study Coach** You made flash cards of questions your teacher might ask on a test. Was this a good strategy for learning the information? Why or why not?

   _____

   _____

End of Section

Organic Compounds

section ② Substituted Hydrocarbons

What You'll Learn

- what aromatic compounds are
- what alcohols and acids are
- some organic compounds you use every day

Mark the Text

Identify Main Ideas As you read this section, highlight the main idea in each paragraph. Read these ideas carefully, then rewrite them in your own words.

✔ Reading Check

1. **Determine** Name one way chemists can change hydrocarbons into substituted hydrocarbons.

_____

_____

● Before You Read

If you're baking a cake you can't substitute baking soda for sugar. It just wouldn't taste the same. On the lines below, tell how you think a cook might come up with a new recipe.

_____

_____

● Read to Learn

Substituted Hydrocarbons

A cook sometimes substitutes one ingredient for another in a recipe. In a similar way, chemists can change molecules by substituting some of their "ingredients." Chemists change hydrocarbons into other compounds with different physical and chemical properties. They may add a double or triple bond. They may substitute different atoms or groups of atoms in a hydrocarbon molecule. A **substituted hydrocarbon** is a hydrocarbon that has one or more of its hydrogen atoms replaced by atoms or groups of atoms of other elements. Chemists first decide what properties they want the new compound to have. Then they choose atoms, groups of atoms, or types of bonds that will give those properties. ☑

What are some substituted hydrocarbons?

Alcohols are one kind of substituted hydrocarbon. They are an important group of organic compounds. They are used as solvents and disinfectants and as pieces to make larger molecules. An **alcohol** forms when a hydroxyl group, –OH, replaces one or more hydrogen atoms in a hydrocarbon. A hydroxyl group contains an oxygen atom and a hydrogen atom.

The first figure shows the structural and chemical formula for ethanol. Notice that the ethanol molecule has –OH on one end. This tells you that ethanol is an alcohol. When the sugar in grains or fruit ferments, it produces ethanol.

Ethanol
C_2H_5OH

What are organic acids?

Another group of substituted hydrocarbons is organic acids. Organic acids form when a carboxyl group, –COOH, attaches to a carbon atom of a hydrocarbon molecule. A carboxyl group contains one carbon atom, two oxygen atoms, and one hydrogen atom. ☑

The second figure shows the structural and chemical formula for acetic acid. Acetic acid is an organic acid found in vinegar. Other organic acids include citric acid, found in fruits such as oranges, and lactic acid, found in sour milk.

Acetic acid
CH_3COOH

What other elements can be added to hydrocarbons?

Hydrogen and oxygen are not the only atoms added to hydrocarbons to make substituted hydrocarbons. Chlorine atoms are used as well. When four chlorine atoms replace four hydrogen atoms on a molecule of ethene, they form a molecule of tetrachloroethene (teh truh klor uh eth EEN). This is a solvent used in dry cleaning. The third figure shows the structural and chemical formula for tetrachloroethene.

Tetrachloroethene
C_2Cl_4

When four fluorine atoms replace four hydrogen atoms in ethylene, they form a compound that can be made into a black, shiny material used for nonstick cookware. Nitrogen, bromine, and sulfur atoms also are used in substituted hydrocarbons. Compounds called thiols are formed when sulfur replaces the oxygen in the –OH group of an alcohol. Thiols also are called mercaptans. Most mercaptans smell very bad. In fact, there is a mercaptan in skunk spray.

Amines An <u>amine</u> forms when an amine group ($–NH_2$) replaces a hydrogen atom in a hydrocarbon. For example, aniline is formed when $–NH_2$ replaces a hydrogen atom on a benzene ring. Aniline is used to make dyes. Amines are also essential for life and help build proteins.

Mercaptans When the group –SH replaces a hydrogen atom in a hydrocarbon, the resulting compound is a thiol, or mercaptans. Most mercaptans have unpleasant odors. This can be useful to animals such as skunks, and strangely enough, to pleasant-smelling grapefruits.

Natural gas has no odor of its own so it is impossible to smell a gas leak. For this reason, gas companies add small amounts of a mercaptan to the gas to make people aware of leaks.

Picture This

2. Use a Diagram In the structural formulas for ethanol and acetic acid, circle the symbols that tell you they are substituted hydrocarbons.

☑ Reading Check

3. Identify What is the symbol for a carboxyl group?

_____

_____

Picture This

4. Infer Look at the diagram. The prefix *tetra-* means "four." How do you suppose tetrachloroethene got its name?

_____

_____

FOLDABLES™

B Classify Make a Foldable like the one below. As you read the section, list the characteristics and examples of aromatic compounds and substituted hydrocarbons.

What is artificial flavoring?

When you mix an acid and a base you get water and a salt. What do you get when you mix alcohol and an organic acid? The answer is water and an ester. An **ester** is a substituted hydrocarbon with a –COOC– group. Esters are used to make soaps, flavorings, perfumes, and fibers for clothing.

How are esters used for flavor?

Many drinks taste like real fruit but do not contain any fruit, only artificial flavors. Most likely, the artificial flavors come from esters. One reaction to prepare an ester involves mixing butyric (byew TIHR ihk) acid, sulfuric acid, and ethyl alcohol. The process removes a hydrogen ion (H+) from the butyric acid and a hydroxide (OH–) ion from the alcohol to make a water molecule. The result is ethyl butyrate which is used in pineapple flavoring. Other esters are used in creating banana, apple, orange, and apricot flavors and odors.

What is polyester?

The reaction between an organic acid and an alcohol produce long, nonpolar chains of many, or poly, esters. The chains are closely packed together making them strong. Polyester fibers can be used alone or are woven or knitted with natural fibers to make fabrics that are strong, water repellant, and colorfast.

Aromatic Compounds

When you chew a piece of gum, you get a pleasant taste in your mouth. You also may notice a smell. Organic compounds produce these tastes and smells. The first figure shows the structural formula for the compound methyl salicylate. Methyl salicylate produces the taste and smell of wintergreen and is used in some gums and candies. Organic compounds also produce the tastes and smells of cloves, cinnamon, and vanilla.

Other organic compounds have unpleasant smells and tastes. The second figure shows the structural formula for acetyl salicylic acid, also called aspirin. Aspirin has a sour taste.

These are aromatic compounds. You might think they get that name because they are smelly. Some of them are, but look at the two structural formulas. What do they have in common? They both have a ring in their structural formula. To chemists, an **aromatic compound** is a compound that contains a benzene structure having a ring made of six carbon atoms.

Structural Formulas

Methyl Salicylate

Acetyl Salicylic Acid

Picture This

5. **Use Math Vocabulary** Look at the structural formulas in the figures. What is the name for a geometric figure with six sides?

● After You Read

Mini Glossary

alcohol: a compound that forms when a hydroxyl group, –OH, replaces one or more hydrogen atoms in a hydrocarbon

amine: a substituted hydrocarbon with a –NH₂ group

aromatic compound: a compound that contains a benzene structure having a ring made of six carbon atoms

ester: a substituted hydrocarbon with a –COOC– group

substituted hydrocarbon: a hydrocarbon that has one or more of its hydrogen atoms replaced by atoms or groups of atoms of other elements

1. Review the terms and definitions in the Mini Glossary. On the lines below, write a sentence that tells how the terms substituted hydrocarbon and alcohol are related.

_____

_____

2. Complete the outline to help you organize what you learned about other organic compounds.

> **Other Organic Compounds**
>
> I.   Substituted Hydrocarbons
>
>   A. A substituted hydrocarbon is _____
>
>   _____
>
>   _____.
>
>   B. In _____, the –OH group is substituted for a _____.
>
>   C. Organic acids contain the group _____.
>
>   D. Substituted hydrocarbons can contain other elements, including _____,
>
>   _____, and _____.
>
> II. Aromatic Compounds
>
>   A. An aromatic compound is _____
>
>   _____.
>
>   B. A _____ contains six atoms bonded together in a ring.
>
>   C. _____ can contain two or more fused rings.

End of Section

Organic Compounds

section ③ Petroleum—A Source of Organic Compounds

What You'll Learn

- how carbon compounds are obtained from petroleum
- how carbon compounds form long chains of molecules
- what polymers are

Study Coach

Make an Outline Make an outline covering the main ideas about petroleum and carbon compounds.

✔ Reading Check

1. **Draw Conclusions** What do we call the process of separating petroleum into fractions?

_____

_____

● Before You Read

Plastics are made from petroleum. Look around you. List and describe two plastic products that have different properties.

_____

_____

● Read to Learn

What is petroleum?

Do you carry a comb in your pocket or purse? It is probably made of plastic. Where did that plastic come from? It probably came from petroleum. Petroleum is a dark, flammable liquid that is often called crude oil. Petroleum exists deep within Earth. Coal, natural gas, and petroleum are all called fossil fuels because they form from fossilized material.

How can thick, dark, liquid petroleum be turned into a hard object like a comb? To answer that question, you have to understand more about petroleum. Petroleum is a mixture of thousands of carbon compounds. First, the petroleum is brought up from its sources underground. Oil wells are used to pump petroleum, or crude oil, to Earth's surface.

Next, chemists and engineers separate the oil mixture into fractions containing compounds with similar boiling points. Within a fraction, boiling points may range more than 100°C. Fractional distillation is the process of separating petroleum into fractions. Fractional distillation happens at petroleum refineries in metal towers called fractionating towers. ✔

What is a fractionating tower?

A fractionating tower can be as tall as 35 m. Inside the tower, metal plates are arranged like floors inside a building. The plates have many small holes to let vapors pass through. Pipes are connected to the outside of the tower at different levels. The tower uses fractional distillation to separate crude oil into fractions.

How does fractional distillation work?

The figure below shows how a fractioning tower works. Crude oil is pumped into the bottom of the tower. The crude oil is heated to more than 350°C. At 350°C, most of the hydrocarbons in the oil mixture turn into vapor and start to rise. The vapors rise up inside the tower. Vapors of the fractions with the highest boiling points only reach the lowest plates in the tower before they condense. Condensed vapor turns into liquid. The first liquid fractions drain out of the tower through pipes connected to its sides and are collected.

The rest of the vapors continue to rise inside the tower. Vapors of fractions with lower boiling points condense when they reach plates in the middle part of tower. These liquid fractions also drain off and are collected.

Vapors of fractions with the lowest boiling points rise to the highest plates before they condense. Some vapors never condense and are collected as gases at the top of the tower. The figure lists some fractions and the temperature at which they condense.

Below 20°C — **Hydrocarbon gases used for fuels and plastics**

40°C – 200°C — Gasoline

175°C – 275°C — Kerosene

250°C – 400°C — Jet fuel and diesel oil

Above 300°C — Lubricating oil

Above 350°C — Asphalt

Heated crude oil

💡 **Think it Over**

2. **Explain** What property do fractions that condense at the lowest part of a fractioning tower have?

_____

_____

Picture This

3. **Interpret a Figure** What do the temperatures in the figure represent?

_____

_____

Applying Math

4. **Use Data** Which fractions condense at a temperature of 300°C or greater?

_____

_____

_____

_____

C Classify Make half-sheet Foldables like the ones below. Use one to explain what petroleum is and then list things that are made from petroleum. On the other tell how monomers and polymers are related.

What is petroleum?

Monomers and Polymers:

Picture This

5. Distinguish Use one color highlighter to circle a monomer in the figure. Use another color to circle the polymer. In the box below, make a key to show what each color represents.

monomer

polymer

Think it Over

6. List Name three things made of polymers.

_____

_____

_____

Uses for Petroleum Compounds

Some petroleum fractions are used for fuels. Butane and propane are some of the lightest fractions. They are taken from the top of the fractioning tower. Molecules of propane have three carbon atoms. Molecules of butane have four carbon atoms.

The molecules of fractions that condense on the upper plates of the tower are larger and heavier than butane and propane. They have five to ten carbon atoms in their molecules. These compounds are used for gasoline and solvents. Fractions that condense on lower plates have 12 to 18 carbon atoms in their molecules. They are kerosene and jet fuel. Most of the bottom fractions are lubricating oil. Whatever is left over is used to make asphalt to pave roads.

Polymers

Have you ever made a paper chain like the one in the figure? A paper chain can be used as a model for the structure of a polymer. A **polymer** is a very large molecule made from small molecules that link together, like links in a chain. The prefix *poly-* means many. A **monomer** is the small molecule that forms a link in the polymer chain. The prefix *mono-* means one. A polymer chain can contain as many as 10,000 monomers.

What are some common polymers?

One common polymer, or plastic, is made from the monomer ethene, also called ethylene. At room temperature, ethylene is a gas. When ethylene combines with itself repeatedly, it forms a polymer, **polyethylene** (pah lee EH thuh leen). Polyethylene is used to make shopping bags and plastic bottles. Another common polymer is polypropylene (pah lee PRO puh leen). Polypropylene is used to make glues and carpets. Two or more different monomers can combine to make one polymer molecule. When this happens, the monomers are called copolymers.

Polymers can be light and flexible. They also can be so strong that they are used to make items such as plastic pipes, boats, and car bodies. Some polymers are used in place of wood and metal in buildings. Some people call this the "Age of Plastics" because so many things we use today are made of plastic polymers.

What determines the properties of polymer materials?

The properties of a polymer depend on which monomers are used to make them. Polymers can have branches in their chains, like hydrocarbons. Also, like hydrocarbons, the amount of branching and the shape of the polymer affect its properties.

Sometimes the same polymer can take two completely different forms. Polystyrene (pah lee STI reen) is an example. Polystyrene is made from the monomer styrene. Clear CD cases are made from polystyrene. When carbon dioxide gas is blown into melted polystyrene as it is molded, it becomes lightweight and opaque. It is used to make foam cups and packing materials. Bubbles remain in the polymer when it cools, so it is a good insulator. ☑

Polymers can be spun into thread. Fabrics made from strong polymer thread can be used for products that are treated roughly, such as suitcases and backpacks. Polymer materials are used to make bulletproof vests. Polymer fibers also can be made to stretch and then return to their original shape. These polymers are used to make exercise clothing.

What are some other petroleum products?

Other petroleum products are made by separating individual compounds from the petroleum fractions. These compounds can be changed into substituted hydrocarbons, as you learned in the previous section. Chemists use these hydrocarbons to make medicines like aspirin, insecticides, printing ink, and flavorings. Dyes made from petroleum have replaced dyes made from plants and other natural sources.

Are there problems with polymers?

Disposing of the many things made of polymers is a problem in this "Age of Plastics." Polymers do not decompose. A piece of plastic will remain a piece of plastic almost forever.

What can we do with all this plastic? One answer is to recycle. Recycling reuses clean plastics to make new products. Another answer is a process called depolymerization. **Depolymerization** uses heat or chemicals to break long polymer chains into monomer fragments. Then the monomers can be used to make other polymers. Today, depolymerization is expensive because a different process must be used to depolymerize each different polymer. There is still much research to do to make depolymerization practical.

✔ Reading Check

7. **Apply** Which is made from polystyrene? (Circle your answer.)

   a. foam cups

   b. aspirin

   c. ink

   d. fabric

💡 Think it Over

8. **List** Name two other petroleum products that are not polymers.

   _____

   _____

● After You Read

Mini Glossary

depolymerization: a process that uses heat or chemicals to break long polymer chains into monomer fragments

monomer: the small molecule that forms a link in the polymer chain

polymer: a very large molecule made from small molecules that link together

1. Review the terms and definitions in the Mini Glossary. Write a sentence using the three terms *monomer*, *polymer* and *depolymerization*.

_____

_____

_____

2. Complete the flow chart to help you organize what you learned about petroleum and its compounds.

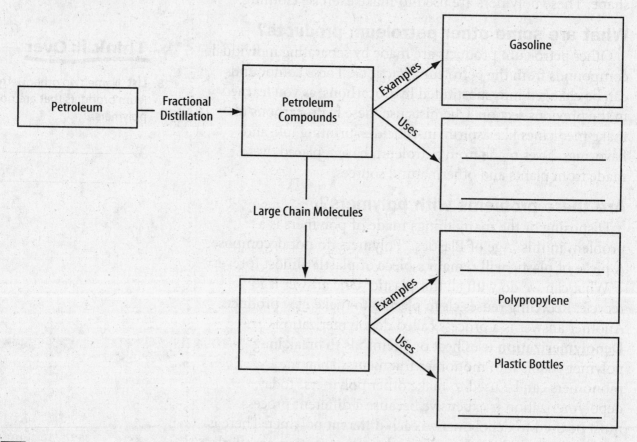

End of Section

chapter 23 Organic Compounds

section ❹ Biological Compounds

● Before You Read

Red meat, poultry, fish, dairy products, some beans, and some vegetables contain protein. List three things you have eaten in the past several days that contain protein.

_____

● Read to Learn

Biological Polymers

Many of the important biological compounds in your body are biological polymers. Biological polymers are huge molecules, just like the polymers you learned about in Section 3. Biological polymers also are made of monomers, but they are larger than monomers in other polymers.

Proteins

Proteins are large organic polymers formed from organic monomers called amino acids. There are only 20 different amino acids. But they can be arranged in so many ways that they make millions of different proteins. Many of the tissues in your body are made of proteins, including muscles, tendons, hair, and fingernails.

What are amino acids?

Amino acids are the monomers that combine to form proteins. The figure shows the structures of two amino acids, glycine and cysteine. Both molecules have an amine group, $-NH_2$. Both also have the group that all organic acids have, the carboxyl group, $-COOH$. All amino acids have an amine and a carboxyl group.

Glycine

Cysteine

Peptide
Glycyl cysteinate

What You'll Learn

- about proteins, nucleic acids, carbohydrates, and lipids
- polymers in food
- biological polymers

Mark the Text

Locate Information As you read about biological compounds in this section, draw lines from the words or phrases in the text to the corresponding parts of the figures.

Picture This

1. **Compare** Highlight the amine group in glycine and cysteine in one color. Highlight the carboxyl group in another color.

A Classify Make a Foldable like the one below. As you read this section, write how your body uses each of these compounds.

Proteins

Nucleic Acids

Carbohydrates

Lipids

✔ Reading Check

2. Infer What kind of organic polymer is DNA?

_____

_____

Picture This

3. Draw To show that there are two bases connected at the slash, draw a slash across each box indicating the base pairs.

The amine group of one amino acid can combine with the carboxyl group of another amino acid. The compound they form is a peptide. The bond joining the two amino acids is a peptide bond. When a peptide contains about 50 or more amino acids, the molecule is called a protein.

What do proteins look like?

Because a protein molecule is such a long chain, it twists around itself. Scientists can identify each protein by the way it twists. Many foods, including meat, dairy products, and some vegetables, contain proteins. When you eat these foods, your body breaks down the large protein molecules into their amino acid monomers. Then, your body uses the amino acids to make new proteins for muscles, blood, and other body tissues.

Nucleic Acids

Nucleic acids are another group of organic polymers that are essential for life. They control the activities and reproduction of cells. One nucleic acid in your body is deoxyribonucleic (dee AHK sih ri boh noo klay ihk) acid, or DNA. **Deoxyribonucleic acid** is in the nuclei of cells, where it codes and stores genetic information. This information is the genetic code. ☑

What is DNA made of?

Monomers called nucleotides make up DNA. **Nucleotides** contain an organic base, a sugar, and a phosphoric acid unit. There are two chains of nucleotides in DNA. These two chains twist around each other and form something like a twisted ladder. This structure is called the double helix. *Helix* means "spiral."

Human DNA contains only four different organic bases, but they can form millions of different combinations. The figure shows how bases on one side of the ladder link with bases on the other side of the ladder to form a base pair. The genetic code of DNA gives instructions for making other nucleotides and proteins needed by the body.

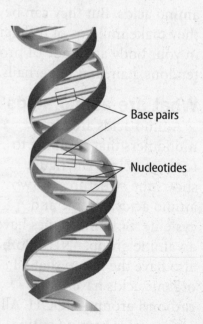

Base pairs

Nucleotides

What is DNA fingerprinting?

Each molecule of DNA in your body has more than 5 million base pairs. Your DNA is different from the DNA of every other person in the world, unless you have an identical twin. Identical twins have exactly the same DNA.

DNA can be used to solve crimes. Scientists can remove DNA from hair, blood, or saliva left at a crime scene. They break up the DNA polymer into its nucleotides. Then scientists use radioactive and X-ray methods to make a picture of the pattern formed by the nucleotides. They compare the DNA pattern from the crime scene to the DNA pattern of a suspect. If the patterns match, they can link the suspect to the crime scene.

Carbohydrates

__Carbohydrates__ are organic compounds made of carbon, hydrogen, and oxygen that have twice as many hydrogen atoms as oxygen atoms. Carbohydrates include sugars and starches. Foods like bread and pasta contain carbohydrates.

What are sugars?

Sugars are one of the main groups of carbohydrates. The figure shows the structural and chemical formulas of sucrose and glucose. Sucrose is also known as table sugar. Your body breaks sucrose down into two simpler sugars, fructose and glucose. Fruit contains the sugar fructose.

Sucrose $C_{12}H_{22}O_{11}$

Glucose $C_6H_{12}O_6$

Glucose is found in your blood. Glucose also is in fruit and honey. You get a quick boost of energy after eating a food rich in sugar.

What is a starch?

A starch is a carbohydrate that is also a polymer. It is made of monomers of the sugar glucose. Your body breaks down starch molecules into monomers of glucose and other sugars. These sugars release energy into your cells.

Picture This

4. **Determine** Look at the structural formula of glucose. How many hydrogen atoms does it contain? _____ How many oxygen atoms? _____ How do these numbers confirm that glucose is a carbohydrate?

_____

_____

_____

_____

_____

_____

Reading Check

6. Determine What are lipids made of? (Circle your answer.)

a. oxygen

b. hydrogen and oxygen

c. carbon and hydrogen

d. oxygen, carbon, and hydrogen

Athletes use starches to give them energy that lasts a long time. The energy from starches can be stored in liver and muscle cells in a compound called glycogen (GLI kuh jun). During a long race, the body releases the stored energy and the runner gets a fresh burst of power.

Lipids

Lipids are organic compounds like fats and oils. Lipids include animal fats, such as butter, and vegetable oils, such as corn oil. Lipids are made of carbon, hydrogen, and oxygen, just as carbohydrates are. But lipids contain these elements in different proportions. For example, lipids have fewer oxygen atoms than carbohydrates have. Also, lipids contain carboxyl groups and carbohydrates do not. ☑

What are some lipids in your diet?

Fats and oils are similar in structure to hydrocarbons. If they have only single bonds between carbon atoms, they are saturated fats. Unsaturated fats have one or more double bonds. An unsaturated fat that has only one double bond is monounsaturated. An unsaturated fat that has two or more double bonds is polyunsaturated.

Fats are lipids that come from animals. Fats usually are saturated and solid at room temperature. Oils are lipids that come from plants. Oils are unsaturated and usually liquid at room temperature. Sometimes hydrogen is added to vegetable oils to saturate their carbon atoms. This makes the oil solid at room temperature. These solid oils are called hydrogenated vegetable shortenings.

Cholesterol Cholesterol (kuh LES tuh rawl) is found in meats, eggs, butter, cheese, and fish. Your body also produces some cholesterol and uses it to build cell membranes. Cholesterol also is found in bile, a digestive fluid. Too much cholesterol in the body may cause serious damage to the heart and blood vessels.

Have you heard that eating too much fat can be unhealthy? Eating too many foods with high amounts of saturated fats and cholesterol may cause heart disease. Some unsaturated fats may protect the heart from some diseases. The body may be more likely to change saturated fats into substances that can block the arteries going to the heart. It is not wise, though, to stop eating all fats. A balanced diet contains some fats as well as proteins and carbohydrates.

● After You Read

Mini Glossary

carbohydrate: an organic compound made of carbon, hydrogen, and oxygen that has twice as many hydrogen atoms as oxygen atoms

deoxyribonucleic acid: a nucleic acid found in the nuclei of cells where it codes and stores genetic information

lipid: an organic compound like fats and oils

nucleic acid: one of a group of organic polymers that are essential for life

nucleotide: a DNA molecule with one of four organic bases, a sugar, and a phosphate unit

protein: a large organic polymer formed from organic monomers called amino acids

1. Review the terms and definitions in the Mini Glossary. Circle the three terms that describe compounds that are in the foods you eat.

2. Write the letter of the term in **Column 2** that matches the example in **Column 1.**

Column 1	Column 2
_____ 1. Made of amino acids	**a.** lipid
_____ 2. Form peptide bonds	**b.** nucleic acid
_____ 3. DNA is one of these	**c.** protein
_____ 4. Gives your body a quick boost of energy	**d.** starch
	e. amino acids
_____ 5. Long-lasting fuel that can be stored in the body	**f.** sugar
_____ 6. Fats and oils	

3. **Mark the Text** You drew a line from the words or phrases in the text to the corresponding part of each figure. How would this strategy help you if you were studying for a test?

_____

_____

_____

End of Section

New Materials Through Chemistry

section ❶ Materials with a Past

What You'll Learn
- the uses of alloys
- how the use of alloys depends on their properties

Outline Create an outline of the section as you read. Include the main ideas and definitions that you read about.

FOLDABLES

A Build Vocabulary
Make the following Foldable to make sure you understand the vocabulary terms in this section.

Alloy
Luster
Ductility
Malleability
Conductivity

● Before You Read

What are some metal objects you have used? How do metals make your life easier?

_____

_____

● Read to Learn

Alloys

For ages, people have searched for things to make their lives easier. Ancient people used stone tools. Later, methods for processing metals were discovered. Today, scientists are still finding ways to blend metals, or make alloys, to make better metal products. An alloy is a mixture of elements that has the properties of metal. Have you ever seen a pewter mug? Pewter is an alloy made of tin, copper, and antimony. Pewter looks and feels like metal. Alloys sometimes have improved properties over the metals they are made of. Alloys can be lighter, stronger, harder, and more durable than the elements they are made of.

How were alloys discovered?

Historians believe the ancient Sumerians discovered bronze in about 3500 B.C. The Sumerians lived in what is now Iraq. By accident, copper and tin ore found in rocks the Sumerians used to keep their campfires from spreading probably melted. The alloy of copper and tin is bronze. Sumerians then used the bronze to make tools and other objects. Bronze was the main alloy used to make tools for about 2,000 years. Historians call this time period the Bronze Age.

Properties of Metals and Alloys

What are the properties of metals and alloys? **Luster** is the ability to reflect light or have a shiny appearance. Ductility (duk TIL uh tee) is the ability to be pulled into wires.

Two other properties are malleability and conductivity. Malleability (mal ee uh BIL uh tee) is the ability to be hammered or rolled into sheets—like aluminum foil. **Conductivity** is the ability to carry heat and electrical charges easily. Copper has good ductility and conductivity so it is often used for electrical wires.

Which properties of an alloy are most important?

The properties of an alloy that are most important depend on how the alloy is used. For example, most gold jewelry is made from an alloy of gold and copper. Pure gold is a bright, expensive metal that is soft and bends easily. Copper is not expensive and it is harder than gold. Gold and copper are melted, mixed, and allowed to cool, forming an alloy. ☑

The properties of the alloy depend on how much of each metal is used. A ring made with a larger percentage of gold would be more expensive, but it would bend more easily. A ring made with a smaller percentage of gold would be less expensive, but it would not bend as easily.

How do you choose an alloy?

A gold and copper alloy would probably not be used to make a drill bit. Why? A drill bit must be much harder than jewelry. The alloys used to make tools, jewelry, or other metal objects depend on how the object will be used. In some tools, such as drill bits, hardness is very important. In jewelry, value and luster are important. In electrical wires, flexibility and conductivity are important. The uses of some alloys and the elements they contain are listed in the table below.

Common Alloys		
Name	**Composition**	**Use**
Bronze	copper, tin	jewelry, marine hardware
Brass	copper, zinc	hardware, musical instruments
Sterling silver	silver, copper	tableware
Pewter	tin, copper, antimony	tableware
Solder	lead, tin	plumbing
Wrought iron	iron, carbon	porch railings, fences, sculpture

☑ **Reading Check**

1. **Conclude** Which metal is softer, gold or copper?

_____

_____

Picture This

2. **Interpret Data** Which metal is used more than any other in the alloys listed in the table?

_____

Uses of Alloys

As you saw in the table on the previous page, alloys are used in many objects. In fact, most metal objects you see around you are made of alloys. Strong alloys are used to make machinery and building materials. The alloys used in car and airplane bodies must be light and strong and resist corrosion.

Alloys are even used in the human body. Steel pins and screws are used to repair broken bones. Steel plates are used to repair damage to the skull. The metal parts used are made from surgical steel. Surgical steel has properties that allow it to resist tissue rejection. Tooth fillings usually are made from silver and mercury.

What is steel and how is it used?

Steel is an alloy of iron, carbon, and other elements. The properties of steel alloys depend on how much carbon and other elements are used in the alloy. The alloys have different properties and different uses. Steel alloys can be very strong and often are used in large buildings and bridges. Ship hulls, bedsprings, and many automobile parts are made from steel. Stainless steel is an alloy that resists corrosion. It is used in surgical instruments and cooking utensils. ☑

New Alloys

Steel is not the only common alloy. Aluminum is a common metal used to make soda cans and cooking foil. Aluminum alloys are strong and light and are used to build large airplanes. Alloys of the metal titanium also are used on aircraft because of their strength and light weight. Strong materials last longer and lighter airplanes are less expensive to fly.

What alloys are used in space?

The original heat shield on the space shuttle was made from ceramic tiles. These tiles resist the large amount of heat generated when a shuttle reenters Earth's atmosphere. The ceramic tiles often crack and break because of the heat and stress of reentry into Earth's atmosphere. Fixing the broken tiles is expensive and time consuming.

Tiles made from a new titanium alloy are being developed. These extremely heat resistant new tiles may be used on space shuttle heat shields. The titanium alloy tiles are larger and easier to attach than the ceramic tiles. The tiles could make the heat shield easier and less expensive to maintain.

✔ Reading Check

3. List What are the two main elements in the alloy steel?

_____

_____

💡 Think it Over

4. Describe What are some properties of aluminum and titanium alloys that make them useful for building aircraft?

_____

_____

● After You Read

Mini Glossary

conductivity: the property of metals and alloys that allows heat and electrical charges to pass through the material easily

luster: the ability to reflect light or have a shiny appearance

1. Review the terms and their definitions in the Mini Glossary. Write a sentence using one of the terms that describes a property of alloys.

_____

_____

2. In the graphic organizer below, list the properties of alloys that might be used to make the listed items. Also give an example of a type of an alloy that has those properties.

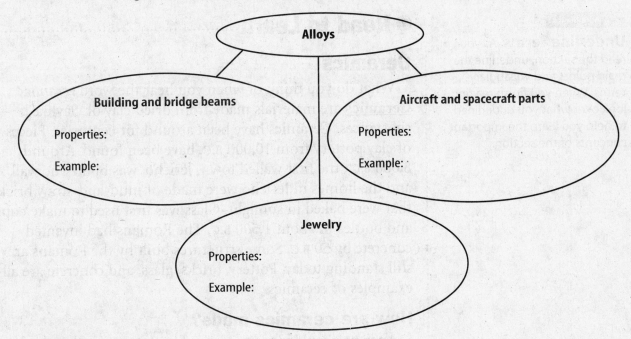

Alloys

Building and bridge beams

Properties:

Example:

Aircraft and spacecraft parts

Properties:

Example:

Jewelry

Properties:

Example:

3. ◀ **Study Coach** You made an outline as you read this section. What did you use for the major categories in your outline?

_____

End of Section

What You'll Learn

- properties of ceramics
- how ceramic materials are used
- what semiconductors are

Underline Terms As you read the section, underline the main points and vocabulary terms. When you finish reading, look over what you underlined to help you learn the important concepts of the section.

● Before You Read

Think about eating a meal at home. You might see metal cooking pans on the stove and metal silverware on the table. What other materials might you see around you?

_____

_____

● Read to Learn

Ceramics

What do you think of when you hear the word *ceramic*? __Ceramics__ are materials made from dried clay or claylike substances. Ceramics have been around for centuries. Pieces of clay pottery from 10,000 B.C. have been found. Around 8,000 B.C., the first walled town, Jericho, was built. The wall and the homes in Jericho were made of mud-and-straw bricks that were baked in sunlight. Glass was first used to make cups and bottles in about 1,500 B.C. The Romans had invented concrete by 50 B.C. Some structures built by the Romans are still standing today. Pottery, bricks, glass, and concrete are all examples of ceramics.

How are ceramics made?

Ancient people made ceramics from materials they could find easily. These were clay, sand, and crystalline rock called feldspar. These materials are still used today to make some ceramics. Newer ceramics might contain metallic compounds and carbon, nitrogen, or sulfur.

Why are ceramics heated?

Ceramic objects are made by molding the processed raw materials into a shape. The ceramic then is heated to a temperature between 1,000°C and 1,700°C. This process, called firing, shrinks the spaces between particles. Ceramic objects are smaller and stronger after firing.

Firing The figures show how firing makes the spaces between the ceramic particles shrink. This makes ceramics strong. They also are resistant to heat. For these reasons, ceramic tiles are used on the heat shield of the space shuttle. But ceramics, even those on the shuttle, are fragile. They will break if dropped or if their temperature changes too quickly.

Not Fired

Fired

How are ceramics used?

Ceramics have some useful properties. They do not react with oxygen, water, acids, bases, salts, or strong solvents. This makes ceramics useful where they might come into contact with these materials. For example, tableware and cookware are often made of ceramics because foods contain acids, salts, and water. Ceramics also make good insulators. You may have seen the ceramic insulators used to connect electric wires to posts. The insulators keep the electric current flowing through the wires instead of through the post and into the ground. ☑

Like alloys, ceramics can be customized, or made to have certain properties. Ceramics are customized by using different raw materials or different processes. The ceramic material in a coffee cup does not need to be the same as the ceramic material in the space shuttle tiles.

What are some new uses for ceramics?

Some new ceramics have been created with properties that most ceramics do not have. For example, most ceramics are insulators. But ceramics made with chromium dioxide are good conductors and some ceramics made with copper are even superconductors. Transparent, conductive ceramics are used to make airplane windshields that stay free of snow and ice.

Ceramics have medical uses in the human body. The properties of ceramics make them resistant to body fluids, which can damage many materials. Surgeons use ceramics to repair or replace joints, such as hips, knees, shoulders, elbows, fingers, and wrists. Dentists use ceramics to repair and replace teeth and to make braces.

Picture This

1. **Describe** Look at the figures showing ceramic material before and after firing. Why is the ceramic denser after firing?

_____

_____

✔ **Reading Check**

2. **Explain** Why are ceramic insulators used to connect electric wires to posts?

_____

_____

B Organize Information
Make the following Foldable to help you organize what you learn about semiconductors.

✔ **Reading Check**

3. **Describe** how the conductivity of a semiconductor can be changed.

_____

_____

Picture This

4. **Explain** In this figure, what is different about the element that was added to the silicon crystal?

_____

_____

Semiconductors

Another group of useful materials is semiconductors. Semiconductors are materials that do not conduct electricity as well as metals but do conduct better than nonmetals, and their conductivity can be controlled. These materials make computers and other electronic devices work.

Look at the periodic table in this book. The elements along the stair-step line between metals and nonmetals are the metalloids. Some metalloids are semiconductors. Recall that metals are good conductors of electricity. Nonmetals are poor conductors of electricity. Semiconductors lie between the two. The metalloid elements silicon (Si) and germanium (Ge) are common semiconductors.

How is the conductivity of a semiconductor changed?

Adding other elements to some metalloids can change their conductivity. For example, you can increase the conductivity of silicon crystals by replacing some silicon atoms with atoms of other elements. Atoms that commonly are added to silicon are arsenic (As) and gallium (Ga). The figure below shows how the conductivity of silicon increases when another element is added. ☑

The added atoms are called impurities. If the impurities have fewer electrons than silicon atoms, the silicon crystals will have areas with fewer electrons. These areas are called holes. Electrons now can move from hole to hole across the crystal. This increases the conductivity of the material. Adding just one atom of an impurity to a million silicon atoms changes the conductivity of the silicon a great deal. By controlling the type and number of atoms added, the conductivity of silicon can vary over a wide range.

Pure silicon crystal   Silicon crystal with impurities

Si atom ●   Electron •   Hole ○   Other element (impurity)

What is doping?

Doping is the process of adding impurities to semiconductors to change their conductivity. If doping increases the number of electrons in a crystal, the semiconductor is called an *n-type* semiconductor. If doping reduces the number of electrons in a crystal, the semiconductor is called a *p-type* semiconductor.

What are integrated circuits?

Electronic devices such as transistors and diodes are made by placing n-type and p-type semiconductor crystals next to each other. Transistors and diodes control the flow of electrons in electric circuits of devices like radios and computers. The figure shows how electrons flow from the arsenic-doped silicon n-type semiconductor to the gallium-doped silicon p-type semiconductor.

Direction of electron flow

n-type   p-type

● Si atom   ● Electron   ○ Hole   ● Free electron

Picture This

5. Classify What is the chemical symbol of the element used in the p-type semiconductor shown in the figure?

_____

In the 1960s, scientists developed very small transistors and other devices. At the same time, the integrated circuit was developed. An **integrated circuit** contains many tiny semiconducting devices. Integrated circuits are sometimes called microchips. The integrated circuit was a great advance in technology. The small electronic devices we use today, such as radios, TVs, cell phones, calculators, and especially computers, would not be possible without integrated circuits. ☑

Integrated circuits are being made smaller and smaller. It takes less time for electric current to travel through smaller circuits, which makes the computers that use them even faster. When computers were first invented in the 1940s, they filled entire rooms. They needed thousands of electric circuits to process information. Today, these thousands of circuits are all found on tiny integrated circuits. In about 60 years, computers have gone from being the size of buildings to being small enough to fit in your hand.

☑ Reading Check

6. Identify What is another name for an integrated circuit?

_____

_____

☻ After You Read

Mini Glossary

doping: the process of adding impurities to semiconductors to change their conductivity

integrated circuit: a circuit that contains many tiny semiconducting devices

1. Review the terms and their definitions in the Mini Glossary. Write a sentence that shows you understand why manufacturers use doping when making some semiconductors.

   _____

   _____

2. Complete the table to describe the properties and uses of ceramics and semiconductors.

Properties and Uses of Ceramics and Semiconductors		
	Ceramics	**Semiconductors**
Materials they are made of		
Properties		
Uses		

3. **Mark the Text** How are semiconductors similar to traffic lights?

   _____

   _____

New Materials Through Chemistry

section ❸ Polymers and Composites

● Before You Read

What was the last plastic item you used and how did you use it?

What You'll Learn

- what a polymer is and the kinds of polymers
- composites and why they are used

● Read to Learn

Polymers

Polymers are natural or manufactured substances that are made of long chains of molecules with small, simple, repeating units. A monomer is one specific molecule that is the repeated unit in polymer chains. For example, polypropylene polymer chains might have 50,000 to 200,000 monomers in each chain. Some examples of manufactured polymers are shown in the table.

Common Polymers	
Polymer	**Uses**
Polyethylene	bottles, garment bags
Polyvinyl chloride (PVC)	pipe, bottles, compact discs, computer housing
Polypropylene	rope, luggage, carpet, film
Polystyrene	toys, packaging, egg cartons, flotation devices

You might know some of the polymers in the table as plastics. Not all polymers are plastics, and not all polymers are manufactured. Proteins, cellulose, and nucleic acids are polymers found in living things. In this section, you will learn mostly about synthetic polymers. **Synthetic** means not naturally occurring but manufactured in laboratories or chemical plants.

Mark the Text

Highlight Main Points As you read this section, highlight the vocabulary words and main points. After you read, look over the sentences you highlighted.

FOLDABLES

❸ Compare and Contrast
Make the following Foldable to help you understand how synthetic and natural polymers are similar and different.

How were synthetic polymers invented?

Humans have used natural polymers for centuries. The ancient Egyptians used polymers in natural resins to help preserve their dead. The first synthetic polymers were made while scientists were trying to improve natural polymers. In 1839, an American inventor named Charles Goodyear found a way to improve natural rubber. He heated it with sulfur. The result was a rubber that did not break when it was cold or get too soft when it was hot. In the 1860s, John Hyatt invented celluloid, a polymer that was used to make billiard balls, umbrella handles, and toys. These early polymers were not as useful as today's polymers. However, they were better than many natural polymers. Now there are many synthetic polymers. They are often divided into groups, such as plastics, synthetic fibers, adhesives, surface coatings, and synthetic rubbers. ☑

What are synthetic polymers made of?

Today, synthetic polymers are made from fossil fuels such as oil, coal, and natural gas. Fossil fuels are called hydrocarbons because they are compounds mostly of hydrogen and carbon. Since synthetic polymers are made from hydrocarbons, synthetic polymers contain mostly hydrogen and carbon.

Why do synthetic polymers have different properties?

The properties of polymers depend on the type of monomers used to make them. The way the monomers are arranged in the chains also affects the properties. There are many different types of synthetic polymers because there are many ways to arrange the chains.

For example, the monomer ethylene is used in several different types of polymers. Polyethylene can be high density or low density, depending on how the molecules are attached to the monomer. Low-density polyethylene (LDPE) and high-density polyethylene (HDPE) differ mostly by the way the monomers are arranged in the polymer chains. The table shows the properties of LDPE and HDPE and some of the ways they are used.

Think it Over

2. **Recognize Cause and Effect** Why could using fewer plastics help save natural resources?

_____

_____

_____

Picture This

3. **Conclude** Would it be better to make plastic buckets from HDPE or LDPE? Explain.

_____

_____

_____

Material	Properties	Typical uses
Low-density polyethlene (LDPE)	Flexible, tough, chemical resistant, low density	Shopping bags, milk jugs
High-density polyethylene (HDPE)	Firm, strong, less translucent, higher density	Toys, outdoor furniture

Plastics One of the groups of synthetic polymers is plastics. Plastics are used in many products because they have useful properties. For example, they are lightweight, strong, impact resistant, waterproof, moldable, chemical resistant, and inexpensive. There are many types of plastics. Some are clear, some are flexible, and some are hard. The properties of plastics depend on the polymers they are made from.

Synthetic Fibers Like plastics, synthetic fibers can be made with almost any properties. Nylon, polyester, acrylic, and polypropylene are examples of synthetic fibers. Nylon is often used in wind- and water-resistant clothing, such as jackets and windbreakers. Polyester is used to make many fabrics. It is often blended with natural fibers, such as cotton. Polyester fibers also are used to stuff pillows and quilts. Polyurethane is the foam used in mattresses and pillows. ☑

Aramids are a family of nylons with special properties. Some aramids are fireproof. They are used to make protective clothing for firefighters, military pilots, and race car drivers. Another aramid fiber is used to make bulletproof vests, race car survival cells, puncture-resistant gloves, and motorcycle clothing. These aramids are light enough for clothing, yet stronger than steel.

Adhesives Adhesives are products that stick things together. Glue is an adhesive. Synthetic polymers are used to make adhesives that can be modified for different uses.

Contact cement is a synthetic polymer adhesive. It is used in the manufacture of car parts, furniture, and leather goods. Contact cement sticks instantly. The bond gets stronger after the cement dries. Structural adhesives are used in construction projects. One structural adhesive is silicon. Silicon adhesive is used to seal the spaces around windows and doors to prevent heat loss. Orthodontists use adhesives to attach braces to teeth. One such adhesive bonds after it is exposed to ultraviolet light.

Surface Coatings and Elastic Polymers Many products that are used to coat and protect surfaces are synthetic polymers. Polyurethane is a polymer that is used to protect wood. Many paints also contain synthetic polymers. Synthetic rubber is an elastic polymer. It is used to make tires, gaskets, belts, and hoses. The soles of some shoes are made from this synthetic elastic polymer.

> **✔ Reading Check**
>
> **4. Identify** Name a synthetic fiber used to stuff pillows.
> _____
> _____

> **💡 Think it Over**
>
> **5. Apply** Suppose water was leaking around a shower door. What kind of synthetic polymer might be used to seal the leak?
> _____
> _____

How do synthetic fibers copy nature?

Spinning long fibers into threads and fabrics is not a new idea. Spiders spun fibers for their webs long before humans copied the idea. Nylon fiber is another idea borrowed from nature. The silkworm produces a fiber that is woven into fabric called silk. Silk is used for blouses and other clothing. Can you imagine how long it takes for a worm to produce enough silk for one blouse? Nylon was produced in a laboratory as a possible substitute for silk.

Composites

The properties of a synthetic polymer can be changed by adding other materials. A **composite** is a mixture of two or more materials—one embedded or layered in the other. Composite materials of plastic have many uses. Some boat and car bodies are made of a composite called fiberglass. Fiberglass is made from small fibers or threads of glass embedded in a plastic. The glass fibers make the plastic stronger. Fiberglass is a strong, lightweight composite. Glass and other fibers are often added to plastics to make them stronger, more flexible, and less brittle. ☑

How are composites used in spacecraft and aircraft?

Some satellites are made from composite materials. These satellites are stronger and weigh less than satellites made of aluminum. Lighter satellites are less expensive to launch.

The figure below shows that aircraft contain composite materials. Just as in satellites, the composites used in aircraft are lighter and stronger than the metals once used. The weight of some commercial aircraft is reduced by more than 2,600 kg by using composite materials and new alloys. Lighter airplanes cost less to operate because they use less fuel.

☑ Reading Check

6. Reflect What effect do the glass fibers have on the plastic they are layered in?

_____

_____

_____

Picture This

7. Observe Look at the figure. In what areas of the aircraft are composites used most often?

_____

_____

_____

■ New composite application

■ Improved composite application

After You Read

Mini Glossary

composite: a mixture of two or more materials—one embedded or layered in the other

synthetic: not naturally occurring, but manufactured in laboratories or chemical plants

1. Review the terms and their definitions in the Mini Glossary. Describe fiberglass using at least one of the terms in the Mini Glossary.

   _____

   _____

2. Fill in the Venn diagram below to compare polymers and composites. Give examples of both polymers and composites.

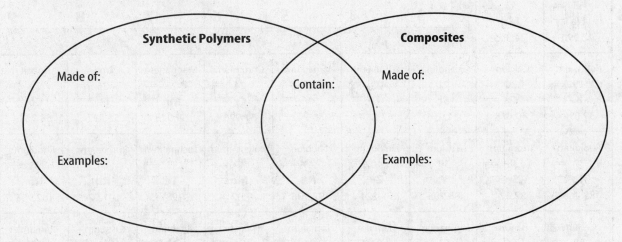

Synthetic Polymers

Made of:

Examples:

Contain:

Composites

Made of:

Examples:

3. Mark the Text As you read this section, you highlighted vocabulary words and main points. After you read, you looked again at the sentences you highlighted. How could you use your highlighting to help you remember the material in this section?

   _____

   _____

   _____

End of Section

PERIODIC TABLE OF THE ELEMENTS

Columns of elements are called groups. Elements in the same group have similar chemical properties.

Element — Hydrogen
Atomic number — 1
Symbol — H
Atomic mass — 1.008

State of matter

Gas
Liquid
Solid
Synthetic

The first three symbols tell you the sta of matter of the element at room temperature. The forth symbol identifies elements that are not preser in significant amounts on Earth. Usefu amounts are made synthetically.

1

1	Hydrogen 1 **H** 1.008	**2**							
2	Lithium 3 **Li** 6.941	Beryllium 4 **Be** 9.012							
3	Sodium 11 **Na** 22.990	Magnesium 12 **Mg** 24.305	**3**	**4**	**5**	**6**	**7**	**8**	**9**
4	Potassium 19 **K** 39.098	Calcium 20 **Ca** 40.078	Scandium 21 **Sc** 44.956	Titanium 22 **Ti** 47.867	Vanadium 23 **V** 50.942	Chromium 24 **Cr** 51.996	Manganese 25 **Mn** 54.938	Iron 26 **Fe** 55.847	Cobalt 27 **Co** 58.933
5	Rubidium 37 **Rb** 85.468	Strontium 38 **Sr** 87.62	Yttrium 39 **Y** 88.906	Zirconium 40 **Zr** 91.224	Niobium 41 **Nb** 92.906	Molybdenum 42 **Mo** 95.94	Technetium 43 **Tc** (98)	Ruthenium 44 **Ru** 101.07	Rhodium 45 **Rh** 102.906
6	Cesium 55 **Cs** 132.905	Barium 56 **Ba** 137.327	Lanthanum 57 **La** 138.905	Hafnium 72 **Hf** 178.49	Tantalum 73 **Ta** 180.948	Tungsten 74 **W** 183.84	Rhenium 75 **Re** 186.207	Osmium 76 **Os** 190.23	Iridium 77 **Ir** 192.217
7	Francium 87 **Fr** (223)	Radium 88 **Ra** (226)	Actinium 89 **Ac** (227)	Rutherfordium 104 **Rf** (261)	Dubnium 105 **Db** (262)	Seaborgium 106 **Sg** (266)	Bohrium 107 **Bh** (264)	Hassium 108 **Hs** (277)	Meitnerium 109 **Mt** (268)

Rows of elements are called periods. Atomic number increases across a period.

The arrow shows where these elements would fit into the periodic table. They are moved to the bottom of the table to save space.

The number in parentheses is the mass number of the longest lived isotope for that element.

	Cerium 58 **Ce** 140.115	Praseodymium 59 **Pr** 140.908	Neodymium 60 **Nd** 144.242	Promethium 61 **Pm** (145)	Samarium 62 **Sm** 150.36	Europium 63 **Eu** 151.965
Lanthanide series						
Actinide series	Thorium 90 **Th** 232.038	Protactinium 91 **Pa** 231.036	Uranium 92 **U** 238.029	Neptunium 93 **Np** (237)	Plutonium 94 **Pu** (244)	Americium 95 **Am** (243)

Legend

- Metal
- Metalloid
- Nonmetal
- Recently observed

The color of an element's block tells you if the element is a metal, nonmetal, or metalloid.

18

Helium
2
He
4.003

13 | **14** | **15** | **16** | **17**

Boron	Carbon	Nitrogen	Oxygen	Fluorine	Neon
5	6	7	8	9	10
B	**C**	**N**	**O**	**F**	**Ne**
10.811	12.011	14.007	15.999	18.998	20.180

Aluminum	Silicon	Phosphorus	Sulfur	Chlorine	Argon
13	14	15	16	17	18
Al	**Si**	**P**	**S**	**Cl**	**Ar**
26.982	28.086	30.974	32.066	35.453	39.948

10 | **11** | **12**

Nickel	Copper	Zinc	Gallium	Germanium	Arsenic	Selenium	Bromine	Krypton
28	29	30	31	32	33	34	35	36
Ni	**Cu**	**Zn**	**Ga**	**Ge**	**As**	**Se**	**Br**	**Kr**
58.693	63.546	65.39	69.723	72.61	74.922	78.96	79.904	83.80

Palladium	Silver	Cadmium	Indium	Tin	Antimony	Tellurium	Iodine	Xenon
46	47	48	49	50	51	52	53	54
Pd	**Ag**	**Cd**	**In**	**Sn**	**Sb**	**Te**	**I**	**Xe**
106.42	107.868	112.411	114.82	118.710	121.757	127.60	126.904	131.290

Platinum	Gold	Mercury	Thallium	Lead	Bismuth	Polonium	Astatine	Radon
78	79	80	81	82	83	84	85	86
Pt	**Au**	**Hg**	**Tl**	**Pb**	**Bi**	**Po**	**At**	**Rn**
195.08	196.967	200.59	204.383	207.2	208.980	208.982	209.987	222.018

Darmstadtium	Roentgenium	Copernicium	Ununtrium	Ununquadium	Ununpentium	Ununhexium		Ununoctium
110	111	112	* 113	* 114	* 115	* 116		* 118
Ds	**Rg**	**Cn**	**Uut**	**Uuq**	**Uup**	**Uuh**		**Uuo**
(281)	(272)	(285)	(284)	(289)	(288)	(291)		(294)

★ The names and symbols for elements 113, 114, 115, 116, and 118 are temporary. Final names will be selected when the elements' discoveries are verified.

Gadolinium	Terbium	Dysprosium	Holmium	Erbium	Thulium	Ytterbium	Lutetium
64	65	66	67	68	69	70	71
Gd	**Tb**	**Dy**	**Ho**	**Er**	**Tm**	**Yb**	**Lu**
157.25	158.925	162.50	164.930	167.259	168.934	173.04	174.967

Curium	Berkelium	Californium	Einsteinium	Fermium	Mendelevium	Nobelium	Lawrencium
96	97	98	99	100	101	102	103
Cm	**Bk**	**Cf**	**Es**	**Fm**	**Md**	**No**	**Lr**
(247)	(247)	(251)	(252)	(257)	(258)	(259)	(262)